Lecture Notes in Comput‹

T0230286

4

Commenced Publication in 1973
Founding and Former Series Editors:
Gerhard Goos, Juris Hartmanis, and Jan v

Max J. Egenhofer   Christian Freksa
Harvey J. Miller (Eds.)

Geographic Information Science

Third International Conference, GIScience 2004
Adelphi, MD, USA, October 20-23, 2004
Proceedings

 Springer

Volume Editors

Max J. Egenhofer
National Center for Geographic Information and Analysis
Department of Spatial Information Science and Engineering
University of Maine, Department of Computer Science
348 Boardman Hall, Orono, ME 04469-5711, USA
E-mail: max@spatial.maine.edu

Christian Freksa
University of Bremen, Department of Mathematics and Informatics
28334 Bremen, Germany
E-mail: freksa@informatik.uni-bremen.de

Harvey J. Miller
University of Utah, Department of Geography
260 S. Central Campus Dr., Room 270, Salt Lake City, UT 84112-9155, USA
E-mail: harvey.miller@geog.utah.edu

Library of Congress Control Number: 2004114080

CR Subject Classification (1998): H.2.8, H.4, H.3, H.2, H.5, J.2

ISSN 0302-9743
ISBN 3-540-23558-2 Springer Berlin Heidelberg New York

Springer is a part of Springer Science+Business Media

springeronline.com

© Springer-Verlag Berlin Heidelberg 2004
Printed in Germany

Typesetting: Camera-ready by author, data conversion by Olgun Computergrafik
Printed on acid-free paper   SPIN: 11331247   06/3142   5 4 3 2 1 0

GIScience 2004
Third International Conference
on Geographic Information Science

Adelphi, MD, USA, October 20-23, 2004

Organizing Committee

General Chair	Michael Goodchild (National Center for Geographic Information and Analysis, Santa Barbara, CA, USA)
Program Co-chairs	Max Egenhofer (University of Maine, Orono, ME, USA)
	Christian Freksa (University of Bremen, Germany)
	Harvey Miller (University of Utah, Salt Lake City, UT, USA)
Co-organizers	Doug Richardson (Association of American Geographers, Washington, DC, USA)
	Mauro Salvemini (Association of Geographic Information Laboratories Europe, Utrecht, The Netherlands)
	John Wilson (University Consortium for Geographic Information Science, Alexandria, VA, USA)

Program Committee

David Abel, CSIRO, Australia
Walid Aref, Purdue University, USA
Claudia Bauzer Medeiros, Universidade Estadual de Campinas, Brazil
Peter Burrough, Universiteit Utrecht, The Netherlands
Barbara Buttenfield, University of Colorado at Boulder, USA
Gilberto Câmara, Instituto Nacional de Pesquisas Espacias, Brazil
Nicholas Chrisman, University of Washington, USA
Anthony Cohn, University of Leeds, UK
Noel Cressie, Ohio State University, USA
Isabel Cruz, University of Illinois at Chicago, USA
Sara Fabrikant, University of California, Santa Barbara, USA
Peter Fisher, University of Leicester, UK
Leila de Floriani, University of Maryland, USA, and Università degli Studi di Genova, Italy
Stewart Fotheringham, University of Newcastle, UK
Andrew Frank, Technische Universität Wien, Austria
Mark Gahegan, Pennsylvania State University, USA
Arthur Getis, San Diego State University, USA
Ralf Güting, FernUniversität Hagen, Germany
Jiawei Han, University of Illinois at Urbana-Champaign, USA

Table of Contents

Contested Nature of *Place*: Knowledge Mapping
for Resolving Ontological Distinctions Between Geographical Concepts 1
   Pragya Agarwal

Geo-Self-Organizing Map (Geo-SOM) for Building
and Exploring Homogeneous Regions 22
   Fernando Bação, Victor Lobo, and Marco Painho

Can Relative Adjacency Contribute to Space Syntax in the Search
for a Structural Logic of the City? 38
   Roderic Béra and Christophe Claramunt

Semi-automatic Ontology Alignment for Geospatial Data Integration 51
   Isabel F. Cruz, William Sunna, and Anjli Chaudhry

Modeling Surface Hydrology Concepts with Endurance and Perdurance 67
   Chen-Chieh Feng, Thomas Bittner, and Douglas M. Flewelling

Procedure to Select the Best Dataset for a Task 81
   Andrew U. Frank, Eva Grum, and Bérengère Vasseur

Floating-Point Filter for the Line Intersection Algorithm 94
   Andras Frankel, Doron Nussbaum, and Jörg-Rudiger Sack

Project Lachesis: Parsing and Modeling Location Histories 106
   Ramaswamy Hariharan and Kentaro Toyama

The SPIRIT Spatial Search Engine:
Architecture, Ontologies and Spatial Indexing 125
   Christopher B. Jones, Alia I. Abdelmoty, David Finch, Gaihua Fu,
   and Subodh Vaid

Comparing Exact and Approximate Spatial Auto-regression Model Solutions
for Spatial Data Analysis 140
   Baris M. Kazar, Shashi Shekhar, David J. Lilja, Ranga R. Vatsavai,
   and R. Kelley Pace

3D GIS for Geo-coding Human Activity in Micro-scale Urban Environments 162
   Jiyeong Lee

Arc_Mat, a Toolbox for Using ArcView Shape Files
for Spatial Econometrics and Statistics 179
   James P. LeSage and R. Kelley Pace

A Predictive Uncertainty Model for Field-Based Survey Maps
Using Generalized Linear Models 191
Stefan Leyk and Niklaus E. Zimmermann

Information Dissemination in Mobile Ad-Hoc Geosensor Networks 206
Silvia Nittel, Matt Duckham, and Lars Kulik

Public Commons of Geographic Data: Research and Development Challenges ... 223
*Harlan Onsrud, Gilberto Camara, James Campbell,
and Narnindi Sharad Chakravarthy*

Alternative Buffer Formation 239
Gary M. Pereira

Effect of Category Aggregation on Map Comparison 251
Robert Gilmore Pontius Jr. and Nicholas R. Malizia

Simplifying Sets of Events by Selecting Temporal Relations 269
Andrea Rodríguez, Nico Van de Weghe, and Philippe De Maeyer

Towards a Temporal Extension of Spatial Allocation Modeling 285
Takeshi Shirabe

Formalizing User Actions for Ontologies................................ 299
Keanhuat Soon and Werner Kuhn

Landmarks in the Communication of Route Directions 313
Elisabeth Weissensteiner and Stephan Winter

From Objects to Events: GEM, the Geospatial Event Model 327
Michael Worboys and Kathleen Hornsby

Author Index 345

Contested Nature of *Place*:
Knowledge Mapping for Resolving Ontological
Distinctions Between Geographical Concepts

Pragya Agarwal

GIS Research Group
School of Geography
The University of Nottingham
Nottingham, NG7 2RD, UK
lgxpa@nottingham.ac.uk

Abstract. In the theoretical literature and geographical models, *place* is defined with reference to other spatial concepts, such as region, neighbourhood and space. The boundaries and distinctions between these concepts and *place* are not clear and the overlap in the semantic fields of these concepts is representative of the vagueness that exists in geographical concepts. This vagueness is a major issue in achieving interoperability in the geographical domain and for the development of a comprehensive geo-ontology. In this paper, the principles of conceptual structures are applied for mapping semantic correspondences between *place* and overlapping geographical concepts. Knowledge mapping is carried out for meaning negotiation from lexical analysis of syntactic proximities between concepts. Experiments with human subjects were performed to elicit the cognitive semantics inherent in conceptual schema of individuals. Distance and proximity measures resolve the cognitive semantics of the concepts. The results provide an indication of the overlap and distinctions between the semantic fields of nearby spatial concepts, and a foundation for defining the relationships and classifications to specify the ontological distinctions between geographic concepts.

1 Introduction

Ambiguous meanings and fuzzy delineation between concepts causes problems in a clear definition and modelling of the geographical domain, and in the specification of a geo-ontology. The semantic heterogeneity of the terms cause problems in developing a consensual ontology for the domain [4, 12] and, therefore, the foremost requirement in the development of a geo-ontology is to clarify the meanings of such geographic terms and concepts that need better definitions [1]. Ontological theories and commitments underlie all forms of cognition, both implicitly as well as explicitly [21]. The University Consortium for Geographic Information Science, within the emerging research theme *Ontological Foundations for Geographic Information Science* [22], has proposed that the process of eliciting ontologies from human subjects, achieved by using standard psychological methods, is a key research area to establish

M.J. Egenhofer, C. Freksa, and H.J. Miller (Eds.): GIScience 2004, LNCS 3234, pp. 1–21, 2004.
© Springer-Verlag Berlin Heidelberg 2004

the conceptual systems that people use in relation to given domains of objects. The process of eliciting ontologies from human subjects can provide guidelines for developing links between three kinds of knowledge domains, namely human attitudes and beliefs, real-world objects and features, and data models. Elicitation work has been previously described [23, 35] where commonsense conceptualisation and categorisation of the real world are investigated through application of prototypical theories within experiments on human subjects [3, 20]. These seminal works provide valuable indications of how cognitive theories and human behaviour can form the basis for ontological development. It is, however, focussed on category explication and on finding the natural categories that result from commonsense view of the world, but its potential for defining concepts and for explication of the semantic content of the categories has not been fully exploited and investigated.

Geographic theories attribute cognitive dimensions to the notion of geographic *place* [6, 13] and the meaning of *place* can be conceptualised as emerging at an embodied interface of mind, body and language. The broad aim in this paper is to apply methods of elicited ontologies to disengage ontological distinctions and commitments for *place* and neighbouring concepts. The underlying principle for the methodology adopted in this paper, developed upon theories of *conceptual spaces* [11], is that the meanings of concepts are imbibed in human cognitive conceptualisations and can be elicited by mapping the concept space existing within the cognitive domain. In this case, *conceptualisation* is used to denote the modelling of the world independent of the technology at the knowledge level, and defined as the idea of the world that an individual has. A combination of semantic proximity and categorical assessments are carried out to elicit the commonsense notions of *place*, where *commonsense* is used to imply the certain core of interconnected beliefs that form the basis for the cognitive notions of reality.

The remainder of the paper is organised as follows. Section 2 provides a background on the contested nature of *place*. A discussion of *place* meanings and its relations with neighbouring spatial concepts based on lexical analysis is presented in section 3. In section 4, a brief summary of the basic principles that form the basis for the experimental work is provided, along with a more detailed discussion of the experimental design, the data collection methods and the results from each stage of the experiment. A summary of the results along with the conclusions from the paper and implications for future work are presented in section 5.

2   The Contested Nature of Geographic *Place*

Place is a key and contested topic in geography. UCGIS has included the study of *place*, conceptualised as a cognitive category, as a priority area of research, because the "continuous physical world is understood in terms of discrete objects and *places*" and because "there are currently inconsistencies in the models for these categorisations and for these *places* that exist" [40]. Since it is shown from experiments that people exchange information and reason about locations based on *places* they conceptualise, the notion of *place* is proposed as a link between human commonsense and the reality in design of wayfinding models [13, 17, 41]. Within information systems and in virtual societies, there is a general concern about the erosion of *sense of place* and the creation of *placelessness* [5, 25, 32]. *Place* as a geographic concept is applied in spatial models in regional geography [7, 24] and for policy and decision-making in geographical environments [36]. Many digital thesauri, such as EDINA GeoXwalk, the Alexandria Digital Library (ADL) and Getty's Thesaurus of Names (TGN), are now using *place*-based mechanisms for query resolutions. Harrison and Dourish [16] have also highlighted the significance of *place*-based information systems for generating behavioural components of space within computer-assisted cooperative virtual and simulated environments.

It is acknowledged that, even within the geographic paradigm, different meanings are attributed to *place*. Theories in human and behavioural geography emphasise the importance of the humanistic and experiential dimensions in shaping the meanings of *place* [37]. The variability in the meanings of *place* gives rise to an inherent vagueness in any integration across this concept in data sources and participatory interfaces. Mapping its meanings from cognitive conceptualisations will help identify the primitives that constitute the primary theory for *place* that can act as a shared resource for interoperability across different domains of applications in GIScience. From a review of commonly used definitions of *place* in different theoretical contexts, and from mapping *place* meanings across a number of ontology and digital libraries, it is seen that the meaning of *place* is linked to meanings of other spatial concepts. The thesaurus [42] states a number of related terms to *place* as synonyms, such as *location, locality, point, spot, space, piazza, plaza, topographic point, area, situation, seat, home, neighbourhood, and landmark*. Although most of the ontologies and theoretical frameworks that were reviewed for this paper have generated a synergy between region and *place*, with regions often forming a *place* type, the discussion in categorisation of region types [28] have also stated *place* to be a sub-type of region. Both *place* and neighbourhood are syntactically linked to notions of distance, such as proximity and closeness. A critical discussion of *place* and its relations to region and neighbourhood raises a number of issues. Few of the questions that are raised are: Is *place* a type of region? Can all regions be *place*s? When does a region become a *place* and vice-versa? Is neighbourhood a sub-type of *place*? And, is a neighbourhood always a *place*? What kind of *place* is a neighbourhood? These questions typify the nature of vagueness within the concept of *place*, which can be resolved once a definitive semantic framework for *place* is defined and relationships to these other concepts explored.

2.1 Mapping Reference Knowledge Domains

WordNet is a lexical reference system that is inspired by current psycholinguistic theories of human lexical memory [27]. The semantic network in WordNet is formed by nodes representing real-world concepts that comprise the synonyms and the definition (gloss) for the concept [29]. *Place* in WordNet is defined as a *concrete entity* functioning as a *location for something else*. The resulting senses from WordNet's semantic taxonomy are sixteen in total and the geographically relevant meanings of *place* are closely related to concepts of *space*, *region*, *location* and *vicinity*. Semantic concordance exercises are helpful in measuring the semantic proximities between *place* and related concepts, and are useful in making the distinctions explicit (Table 1). Semantic similarity measures [29] were employed for measuring proximity between *place* and related concepts. The results from the analysis using WordNet-similarity are shown in Table 2. For Hirst and St-Onge, the highest value is 16; for Path length, Lin and Wu Palmer, the highest value is 1. In all cases, higher values indicate a higher measure of semantic similarity. These measures use different parameters for estimating semantic relatedness and, therefore, a cumulative figure from all the measures gives an effective indication of the relative semantic similarities between concepts.

Table 1. Semantic correspondence scores for *place* and related concepts

	Place	Region	Area	Location	Neighbourhood	Space	Cognition
Place	1.000	1.000	1.000	1.000	1.000	1.000	1.000
Region	x	1.000	x	0.750	x	x	-0.139
Area	x	1.000	1.000	0.250	x	x	0.250
Neighbourhood	x	0.111	0.250	0.062	1.000	x	x
Space	x	1.000	1.000	x	x	1.000	x
Cognition	x	x	x	x	x	x	x

Table 2. Similarity relatedness measures for *place* with correspondent concepts

	region	neighbourhood	space	location	area	district
Resnick	3.7552	6.0831	11.9046	3.2232	8.9089	3.7522
Path length	0.3333	0.5	1.000	0.3333	0.5	0.25
Leacock & Chodorow	2.4849	2.8904	3.5835	2.4849	2.8904	2.1972
Wu & Palmer	0.800	0.9333	1.000	0.75	0.9333	0.7273
Hirst & St-Onge	6.00	4.00	16.00	6.00	6.00	4.00
Jiang & Conrath	0.3382	0.2716	29590099	0.3033	0.3338	0.2558
Adapted Lesk	37.00	24.00	448.00	91.00	110.00	31.00
Lin	0.7175	0.7677	1.00	0.6616	0.8561	0.6576
Cumulative	**51.4291**	**39.4461**	**29505982**	**104.756**	**130.42**	**42.8401**

Place and *space* share maximum semantic relatedness based on the WordNet lexical database and the corpus of concepts created from it. Results from this analysis also show that, lexically and semantically, *place* and *region* are more closely related than *place* and *neighbourhood*. A high measure of semantic relatedness is also seen with *location* and *area*. It is worth considering that these measures are dependent on concept definitions and the nature of hierarchical is-a relationships, specified in WordNet. However, WordNet is constructed based on the distinctions determined from human reasoning and knowledge [9] and can be viewed as a reliable source of ontological distinctions based on commitments for the concepts in the real world. The lexical analysis provides important indications for the way these concepts are distinguished semantically. It also facilitates a framework for aligning and coordinating different ontologies by defining points of articulation based on the concepts that are semantically related to *place* in cases where *place* does not exist naturally in a particular knowledge base. A comparison of the proximity measures generated from syntactic space in WordNet, with the results from the experimental work with human subjects, will also be indicative of the discrepancies between commonsense semantics and that conceptualised in reference systems.

3   Theoretical Framework and Hypotheses

Meaning is a much discussed subject in psychology as well as philosophy, and several theories have been proposed to explain meanings, concepts and their formation in the real world [2, 19, 30, 31]. The principal idea in the theory of *conceptual space* is the existence of a conceptual structure that facilitates the grounding of meanings in the cognitive models formed in the real world [10, 11]. The semantics are primarily dependent on the individual nature of internalised spatial representations and cognitive structures. The meanings for concepts can, therefore, be realised from an understanding of the relations between the conceptual structures and the real world. *Similarity* is one of the most fundamental notions in concept formation, and is defined simply as "concepts group together things that are similar" [11 p. 109]. The notions of proximity and similarity are also applied in prototypical theories that are used as a basis for elicitation experiments in the geographic context. The theory of conceptual spaces extends the theory of prototypical effects by introducing the notion of properties that have a partitioning effect within the concept space [10]. The meaning of the concept emerges from an inter-relationship of the different domains and dimensions in it, evolves and changes in different contexts, and is represented as a measure of the salience of the different domains that act in the determination of the meanings.

   In the methodology used here, salience and correlation factors are applied to categorisation experiments for similarity assessment between concepts. The key idea is that a *core* (essential) property of a concept is a property that belongs to a dimension with high salience, while a *peripheral* property is associated with a dimension with lower salience. Extreme salience is attributed to the essential dimensions while determining the content of the concept. The partitioning of the concept space, from the different dimensions, results in the discretisation of space using a finite number of classes that define the concept. The classes are psychologically determined and the classification metrics are context-determined. Therefore, the properties are represented as regions in a conceptual space, and the metrics for representation are externally imposed. Previous theories [14, 26] proposed that similarity judgement is based either in shared properties or distance assessment in the conceptual space. The notion of shared properties assumes that a set of properties are identifiable for a concept [14]. This can be problematic in the case of a cognitive concept, where the properties of the concept are often linked intrinsically to the meaning of the concept itself that creates an "inherent circularity" in this argument [11, p. 111]. There is also the problem of identifying which properties or instances are significant in determining a concept. On the other hand, the use of distance assessment, employed in the methodology developed for semantic similarity in this paper, does not suffer from these limitations. Compared to other semantic theories, the use of the principles of conceptual space to explain the variability and context-dependence of meanings of concepts, requires less dependence on perceptual similarity in shared properties for explaining the distinctions between concepts, and more on the distances from the core set of properties, forming a primary theory in the domain.

Two primary and other subsidiary hypotheses are defined for the experimental setup that is described in the next section.

H1: The ontological distinctions between spatial concepts can be generated from empirical tests with human subjects.

1. Distance metaphors can be employed to explicate the relative semantic relatedness of geographic concepts.
2. *Place* is semantically closer to neighbourhood than to region in meaning.
3. Hyponymn and hypernymn taxonomic relations for *place* can be generated from tests of cognitive semantic relatedness.

H2: Semantic heterogeneity is context-dependent.

1. Semantic relatedness assessed based on domain specification in similarity matrix varies significantly from that carried out solely as distance measures between concepts.
2. Domain saliency influences the meanings of concepts in the conceptual space.

4 Experimental Design

The questionnaire-based experimental design is aimed at finding the similarity relations between concepts to define (1) ontological distinctions based on semantic proximity and (2) the relations elemental in determining the meaning of *place*. Semantic priming relates terms that are semantically close, and the notion of similarity and distance is used to judge categorical distinctions and boundaries between concepts [33]. Methods of semantic similarity assessment have faced criticism that symmetrical relationships are assumed in either direction for similarity judgements [38]. On the other hand, it has also been shown that for semantic similarity assessment, the asymmetry effect is less prominent when the set of stimuli provided in the experiment is devoid of any concept that is a super-ordinate concept for many others in the set

[34]. There is no strongly explicit hierarchical ordering for the concepts that constitute the experiment setup (*place, region, neighbourhood, location, space, district* and *area*) and, therefore, no triangular inequality and asymmetry effects are prominent in applying methods of semantic proximity. These concepts were generated in a background study from mapping *place* conceptualisation in theoretical and modelling frameworks, and from a discussion of knowledge domains in ontologies, such as SUMO, Cyc, DAML and DOLCE. Since critics of distance-based similarity assessment [38] pointed out that the relations are not symmetrical, the similarity assessment was carried out in either direction. Experimental control on the nature of stimuli was provided by specifying that the similarity is assessed based on "meaning as a geographic concept."

4.1 Data Collection Methods

A pilot study was undertaken as a pre-test to develop viable methods of data collection and test the nature of instructions, set of stimuli, control setting, appropriate rate of response, and the identification and verification of suitable analysis and inference methods. Although it has been shown by psychological studies that there is a difference in cognitive processing between sexes, inter-group analysis was not considered relevant within the remits of this study. The length of residence was excluded from the final experiment design since the analysis in the pilot study showed that it did not yield conclusive results regarding the meaning and conceptualisation of *place*. Theoretical and empirical studies [8, 39] have shown that cross-cultural factors can have an influence on the perception of *place*. However, this did not appear to be significant in this experiment.

In the main experimental study, the total number of respondents N=50, where the number of males N_1=28, and the number of females N_2=22. The subjects were predominantly white, with 92% White and 8% British Asians. All subjects were native English speakers. The respondents constituted undergraduate and graduate students, and academic staff members from The University of Nottingham. Among the subjects, 64% (N=32) were trained geographers. Random assignment was carried out for selecting the sample population. Out of the 50 respondents used in the experimental design, there were undergraduates (N_1=28), graduate students (N_2=11), and academic staff members (N_3=11). In addition, the experimental framework required a concentrated locational reference for the subjects to avoid locational biases, and to minimise the effects that any extraneous factors specific to certain locations might have on the results. Out of the 50 subjects, a number of respondents lived on the University Campus (N_1=28), and a number of respondents lived in an area just outside the campus, on the west side of the University (N_2=22). None of the subjects received any payments for their participation. The total number of geographers was 32 (20 were students on geography undergraduate courses, eight were faculty members in the School of Geography at the University of Nottingham, and four were geography postgraduate students). From the rest of the subjects, eight were undergraduate students and seven postgraduate students in such courses as law, medicine, English and social

studies, and three were faculty members from English, Art History and Music de-
partments in the University.

Each subject was provided with a paper-based questionnaire, and groups of ten
subjects were tested at the same time. The response time was 60 seconds for each part
of the questionnaire. It was found that the subjects were able to respond within this
time and no problems were noted. The subjects were instructed to make similarity
judgements based on their own individual notions and mark their responses in a clear,
legible manner on the questionnaire itself. In all the experiments, instructions were
provided with the questionnaire, and the verbal instructions were limited to maintain
constancy and reduce the chances of any bias creeping in between different groups
due to difference in instructions. All 50 questionnaires were found to be suitable and,
therefore, none of the questionnaires from the sample was excluded from the analysis.

4.2 Stage One

Since the similarity measures were assessed in both directions, an asymmetrical matrix resulted for the 50 subjects. Table 3 shows the raw counts resulting from the
experiment for each pair of concepts in the matrix. The data were normalised based
on the total number of counts. Cronbach's alpha (α) was used to test the reliability
of the questionnaire. For all eight variables in the matrix, α was relatively low, at
0.3991. However, it was expected that the variables were multi-dimensional and a
Principal Component Analysis (PCA) was performed to investigate the latent dimensionality of the variables. From a rotated Varimax Kaiser Normalisation, subsets of
items were identified that shared more dimensionalities than others did. The subsets
q1 (*place, region, district, space*), q2 (*neighbourhood, area*), q3 (*location*), and q4
(*time*) were then checked for reliability, and for q1, $\alpha = 0.7104$, and for q2, $\alpha =
0.8106$. *Location* and *time* showed a negative covariance. The reliability for the two
individual subsets was higher than when all variables were used together. It also indicated that the variables in the subset q2 were more correlated than the ones in q1,
which is validated by the correlation matrix. The t-test shows significance levels for
all classes except *time*, where p>0.05. The test data were also normally distributed,
with 2-tailed p<0.05 for Kolmogorov-Smirnov test.

A factor analysis, carried out to test for any natural clusters, resulted in a three-
factor solution. The factor loadings for the variables are presented in Table 4, where
the first factor explains the most variance in the model (29.9%), and highest overall
correlation to concepts. A three-dimensional rotated Varimax component plot was
generated for the matrix showing the loadings in the component space (Figure 1). A
negative correlation was shown imposing constraints on the generation of any distinct
clusters and classes from the factor analysis. In addition, the factor analysis did not
provide adequate consideration of the dimensionalities inherent in the common space
or indications for the proximity measures between the concepts.

A multi-dimensional scaling (MDS), using the PROXSCAL algorithm, returned
the stress values, conceptual space and the proximities between the variables. A two-
dimensional scatter plot of the objects in the proximity space is shown in Figure 2,

with the axes denoting the internal metrics generated by MDS based on the dimensionality of the dataset, and showing the relative proximities between the concepts. From a ratio transformation, the distance of *place-neighbourhood* was 0.571, *place-region* was 0.408, *place-area* was 0.888, *place-location* was 0.879, *place-district* was 0.837, *place-space* was 0.513, and *place-time* was 1.327. *Time* is furthest in meaning to other concepts in the common space. As shown by this experiment, *place* is close in meaning to *neighbourhood* and *space*, and nearest to *region* in the semantic space.

Table 3. Response matrix for *closeness* in meaning between concepts using raw counts of data

	Place	*Neighbour-hood*	*Region*	*Area*	*Location*	*District*	*Space*	*Time*	TOTAL
Place	50	31	43	11	28	14	24	13	214
Neighbourhood	41	50	19	19	32	46	44	21	272
Region	32	16	50	8	12	42	37	8	205
Area	36	42	18	50	13	21	32	2	214
Location	27	17	16	9	50	3	24	0	146
District	38	45	38	33	5	50	11	1	221
Space	39	12	27	18	9	16	50	6	177
Time	15	8	8	3	0	0	2	50	86
TOTAL	278	221	219	151	149	192	224	101	1,535

Table 4. Factor loadings showing correlations between the three factors and the total variance explained by each factor

	Factor1	*Factor2*	*Factor3*
Place	-0.7397796	0.423689	-0.02171
Neighbourhood	0.8407402	-0.1393	-0.17742
Region	-0.7772876	0.056624	1.670868
Area	1.5607931	0.126885	-0.43021
Location	-0.1981053	1.082207	-1.41041
District	1.064032	-0.50047	1.088614
Space	-0.8414053	1.025873	0.168255
Time	-0.9089875	-2.0755	-0.88799
Total	**0**	**-1E-07**	**1E-07**
Explained Variance	**2.392046195**	**2.059576**	**1.556147**

The normalised dataset was then analysed using a self-organising map (SOM) algorithm. Although the 8-by-8 matrix of the input dataset was small for a SOM algorithm, the number of iterations were sensitised accordingly to cope with the limited number of dimensions. The DOS-based freeware package SOM_PAK (http://www.

Component Plot in Rotated Space

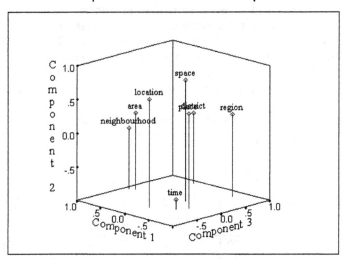

Fig. 1. Rotated three-dimensional varimax scatterplot showing factor loadings for each concept in the component space

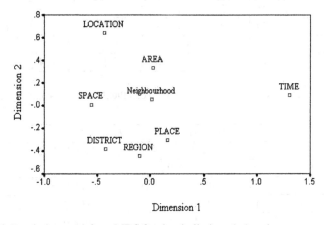

Dimension 1

Fig. 2. Proximity space from MDS for the similarity relations between concepts

cis.hut.fi) and a Windows-based trial version of Viscovery SOMine were used for this analysis. A variance scaling was applied and with a tension of 0.05, the results were produced for hierarchical clustering. Figures 3a-c show the conceptual distances between the concepts for three, five and seven clusters. The shaded areas show the cluster centroids and the visualisation uses *flat clusters* for ease of interpretation, rather than a U-Matrix, which is more commonly employed but not as easy to interpret [18]. *Place* and *region* lay in the same cluster when the number of clusters was constrained to three, and the distances between nodes indicated that *place* and *region* are closer in the conceptual similarity space than *neighbourhood* and *place*. Kendall's concordance test was applied to test the significance of the results and the probability of them occurring by random chance. The analysis (df =7) resulted in Kendall's Concordance, W=0.296, chi-square=16.550, p=0.02, thereby showing a high significance for the results and indicating that the results were not random and were statistically valid.

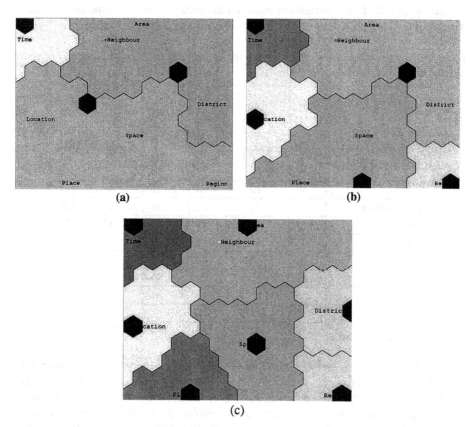

Fig. 3. Hierarchical clustering of the concepts in the similarity space showing (a) 3 clusters, (b) 5 clusters, and (c) 7 clusters

4.3 Stage Two

In stage two, the experiment is designed to generate a set of properties that are salient in determination of the concept, instead of allowing the similarity judgement to be based on a set of pre-determined properties. The theoretical foundation in the context-dependence of semantic space means that the context determines the salience of the domains that explain the concept, thereby making the concepts and similarity judgements dynamic [15]. Introducing a set of domains in similarity judgements addresses the context-variability and dynamic nature of semantics. The domains used in this experimental set-up were *extent, physical attributes, cartographic representations, boundaries* and *dynamic nature*. These variables were identified as significant factors in controlling meanings of geographic concepts and for specifying ontological distinctions between different spatial concepts. The domain specification was kept simple to reduce the error in interpretation of these terms. A brief description for the terms was also provided to reduce the chances of individual interpretations. The responses were scaled and normalised based on the total number of responses for each domain (total from each column) and the raw counts are shown in Table 5.

Table 5. Response matrix showing raw counts for similarity assessment based on pair of concept and on basis of domains

	Extent	Physical Attributes	Cartographic Representation	Boundaries	Dynamic	Total
Place	36	19	18	12	46	**131**
Neighbourhood	47	41	32	39	41	**200**
Region	45	28	29	43	34	**179**
Area	40	31	38	42	28	**179**
Location	14	43	48	8	11	**124**
District	47	31	24	42	18	**162**
Space	35	17	21	46	49	**168**
Time	0	2	3	8	50	**63**
Total	**264**	**212**	**213**	**240**	**277**	**1,206**

Cronbach's alpha (α =-0.07) indicated a high degree of latent dimensionality for the different variables in this matrix. A PCA showed a grouping of variables as q1 (*place, space, time*), q2 (*region, area, district, neighbourhood*), and location showed a high negative covariance with all the other variables (except *area*, where r=0.371), which explained the overall negative value for α. Reliability analysis on the subsets, with spatial concepts as primary variables and the domains as determinants, gave values of α_{q1} =0.7031 and α_{q2} =0.7992, which was more than the results from the whole set of variables, indicating stronger correlation between the second set of variables. Two components were extracted from the Varimax rotation, and the rotated component matrix mapped in component space is shown in Figure 4.

The goodness of fit tests showed low stress values and high dispersion, and Tucker's coefficient showed a good fit for the model (Table 6). The output from MDS (transformation=ratio, initial configuration=simplex, stress convergence=0.001, no restrictions) showed *place* to be closer to *region* (distance=0.751) than to *neighbourhood* (distance=0.825). *Place* and *space* are closest in the semantic space (distance=0.656) and *place* and *time* are not that close (distance=0.878), as demonstrated by cluster analysis in PCA. *Location* showed least proximity to *place* (dis-

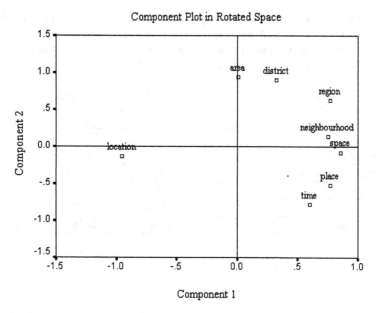

Fig. 4. Rotated Varimax plot showing factor loadings in the component space

Table 6. Stress Measures showing low stress and high values for DAF and Tucker's coefficient

Stress and Fit Measures

Dimensionality: 3

Normalized Raw Stress	.00237
Stress-I	.04870[e]
Stress-II	.11441[e]
S-Stress	.00444[f]
Dispersion Accounted For (D.A.F.)	.99763
Tucker's Coefficient of Congruence	.99881

e. Optimal scaling factor = 1.002.

f. Optimal scaling factor = .995.

tance=1.160) and confirmed the analysis from PCA. Figure 5 shows the distances for the concepts mapped out in a two-dimensional proximity space. Factor analysis results indicate that *cartographic representations* and *boundaries* are the domains that are prominent in explaining the variance in the similarity matrix, with *extent* and *dynamic nature* having less influence and *physical attributes* explaining the least variance in the model. The factor loadings for the domains used as senses for semantic discriminability in this experiment are shown in Table 7. A Chi-square test, however, showed that none of the domain variables were becoming significantly more salient than others in the model (Kendall's W=0.194, Chi-square=6.788, p=0.450). Correlation results showed that the only significant correlations were between *physical attributes* and *cartographic representation* (r=0.810, p=0.015), and the domain of *dynamic nature* showed a significant negative correlation with that of *physical attributes* (r=-0.833, p=0.010), and with *cartographic representation* (r=-0.833, p=0.010). K-related sample testing showed that although the mean ranks for the different domain variables differed, these were not markedly different.

Table 7. Factor loadings showing total variance explained

	Factor 1	Factor 2	Factor 3
Extent	-0.18275	0.57288	1.11386
Physical Attributes	-0.24393	-1.05951	0.96727
Cartographic Representation	0.79652	-1.04394	-0.9332
Boundaries	1.08083	1.14483	-0.19058
Dynamic	-1.45067	0.38575	-0.95735
Explained Variance	**4.059089834**	**2.42816**	**1.186526**

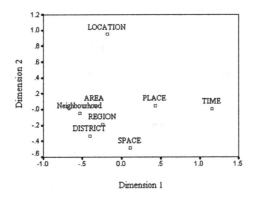

Fig. 5. Scatterplot showing distances between concepts in the similarity matrix

As in stage one, a SOM analysis was carried out and the output was visualised in the form of flat clusters for three, five and seven clusters. The shaded areas on the SOM map (Figure 6) indicate the cluster centroids and the mapped clusters show proximity between *place* and *area*. *Place*, *region* and *neighbourhood* are not indicated to be close in meaning based on the domains introduced in the experiment. However, on a comparative basis, *place* is closer to *neighbourhood* than to *region*. This contradicted the results produced from the MDS analysis for the same dataset that showed *place* was closer to *region* in conceptual space than to *neighbourhood*. The results from MDS, PCA and SOM show consistency in all other semantic relations between concepts, except for the semantic distances from *place* to both *region* and *neighbourhood* that showed conflicting results. Significance testing revealed that there was a high level of agreement between subjects for the general trends emerging in the model (p<0.05). This, along with the other multivariate testing, has shown that the introduction of domains has influenced the meanings of concepts and the proximity measures of semantic relatedness between them.

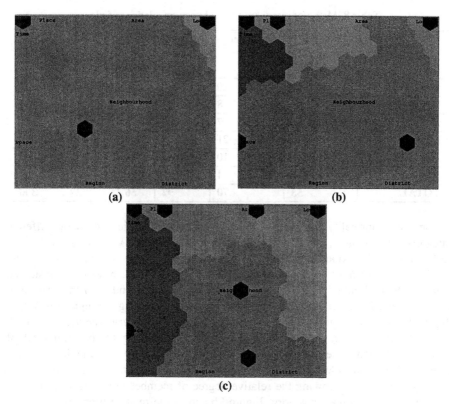

(a)

(b)

(c)

Fig. 6. Hierarchical clustering of the concepts in the domain-dependent similarity space showing (a) three clusters, (b) five clusters and (c) seven clusters.

4.4 Stage Three

Category norms and category scaling proposed for human subject experiments [3] have been previously tested in geographic contexts [20, 21, 35]. In this stage, category norms were employed to generate a conceptual schema for the relation is-a-kind-of-place, to extract a set of proximity measures in a specific direction. The concepts used in the first two stages *(region, neighbourhood, area, location, district, space)* were considered to be instances for the overall class *place* and the experiment was aimed at finding the degree of membership. Similar to Battig and Montague's [3] experimental framework, a scale of 1 to 5 was designed, denoting a range from excellent to poor membership classes (1=very excellent, 2=excellent, 3=good, 4=poor, 5=very poor). A matrix of the total number of responses was created for each category of membership, corresponding to each of the concepts included in the experimental setup. For example, out of 50 subjects, eleven indicated that neighbourhood was a *very excellent* example of *place*, while only two respondents indicated that *region* was a *very excellent* example of *place*. The response matrix resulting from the experiment is shown in Table 8 (N=50).

Table 8. Matrix of responses for the hyponym relations of *place*

	Very Excellent (1)	Excellent (2)	Good (3)	Poor (4)	Very Poor (5)	Total
Neighbourhood	11	20	18	0	1	50
Region	2	8	27	11	2	50
Area	4	7	20	13	6	50
Location	19	21	5	3	2	50
District	14	10	11	14	1	50
Space	3	4	11	15	17	50
TOTAL	**53**	**70**	**92**	**56**	**29**	**300**

The test of normality was valid at the lower level of significance for the different categories of responses, $p < 0.05$. It is seen that *neighbourhood* was categorised as the best example of "a kind of *place*" with 62% of subjects agreeing on scales of *very excellent* and *excellent* (1 and 2). Only 20% of the subjects, however, considered *region* to be a "kind of *place*" in the membership grades 1 and 2 on the scale provided to them. *Location* was shown, from the results, as having the highest membership to the overall category of *place* with 80% responses for *very excellent* and *excellent* categories. The raw counts were normalised according to the sum total of responses on each scale to find the proportional membership of variables for each category. Figure 7a shows the membership of each concept to the different categories on the scale, thereby showing the relative degree of membership of each concept to *place* as a super-ordinate category. Figure 7b shows the relative proportion of variables corresponding to each of the categories.

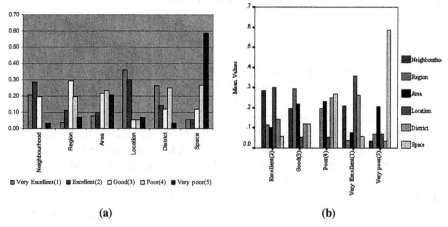

(a)   (b)

Fig. 7. (a) Relative membership of each concept to the overall category of *place* and (b) Relative proportion of variables to each category as an example of *a kind of place*

Location and *district* show the highest membership to *place*, while *space* has the lowest membership, with the variation from highest to lowest membership category being 0.28525 and 0.229688 for *location* and *district* respectively, and -0.529603 for *space*. The membership for *neighbourhood* is 0.173064, for *region* it is -0.03123, and for *area* it is -0.131425. It was inferred from these results that the degree of membership to *place* for the relation is-a-kind-of-*place* is shown in descending order to be *location, district, neighbourhood, region, area* and *space*. Reliability analysis showed negative correlation and covariance between the set constituting *region, area and space*, to that including *neighbourhood, district and location*. Two distinct subsets q1 (*neighbourhood, location, district*) and q2 (*region, area, space*) emerged. The reliability analysis on the separate subsets showed a high degree of reliability, with $\alpha q1=0.6918$ and $\alpha q2=0.3516$. When reliability testing for the set q2 was carried out without *space* included in it, only for *region* and *area*, the results show that $\alpha=0.7604$, thereby indicating a high relatedness in the dimensionalities of *region* and *area*. The component plot in Figure 8 demonstrates the inverse nature of dimensionality in *neighbourhood, location and district* to *region* and *area*. The membership nature of *space* was not clearly indicated from the results and although it had a positive dimensionality, it was not high enough either for it to belong distinctly to subsets or for its relation to *place* to be defined clearly.

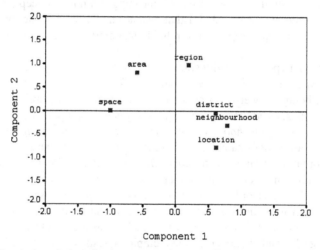

Fig. 8. Proximity relations for the concepts in a two-dimensional component space

A variation between the observed count and expected count was seen, and the residuals showed that the results were not random. Chi-square tests showed that the variation in data was not random (chi-square=130.379, p=0.000, df =20). Significance testing was performed to test the validity of applying the categories in the experimental design and results indicated that there is a significant effect of the categories in determining the overall model and on the overall trends for each concept in the model, p=0.000. The median tests showed a significant variation in the frequencies for the different concepts across the different classes, p=0.000. These results validate the use of category norms for testing taxonomic relations between concepts in the semantic space.

5 Conclusions and Future Directions

The results from the experiments provide meaningful indications for semantic distinctions between concepts and, thereby, corroborate hypothesis H1 that meanings for concepts can be made explicit from human conceptual schemas by treating semantic relatedness as distances. The experimental results indicate that *place* is conceptualised to be semantically closer to *region* than *neighbourhood* is. *Time* was tested for its semantic equivalence to *place* and a negative covariance was shown, thereby indicating that, within commonsense notions, *time* is not related to *place* in meaning or to the other concepts used as control variables in the experiment (i.e., *region, neighbourhood, space, location, area* and *district*). On testing for the measure of semantic relatedness between concepts based on domains introduced within the experimental framework, *place* is once again shown to be more proximate to *region* than to *neighbourhood*, thereby refuting hypothesis H2 that *neighbourhood* is closer in meaning to *place*. Statistical validation showed a low significance for the correlation and proximity vectors obtained from the tests, but the results from MDS and SOM validate hypothesis H2, that cognitive semantics are context-dependent and that domains influence similarity measures and meanings within a conceptual space. The results also corroborate hypothesis H1 that distance and proximity measures using the notion of conceptual spaces provide a reliable technique for semantic explication of concepts.

On testing for hyponym relationships with *place*, using categorical scaling methods, it is shown that *neighbourhood*, along with *location* and *district*, have a higher probability of being a lower-order concept to *place* than *region, area* or *space*. From the results, it is suggested that *region, area* and *space* have a positive dimensionality, as compared to *neighbourhood, location* and *district*. Since *location, district* and *neighbourhood* are better examples of "a kind of *place*," they can be axiomatised as having a hyponym relationship with *place*. On the other hand, *region, area* and *space* are shown to have an inverse dimensionality to these concepts and it can be inferred that they form the higher-order concepts for *place*, with *place* closer to *region* than to *neighbourhood* in the other direction. Therefore, it can be inferred that *region* is a super-type for *place* and that *place* is conceptualised as a sub-set of *regions*. Hence, both hypotheses H1 and H2 are confirmed: empirical tests provide reliable indications for boundaries and ontological distinctions between geographic concepts, to define the semantic variability between them in context-specific conditions, and can be used to define hyponym and hypernym relations for geographic concepts.

The results are indicative of the ontological theory that is embodied in human cognition and representative of the properties and relations specified in the real world for conceptualisation of a *place*. The method of eliciting ontologies from human subject experiments is a reliable method for extracting the hierarchical and categorical relations. As compared to the prototypical theories, theories in conceptual space, by specification of similarity measures from cognitive schemas, provide mechanisms for effectively extracting meanings of concepts. This paper has also demonstrated that conceptual structures and cognitive metrics can be effectively employed for resolving ontological distinctions in geographical concepts and for defining ontological theories in the geographical domain. Theories in conceptual space and cognitive semantics propose that meanings and conceptual associations are context-dependent, based on the saliency of the different domains that act in different contexts. The results indicate that the introduction of domains cause a shift in the conceptual space and mapping of locations of the concepts in it.

The results in this paper have notable implications in defining the ways that *place* and its related concepts can be mapped in conceptual space using the associations and distances that are generated from the cognitive schema. These results provide significant foundations for resolving the semantic heterogeneity in *place* and for definition and development of a reference ontology that can form the basis for inter-operability in *place*-based information systems. The similarity and distance assessment between *place* and semantically overlapping concepts of region, neighbourhood and location shows that such categories can form the basis for *place*-based information systems that are grounded in cognitive conceptualisations as well as the real world, thereby resulting in semantically inter-operable environment for information sharing and user-based applications. Although the experiments described in this paper indicate usefulness of such categories and the hierarchical relations to the notion of *place*, the boundaries of these notions and representational aspects are not clear yet and are being investigated as part of further work. Also, as further work in this direction, the methods developed are being tested and applied to a range of contexts and domains to form a comprehensive set of rules for negotiating meanings of *place* in the geographic context.

Acknowledgments

I am grateful to Prof. Dan Montello for his constructive comments on the experimental framework described in this paper and to Dr. Paul Aplin for proof-reading the initial draft. Thanks are also due to the three anonymous reviewers and to Prof. Max Egenhofer for their feedback that helped improve the quality of this paper.

References

1. Agarwal, P.: Ontological Considerations in GIScience: A Review, *International Journal of Geographical Information Science* (in press).
2. Barsalou, L., Yeh, W., Luka, B., Olseth, K., Mix, K., and Wu, L.: Concepts and meaning. In: K. Beals, G. Cooke, D. Kathman, K. McCullough, S. Kita, and D. Testen (eds.), *Chicago Linguistics Society 29: Papers from the Parasession on Conceptual Representations*. University of Chicago:Linguistics Society (1993) pp. 23-61
3. Battig, W. and Montague, W.: Category Norms for Verbal Items in 56 Categories: A Replication and Extension of the Connecticut Norms, *Journal of Experimental Psychology Monograph Suppl.*, 80(3, Part 2): 1-46 (1968)
4. Bishr, Y.: Semantic Aspects of Interoperable GIS, ITC Publication No. 56, Enschede, NL. (1997)
5. Boyer, C.: Cyber Cities, New York: Princeton Architectural Press (1996)

6. Canter, D.: The Facets of Place, in G. Moore and R. Marans (eds.) *Advances in Environment, Behavior, and Design*. 4. New York. Plenum. (1997) pp. 109-148

7. Christaller, W.: How I Discovered the Theory of Central Places: A Report about the Origin of Central Places. In: P. English and R. Mayfield (eds.), *Man Space and Environment*. Oxford Univ. Press. (1972) pp. 601-610.

8. Duncan, J. and D. Ley, Eds.: *Place/Culture/Representation*. London, England: Routledge. (1993)

9. Felbaum, C.: *WordNet: An Electronic Lexical Database*, MIT Press, Cambridge, MA. (1998)

10. Gardenfors, P.: Conceptual Spaces as a Framework for Cognitive Semantics. In: A. Clark (ed.), *Philosophy and Cognitive Science*. Kluwer, Dordrecht. (1996) pp. 159-180.

11. Gärdenfors, P.: *Conceptual Spaces: The Geometry of Thought*. MIT Press, Cambridge, MA. (2000)

12. Goh, C.: *Representing and Reasoning about Semantic Conflicts in Heterogeneous Information Sources*. PhD Thesis, MIT. (1997)

13. Golledge, R.: Place Recognition and Wayfinding: Making Sense of Space. *Geoforum*, 23(2): 199-214. (1992)

14. Hahn, U. and Chater, N.: Concepts and Similarity. In: K. Lamberts and D. Shanks (eds.), *Knowledge, Concepts and Categories*. Psychology Press, East Sussex. (1997) pp. 43-92.

15. Hardt, D.: Sense and Reference in Dynamic Semantics. In: P. Dekker and M. Stokhof (Eds.) *Proceedings of the Ninth Amsterdam Colloquium*. (1994) pp. 333-348.

16. Harrison, S. and Dourish, P.: Re-place-ing Space: The Roles of Place and Space in Collaborative Systems, In: M. Ackerman (ed.) *Proceedings of Computer Supported Collaborative Work*. ACM, Cambridge, MA. (1996) pp. 67-76.

17. Jordan T, M. Raubal, B. Gartrell and M. Egenhofer: An Affordance-Based Model of Place in GIS,. In: T. Poiker and N. Chrisman (eds.), *Proceedings of the Eighth International Symposium on Spatial Data Handling*, International Geographical Union. (1998) pp. 98-109.

18. Kohenen T.: *Self-Organising Maps*, Second Edition, Springer, Berlin. (1997)

19. Lakoff, G.: 1988. Cognitive Semantics. In: U. Eco, M. Santambrogio, and P. Violi (eds.) *Meaning and Mental Representations*, Bloomington, IN: Indiana University Press. (1988) pp. 119-154

20. Lloyd, R., Patton, D., and Cammack, R.: Basic-Level Geographic Categories, *Professional Geographer*, 48(2): 181-194. (1996)

21. Mark, D., Smith, B. and Tversky, B.: Ontology and Geographic Objects: An Empirical Study of Cognitive Categorization. In: C. Freksa (ed.), *Spatial Information Theory. International Conference COSIT '99*. Lecture Notes in Computer Science 1661 (1999), pp. 283-298.

22. Mark, D., Egenhofer, M., Hirtle, S. and Smith, B.: Ontological Foundations for Geographic Information Science, UCGIS Emerging Research Theme. In: R. McMaster and L. Usery (eds.), Research Challenges in Geographic Information Science. John Wiley and Sons, New York. (in press)

23. Mark, D., Skupin, A. and Smith, B.: Features, Objects, and Other Things: Ontological Distinctions in the Geographic Domain. In: D. Montello (ed.), *Spatial Information Theory*, Lecture Notes in Computer Science 2205. Springer, Berlin, (2001) pp. 488-502.

24. Massey, D.: *Space, Place and Gender*. Cambridge: Polity. (1994)

25. McCullough, D.: Images of America: On the Road with Photographer David Plowden, *Historic Preservation*, July/Aug.1982, 24-33 (1982)

26. Medin, D.L.: Concepts and Conceptual Structure. *American Psychologist*, 44(12): 1469-1481. (1989)
27. Miller, G., Beckwith, R., Fellbaum, C., Gross, D. and Miller, K.: *Introduction to WordNet: An On-line Lexical Database*, ftp://ftp.cogsci.princeton.edu/pub/wordnet/ 5papers.pdf. (1993)
28. Montello, D.: Regions in Geography: Process and Content. In M. Duckham, M. Goodchild, and M. Worboys (Eds.), *Foundations of Geographic Information Science*, London: Taylor & Francis. (2003) pp. 173-189
29. Patwardhan, S., Michelizzi, J. and Pedersen, T.: *WordNet-Similarity*., http://search.cpan.org/dist/WordNet-Similarity/./ (2003)
30. Peacocke, C.: Interrelations: Concepts, Knowledge, Reference and Structure. *Mind and Language*, 19(1): 85-98. (2004)
31. Putnam, H.: The Meaning of "Meaning." In: H. Putnam (ed.), *Mind, Language, and Reality*. Cambridge University Press, London. (1975) pp 215-271
32. Relph, E.: *Place and Placelessness*, Pion Press, London. (1976)
33. Schvaneveldt, R. and Meyer, D.: Retrieval and Comparison Processes in Semantic Memory. In: S. Kornblum (ed.), *Attention and Performance IV*. Academic Press, New York. (1973), pp. 395-409
34. Schvaneveldt, R. and McDonald, J.: Semantic Context and the Encoding of Words: Evidence for Two Modes of Stimulus Analysis. *Journal of Experimental Psychology: Human Perception and Performance*, 7: 673-687. (1981)
35. Smith, B. and Mark, D.: Geographic Categories: An Ontological Investigation *International Journal of Geographical Information Science*, 15(7): 591-612. (2001)
36. Starrs, P.F.: The Importance of Places, or, a Sense of Where You Are. *Spectrum: The Journal of State Governments* 67 (3): 5-17 (1994)
37. Tuan, Y.-F.: *Space and Place: The Perspective of Experience*, University of Minnesota Press. (1990)
38. Tversky, A. and Gati, I.: Similarity, Seperability and the Triangle Inequality. *Psychological Review*, 89(4): 123-154. (1982)
39. Twigger-Ross, C. and D. Uzzell: Place and Identity Processes. *Journal of Environmental Psychology* 16(3):205-220. (1996)
40. UCGIS: Cognition of Geographic Information, http://www.ucgis.org/priorities/research/ research_white/1998%20Papers/cog.html (1998)
41. Vögele, T., Schlieder, C., and Visser, U.: Intuitive Modelling of Place Name Regions for Spatial Information Retrieval, In: W. Kuhn, M. Worboys, and S. Timpf (eds.) *Conference on Spatial Information Theory COSIT'03*, Lecture Notes in Computer Science2825, Springer. (2003) pp. 239-252
42. VT: *The Visual Thesauras Ontline*, trial version, http://www.visualthesaurus.com/_online/index.html. (2003)

Geo-Self-Organizing Map (Geo-SOM) for Building and Exploring Homogeneous Regions

Fernando Bação[1], Victor Lobo[1,2], and Marco Painho[1]

[1] Instituto Superior de Estatística e Gestão de Informação Universidade Nova de Lisboa,
Campus de Campolide 1070-312 Lisboa, Portugal
{bacao,vlobo,painho}@isegi.unl.pt
[2] Academia Naval, Alfeite, 2810-001 Almada, Portugal

Abstract. Regionalization and uniform/homogeneous region building constitutes one of the most longstanding concerns of geographers. In this paper we
explore the Geo-Self-Organizing Map (Geo-SOM) as a tool to develop homogeneous regions and perform geographic pattern detection. The Geo-SOM presents several advantages over other available methods. The possibility of "what-
if" analysis, coupled with powerful visualization tools and the accommodation
of spatial constraints, constitute some of the most relevant features of the Geo-
SOM. In this paper we show the opportunities made available by this tool and
explore different features which allow rich exploratory spatial data analysis.

1 Introduction

Small area census data constitute a major data source in Geographic Information
Science (GISc). The advent of digital census boundaries and the consequent assembly
of Geographic Information Systems (GISs) and census data made available huge
databases of small administrative geographical features characterized by high dimensional vectors of socio-demo-economic information.

This fact created opportunities for developing an improved understanding of a
number of socioeconomic phenomena that are at the heart of GISc. Nevertheless, it
also shaped new challenges and raised unexpected difficulties for the analysis of multivariate spatially referenced data. Today, the availability of methods able to perform
sensible reduction on huge amounts of high dimensional data, is a central issue in
science generically and GISc is no exception.

The urgency of transforming into information the massive digital databases that result from decennial census operations has motivated work in a number of research
areas. Geodemographic typologies (Openshaw and Wymer 1995; Openshaw et al.
1995; Birkin and Clarke 1998; Feng and Flowerdew 1998), identification of deprived
areas (Fahmy et al. 2002), and social services provision (Birkin et al. 1999) constitute
a few examples of subjects where private and public organizations can benefit from
techniques that can isolate important trends and patterns from such large datasets,
although many more can benefit (i.e., research on suburbanization, residential segregation, immigrant settlement patterns, rural depopulation, etc.).

The challenge is to take advantage of new computationally intensive tools made
available in research areas like knowledge discovery (Fayyad et al. 1996; Han and
Kamber 2000; Miller and Han 2001), which are particularly adapted to process large

M.J. Egenhofer, C. Freksa, and H.J. Miller (Eds.): GIScience 2004, LNCS 3234, pp. 22–37, 2004.

quantities of data. Nevertheless, it is fundamental to introduce some kind of spatial reasoning into these methods (Openshaw *et al.* 1995), as spatial data comprises special characteristics (Anselin 1990) and spatial analysis should not ignore some important paradigms (e.g., the 1st Law of Geography (Tobler 1970)).

In this paper we put forward a new tool, based on the Self-Organizing Map (Kohonen 1982; Kohonen 2001), for the development of homogeneous regions and spatial pattern detection. Here the zone design problem is approached as a tool for discovery and spatial data exploration and reduction. The idea is to provide a tool that enables the user to interact and explore spatial data, emphasizing the fuzzy nature of most classifications, allowing for "what-if" analysis and providing rich visualization context for exploratory analysis.

The improvements pursued here lie on the capability of introducing more geographical knowledge within the classification process. The option in this work is to emphasize the importance of the geo prefix in small area analysis. In this paper we develop the Geo-SOM which consists of variants of the original Self-Organizing Map (SOM), and is particularly adapted to process and deal with specific features of spatial data, such as geographic location.

2   Discovery and Exploration Through Zone Design

In the context of GISc the zone design problem constitutes a paradox as it can be seen both as a problem and an opportunity. On one hand the detrimental effects of arbitrary zoning in administrative and statistical area reporting have plagued geographic-based research (Openshaw 1984; Fotheringham and Wong 1991; Amrhein 1995). On the other hand, computational and algorithmic evolution created the opportunity to transform the problem into a valuable exploratory tool in spatial analysis (Openshaw 1984; Wise *et al.* 1997; Guo *et al.* 2003).

Haggett *et al.* (1977) propose three different types of regions which they classified as uniform regions; nodal regions; and planning or programming regions. Usually, planning or programming regions are developed with a specific and well-defined purpose in mind. They tend to result from explicit needs of institutions that have to manage territorial dispersed activities, providing a management perspective on zone design. Examples of this are the problem of electoral districting (Horn 1995; George *et al.* 1997; Mehrotra *et al.* 1998; Macmillan and Pierce 1994), sales territories (Leischmann and Paraschis 1998), police reporting areas (Sarac *et al.* 1999), and census output Areas (Martin 1997; Martin 1998). Uniform and nodal regions can be viewed as having a fairly exploratory nature, as they usually assist research and discovery processes. Although it is difficult to produce a clear-cut differentiation between zone design as a management tool and zone design as a discovery tool, some distinctions can be made.

Typically, zone design as a management tool assumes restrictive constraints on the geographic configuration of the resulting regions. Thus, the contiguity constraint is usually present and compactness is generally considered a desirable characteristic. The rationale behind the construction of the algorithms is to produce highly efficient procedures that make use of computational resources and improved optimization techniques to eventually arriving at global optimum solutions. An optimality criterion is previously defined and needs to be encoded into the optimization procedure.

In the case of zone design as a discovery and exploratory tool the aim is more focused on evaluating different possibilities, probably using fewer and less restrictive constraints on the geographic configuration of the resulting zonal systems. Here the main objective is to, "let data speak for themselves" (Gould 1981). This way, strict geographic constraints, such as contiguity, and the need to define in advance the number of regions can be interpreted as restrictive factors that might obscure the identification of interesting patterns. As a discovery tool, regionalization can be seen as a pre-processing method, which is used to generate the basic units for subsequent analysis (Haining *et al.* 1994). Additionally it can be used to detect particular areas with specific characteristics.

Logically, the differences expressed above have a major impact on the specifications of the algorithms developed to deal with zone design. Exploratory zone design algorithms should allow for user interaction, because the objectives are somewhat fuzzy, and an adequate human/system interaction can guide an otherwise "black box" clustering process (Guo *et al.* 2003). The critical analysis of the results, which can be provided by skilled analysts, can be a valuable contribution, and algorithms should provide adequate means for result interpretation and diagnosis. Additionally, the possibility of "what-if" analysis can be a helpful addition in the sense that it allows probing based on prior knowledge and prompt evaluation between different design options. Finally, visualization capabilities of the algorithms should take advantage of GIS technology and, if possible, push forward analysis based on visualization tools.

3   Characteristics of Typologies Based on Small Area Census Data

We briefly review some of the major characteristics of small area census data typologies. These characteristics constitute the motivation for the development of the Geo-SOM.

3.1   The Fuzziness Associated with Small Area Typologies

According to Feng and Flowerdew (1998) there are two different types of fuzziness in typologies developed from small area census data: one related with the attribute space, the other associated with the geographical space. The first kind of fuzziness is due to the "all or nothing nature of the classification assignment" (Openshaw *et al.* 1995), which conceals the fact that enumeration districts (EDs) may be close to more than one neighbourhood type in the attribute space.

The geographical fuzziness comes from the arbitrary nature of the EDs, which serve as the elementary units on which typologies are based, the well known problem of the modifiable areal unit problem (Openshaw 1984). In fact, the nature of these units reflects the operational needs that steer data collection, and for that matter should not be assumed to have any type of homogeneity in terms of socioeconomic characteristics (Morphet 1993). Although near things are more related than distant things (Tobler 1970), the problem lies with the fact that the scale on which this empirical regularity can be observed does not have to coincide with the scale represented by the EDs. The neighbourhood effects might be expressed at different scales in different areas of the study region and only fortuitously (probably only miraculously) will coincide with the scale denoted by the EDs.

3.2 Resolution and Precision Issues

Openshaw *et al.* (1995) point out problems related with the variability in the size of EDs, which influence the precision and resolution of data. This variation is not random and often reflects the duality of urban-rural areas. In urban areas, EDs tend to be more densely populated, which results in mixed characteristics but with accurate values. On the other hand, rural EDs tend to be more homogeneous, but also present more extreme results due to their sizes. Conventional classifiers give equal importance to each ED, this will result in a better representation of extreme results and a poor representation of more mixed ED, this way the least reliable results will benefit from a better representation (Openshaw *et al.* 1995).

Another relevant issue that needs to be pointed out is the reliability and statistical significance of relations found at the ED level. For instance, when linking health data and census data it is fundamental that areas have large enough populations to ensure that rates are reliable; additionally, these areas should be homogeneous with respect to relevant socioeconomic attributes (Haining *et al.* 1994).

3.3 The Relevance of Providing Geographical Context

A lot of work in this area is related to the field of geodemographics, where the main focus has been on developing highly efficient variance optimizers, and for this reason little if any attention has been given to the geographical context of the data. In fact, it is clear that the inclusion of contiguity restrictions (Openshaw and Wymer 1995) or geographic references (Lobo *et al.* 2004) will increase the variance of the resulting clusters. Nevertheless the question is: will this variance reduction frenzy improve typologies? Probably no one can answer this question. Nevertheless, we argue that it is geographically coherent to provide a geographical framework in small area based typologies. This is especially pertinent in the light of the characterization provided above, which points to all sorts of fuzziness and uncertainty when dealing with small area census data. The geographical framework can allow the identification of regularities, the detection of unusual EDs, and the discovery of boundaries where major shifts in the phenomena under study occur. Within its geographic context, the possibility of sorting out and understanding the fuzziness in data is much higher. Clearly the assumption here is that the studied phenomena happens at a smaller scale than the scale that characterizes the EDs.

4 Self-Organizing Maps (SOM)

Although the term "Self-Organizing Map" could be applied to a number of different approaches, we use it as a synonym of Kohonen's Self Organizing Map, or SOM for short. The basic idea of a SOM is to map the data patterns onto an n-dimensional grid of neurons or units. That grid forms what is known as the output space, as opposed to the input space, which is the original space where the data patterns are. This mapping tries to preserve topological relations (i.e., patterns that are close in the input space will be mapped to units that are close in the output space, and vice-versa). The SOM algorithm for training a 2-dimensional map may be defined as follows:

Let

    X be the set of n training patterns $x1, x2, ..xn$

    W be a $p \times q$ grid of units wij where i and j are their coordinates on that grid

    α be the learning rate, assuming values in $]0,1[$, initialized to a given initial learning rate

    r be the radius of the neighborhood function $h(wij,wmn,r)$, initialized to a given initial radius

1 Repeat

2   For k=1 to n

3   For all $wij \in W$, calculate $dij = || xk - wij ||$

4   Select the unit that minimizes dij as the winner $wwinner$

5   Update each unit $wij \in W$: $wij = wij + \alpha\, h(wwinner, wij, r) \; ||xk-wij||$

6   Decrease the value of α and r

7 Until α reaches 0

To visualize the results of a SOM, we may use U-Matrices (Ultsch and Siemon 1990). This is a representation of a SOM in which distances, in the input space, between neighbouring neurons are represented, usually by a colour code. If distances between neighbouring neurons are small, then these neurons represent a cluster of patterns with similar characteristics. If the neurons are far apart, then they are located in a zone of the input space that has few patterns and can be seen as a separation between clusters. Thus, a visual inspection of the U-Matrices allows the user to identify different clusters of data with variable "similarity resolution."

For a thorough review the reader is referred to Kohonen (2001). SOMs have been used in many different areas, and in geographical problems they have been used to perform nonlinear mappings (Skupin 2003), clustering (Openshaw *et al.* 1995; Painho and Bação 2000), and in localization problems (Gomes *et al.* 2004; Hsieh and Tien 2004). Most of these applications focus either on the geographical coordinates or on the other features. However, to address the issues of homogeneous and geographically coherent region design, it is necessary to give special attention to geographical coordinates and at the same time use the non-geographical features for clustering. We identified two distinct ways of doing so (Figure 1). The first consists of including the geographical coordinates in the pattern vector, and giving them increasing importance (Lobo *et al.* 2004). The second is a new SOM architecture, which we named Geo-SOM, and in this paper we discuss its relation with other well known architectures.

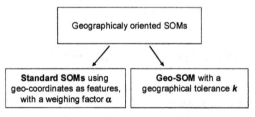

Fig. 1. Types of geographically oriented SOMs

4.1 Geographically Oriented SOMs

In the SOM training algorithm, the most important step in establishing which patterns are clustered together is the one where we choose the Best Matching Unit (BMU). In the basic SOM this is done by comparing all components of the input pattern with all components of each unit (and normally calculating the distance between these two vectors). By changing the way the BMU is selected, we can give greater importance to the geographical coordinates.

One way of doing this is simply to include these coordinates in the pattern vector, and scaling them by a parameter α. If $\alpha=0$ the geographical coordinates have no importance whatsoever, whereas if $\alpha=\infty$, only these are relevant. In this latter case, the BMU will always be the unit geographically closer, and the update process will simply calculate local averages of the other parameters. By observing the way in which these local averages differ (normally with a U-Matrix) we may establish regions with the desired regularity.

The same basic approach of including the geographical coordinates in the data pattern can be used with most other SOM variants, such as Hierarchical SOMs (Ichiki *et al.* 1991; Behme *et al.* 1993) or ASSOM (Kohonen 2001).

4.2 Geo-SOM

Another way of forcing the BMU to be in the geographical vicinity of the input pattern is to explicitly divide the search for the BMU in two phases: first establish a geographical vicinity where it is admissible to search for the BMU, and then perform the final search using the other components. The vicinity where we search for the BMU can be controlled by a parameter k, defined in the output space[1]. If we choose k=0, then the BMU will necessarily be the unit geographically closer. If we allow k to grow up to the size of the map then we will ignore the geographical coordinates altogether.

When k=0, the final locations in the input space of the units will be a quasi-proportional representation of the geographical locations of the training patterns (for a discussion on the proportionality between units and training patterns see (Cottrell *et al.* 1998)), and thus the units will have local averages of the training vectors. Exactly the same final result may be obtained by training a standard SOM with only the geographical locations, and then using each unit as a low pass filter of the non-geographic features. The exact transfer function (or kernel function) of these filters depends on the training parameters of the SOM, and is not relevant for this discussion.

As k (the geographic tolerance) increases, the unit locations will no longer be quasi-proportional to the locations of the training patterns, and the "equivalent filter" functions of the units will become more and more skewed, eventually ceasing to be useful as models.

---

[1] The geographical tolerance k could be defined in the input space. This would lead to a fixed geographical radius where clustering would be allowed to occur. Choosing k in the output is preferable since it allows a finer resolution in areas with greater pattern density and a coarser resolution in the rest of the space.

Formally, the Geo-SOM may be described by the following algorithm:

```
Let
            X be the set of n training patterns x1,
            x2,..xn, each of these having a set of
            components geoᵢ and another set ngfᵢ.

            W be a p×q grid of units wij where i and j are
            their coordinates on that grid, and each of
            these units having a set of components wgeoᵢⱼ
            and another set wngfᵢⱼ.

            α be the learning rate, assuming values in
            ]0,1[, initialized to a given initial
            learning rate
            r be the radius of the neighborhood function
            h(wij,wmn,r), initialized to a given initial
            radius
            k be the radius of the geographical BMU that
            is to be searched
            f be a logical variable that is true if the
            units are at fixed geographical locations.
1 Repeat
2           For m=1 to n
3               For all wij∈W,
4                   Calculate dij = ||wgeok - wgeoij||
5                   Select the unit that minimizes dij as the
                      geo-winner wwinnergeo
6                   Select a set Wwinner of wij such that the
                        distance in the grid between wwinnergeo and
                        wij is smaller or equal to k.
7                   For all wij∈Wwinner,
8                       calculate dij = ||xk - wij||
9                       Select the unit that minimizes dij as the
                          winner wwinner
10                  If f is true, then
11                      Update each unit wij∈W: wngfij = wngfij +
                          α h(wngfwinner,wngfij,r) ||xk - wij||
12                  Else
13                      Update each unit wij∈W: wij = wij +
                          α h(wwinner,wij,r) ||xk - wij||
14          Decrease the value of α and r
15 Until α reaches 0
```

4.3 Comparison Between Standard SOM and Geo-SOM

In both approaches (Standard and Geo-SOM), geographical coordinates can be the only relevant feature (when $\alpha=\infty$ or $k=0$), or they may be irrelevant (when $\alpha=0$ or k=maximum size of map), and thus in both limits the approaches produce the same result (Figure 2). In practical terms it is easier and more efficient to use the Geo-SOM when geographical coordinates are very important, and the standard SOM otherwise.

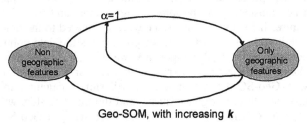

Fig. 2. Different models for standard SOM and Geo-SOM

The Geo-SOM architecture is closely related to the hypermap architecture (Kohonen 1991), the spatio-temporal SOMs (Chandrasekaran and Palaniswami 1995; Chandrasekaran and Liu 1998; Euliano and Principe 1996; Euliano and Principe 1999), and in the kangas map (Kangas 1992) adapted to spatial coordinates (Lobo *et al.* 2004). However, a thorough discussion of these relationships is outside the scope of this paper.

Finally, when using the Geo-SOM, we may include the geographical coordinates in the final search for the BMU, thus obtaining a continuum of models between the pure Geo-SOM and the pure standard SOM. For the sake of simplicity we call all these models Geo-SOM, and in our experimental tests we used the geographical coordinates in the final search for the BMU, with $\alpha=1$.

5   Some Experimental Results

In order to test the Geo-SOM, we used two datasets. One of them is a very simple artificial problem with only one non-geographic feature. It was used to understand some basic properties of the Geo-SOM. The other dataset refers to EDs of the Lisbon Metropolitan Area and includes 3,968 EDs, which are characterized based on 65 variables. The Geo-SOM was implemented in Matlab® compatible with Somtoolbox (Vesanto *et al.* 2000), and is available at http://www.isegi.unl.pt/docentes/vlobo/projectos/programas.

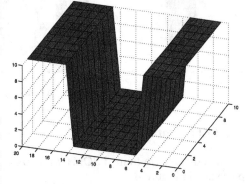

Fig. 3. The artificial dataset

5.1   The Artificial Dataset

For this example we used a set of 200 data points evenly spaced on a surface with coordinates $x \in [0,1]$, $y \in [0,2]$. Each point is associated with a single feature z, which is 0 whenever $0.5 < y < 1.5$ and 10 otherwise (Figure 3).

If we cluster the data based on non-geographical features, then we will have two very well defined clusters: one where z=10, another where z=0 (Figure 4). If we consider only geographical coordinates, then we will have no well defined clusters, since the points are evenly spaced. If we consider all three components, we may or may not obtain well defined clusters. If no pre-processing is done, and since in this case the geographical features have a very small scale when compared to the other feature, we will basically obtain only two clusters. If we pre-process the data points to have approximately the same scale in all components, we will obtain rather fuzzy clusters. Depending on the different scalings, we may obtain 1, 2, or 3 clusters, but never clear-cut separations. A Geo-SOM with 0-tolerance will simply calculate local averages, and thus will just smooth the original dataset, and the three clusters will still appear clearly in the U-matrix. The best results are obtained using a Geo-SOM with k=2 (Figure 5). It is interesting to note that a 0-tolerance in the Geo-SOM produces blurred clusters, while relaxing this constraint will allow the clusters to define themselves better without loosing their geographic localization.

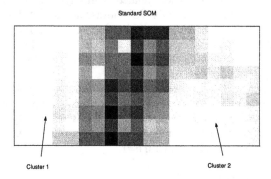

Fig. 4. U-matrix obtained for the artificial dataset with the standard SOM

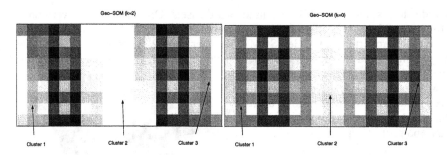

Fig. 5. U-Matrix obtained for the artificial dataset with Geo-SOMs

5.2 Lisbon Dataset

For this dataset we trained SOMs with 20x30 units, and the U-matrices obtained are presented in Figure 6.

The connection between the U-Matrices and the geographic map is a key issue in using SOM-based methods. The objective is to provide an interactive exploration environment that interconnects both spaces. At this time, this interaction is provided by ArcView, where the U-Matrix is geocoded, and linked to the geographic map. This way the selection of a unit on the U-Matrix automatically highlights the geographical areas that are classified in that specific unit. Through this mechanism, one can analyse the U-Matrix, define clusters in the data and, by selecting them in the U-Matrix, automatically get a "picture" of their geographic location.

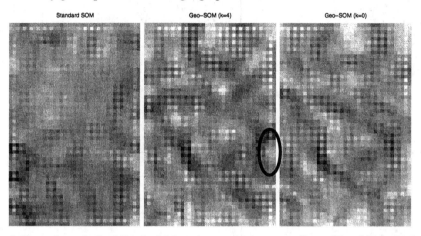

Fig. 6. U-matrices obtained for the Lisbon dataset. In the center figure (Geo-SOM with k=4) one of the clearly separated clusters, which we later use as example, is marked with an ellipse

A rough analysis of these U-matrices allows us to see that the standard SOM clusters most units in a single cluster (top half of the map), and separates, although not very clearly, a few clusters in the bottom half. An analysis of the EDs mapped onto these clusters shows, as expected, that while these EDs do in fact have similar characteristics, they are not geographically close. The Geo-SOMs lead to a very different clustering.

We would like to emphasize the exploration possibilities provided by the Geo-SOM. After training the Geo-SOM the units were georeferenced in Lisbon's map. The first aspect that is important to highlight is related with the geographic distribution of the units of the Geo-SOM. As can be seen in Figure 7 there is an important difference between the distribution of the units in the standard SOM and the Geo-SOM. In the Geo-SOM (with k=0 and 4) the units are geographically spread, mimicking the density of the centroids of the EDs. This way, the more densely populated areas will receive more classifying resources (i.e., more units), but sparsely populated areas will also receive some resources. It is quite clear from the analysis of Figure 7 that the unfolding of the Geo-SOM is ruled by the spatial distribution of the EDs. As would be expected the standard SOM distributes its units in the centre of the region, as it is there that the global variance can be minimized.

Once the units are geocoded, the next step consists on defining to which unit is each ED centroid associated. This was done by generating Thiessen polygons based on the units and assigning each ED centroid to the nearest unit. Figure 8 shows the assigning process. In this particular case Lisbon's downtown area is depicted. As it can be seen several Geo-SOM units are placed in the river, due to the fact that it separates two high-density areas. This is natural as the Geo-SOM produces a surface for the whole of the study area.

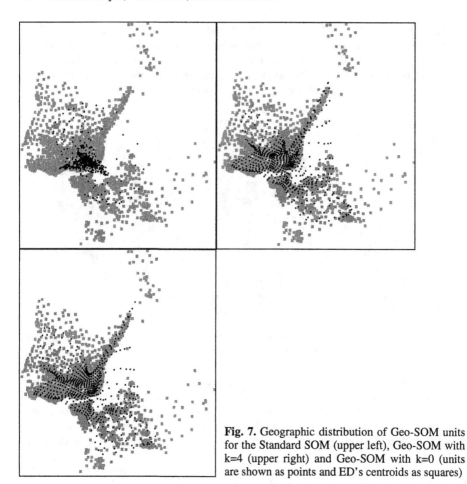

Fig. 7. Geographic distribution of Geo-SOM units for the Standard SOM (upper left), Geo-SOM with k=4 (upper right) and Geo-SOM with k=0 (units are shown as points and ED's centroids as squares)

Building homogeneous regions based on the Geo-SOM can be done in a number of ways. One of them is to define thresholds based on which homogeneous regions will be built. In Figure 9, three different thresholds are tested and as the threshold grows the same happens to the number of homogeneous regions.

This type of visualization is in effect a visualization of the U-Matrix in the geographic sub-space of the input space, as opposed to the traditional visualizations of that matrix in the output space (Figures 4-6). The mapping of clusters identified in the traditional visualization of the U-Matrix to this geographical representation, although not linear, is quite simple. As an example, the cluster detected in the U-Matrix of Figure 6 can be readily identified in Lisbon's geographical map (Figure 10).

Similarly, if a surface is built based on the similarities of each unit and its neighbours, an elevation model can be developed (Figure 11). A progressive flooding of this surface will indicate which areas should be aggregated first. Additionally, the ridges indicate areas of change, where EDs present important dissimilarities. For instance, areas, such as the Lisbon International Airport and Monsanto (a big green area with no housing within the city limits), constitute obvious transition areas, which are well depicted by the Geo-SOM.

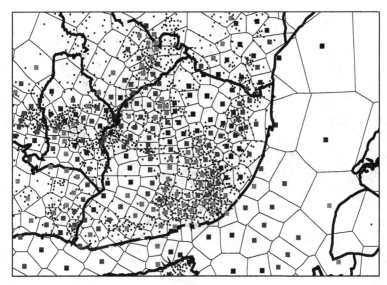

Fig. 8. Geographic distribution of Geo-SOM units and the EDs that are mapped to them in the center of Lisbon (units are shown as squares and EDs centroids as points)

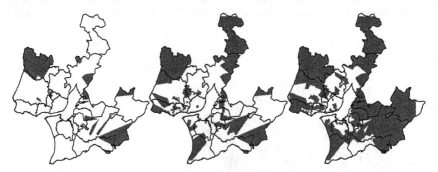

Fig. 9. Identification of homogeneous regions (darker areas represent homogeneous regions). The leftmost figure was obtained using a low threshold (forcing very homogeneous regions), and the other two were obtained using increasing thresholds

6 Conclusions

An overview of different techniques for using SOMs as tools for designing homogeneous geographical regions and pattern detection was presented, and a new SOM-based architecture, named Geo-SOM, was proposed. It was shown, both in an artificial problem and in a real problem, that this new architecture can provide better insights for the region design problem. One of the advantages of using the Geo-SOM is related with its exploratory nature. Various ways of exploring the information provided by the Geo-SOM were explained. Finally, it was shown that this new architecture provides a meaningful clustering of the Lisbon Metropolitan area given a large set of census data. The idea of creating a bridge between the geographic space and the

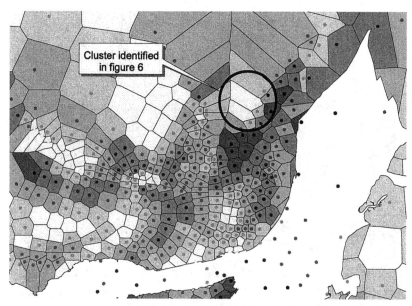

Fig. 10. Map identification of the cluster detected in figure 6 U-Matrix, here with a set of five similar units (light shade) near a well defined "boundary" of very dissimilar units (dark shade)

Fig. 11. Elevation map where ridges represent transition areas

feature space is particularly appealing, as it allows processing features and subsequent visualization with geographical context. The result is a partition of space, which is primarily ruled by the density of geographic occupation and secondly by the similarity of the patterns. Further work needs to be done in two major areas. The first one is related with extending the exploratory tools provided by the Geo-SOM. The second one is rather theoretical and regards the relations between the Geo-SOM and geographical concepts like spatial autocorrelation and spatial heterogeneity.

References

Amrhein, C. G. (1995). Searching for the elusive aggregation effect: evidence from statistical simulation. *Environment and Planning A* 27: 105-120.

Anselin, L. (1990). What is special about spatial data? Alternative perspectives on spatial data analysis. In D. Griffith (ed.) *Spatial Statistics, Past, Present and Future*. Ann Arbor, MI, Institute of Mathematical Geography: 63-77.

Behme, H., W. Brandt and H. Strube. (1993). Speech Recognition by Hierarchical Segment Classification. In: S. Gielen and B. Kappen (eds), *Proceedings of the International Conference on Artificial Neural Networks*, London, UK. 416-419.

Birkin, M. and G. Clarke. (1998). GIS, geodemographics and spatial modeling in the UK financial service industry. *Journal of Housing Research* 9: 87-111.

Birkin, M., G. Clarke and M. Clark. (1999). GIS for Business and Service Planning. In M. Goodchild, P. Longley, D. Maguire and D. Rhind (eds.) *Geographical Information Systems*. Cambridge, Geoinformation.

Chandrasekaran, V. and Z.-Q. Liu. (1998). Topology Constraint Free Fuzzy Gated Neural Networks for Pattern Recognition. *IEEE Transactions on Neural Networks* 9(3): 483-502.

Chandrasekaran, V. and M. Palaniswami. (1995). Spatio-Temporal Feature Maps using Gated Neuronal Architecture. *IEEE Transactions on Neural Networks* 6(5): 1119-1131.

Cottrell, M., J. Fort and G. Pagès. (1998). Theoretical Aspects of the SOM Algorithm. *Neurocomputing* 21(1): 119-138.

Euliano, N. and J. Principe. (1996). Spatio-Temporal Self-Organizing Feature Maps. *Proceedings of the International Conference on Neural Networks*, Washington, DC. 1900-1905.

Euliano, N. and J. Principe. (1999). A Spatio-Temporal Memory Based on SOMs with Activity Diffusion. In E. Oja and S. Kaski (eds.) *Kohonen Maps*. Amsterdam, Elsevier: 253-266.

Fahmy, E., D. Gordon and S. Cemlyn. (2002). Poverty and Neighbourhood Renewal in West Cornwall. *Social Policy Association Annual Conference*, Nottingham, UK. http://www.bris.ac.uk/poverty/cornw/02SPA.doc

Fayyad, U., G. Piatetsky-Shapiro, P. Smyth and R. Uthurusamy. (1996). *Advances in Knowledge Discovery and Data Mining*. AAAI/MIT Press, Cambridge, MA.

Feng, Z. and R. Flowerdew. (1998). Fuzzy Geodemographics: A Contribution from Fuzzy Clustering Methods. In S. Carver (ed.) *Innovations in GIS* 5. London, Taylor & Francis: 119-127.

Fotheringham, A. S. and D. Wong. (1991). The Modifiable Areal Unit Problem in Multivariate Statistical Analysis. *Environment and Planning A* 23(7): 1025-1044.

George, J., B. Lamar and C. Wallace. (1997). Political District Determination Using Large-Scale Optimization. *Socio-Economic Planning Sciences* 31(1): 11-28.

Gomes, H., V. Lobo and A. Ribeiro. (2004). Application of Clustering Methods for Optimizing the Location of Treated Wood Remediation Units. *XI Jornadas de Classificação e Análise de Dados*, Lisbon, Portugal.

Gould, P. (1981). Let Data Speak for Themselves. *Annals of the Association of American Geographers* 71: 166-176.

Guo, D., D. Peuquet and M. Gahegan. (2003). ICEAGE: Interactive Clustering and Exploration of Large and High-Dimensional Geodata. *GeoInformatica*, 7(3): 229-253.

Haggett, P., A. D. Cliff and A. E. Frey. (1977). *Locational Analysis in Human Geography*. Second Edition. London, Arnold.

Haining, R., S. Wise and M. Blake. (1994). Constructing Regions for Small Area Analysis: Material Deprivation and Colorectal Cancer. *Journal of Public Health Medicine* 16(4): 429-438.

Han, J. and M. Kamber (2000). *Data Mining: Concepts and Techniques*. New York, Morgan Kaufmann.

Horn, M. (1995). Solution Techniques for Large Regional Partitioning Problems. *Geographical Analysis* 27(3): 230-248.

Hsieh, K.-H. and F. Tien. (2004). Self-Organizing Feature Maps for Solving Location–Allocation Problems with Rectilinear Distances. *Computers and Operations Research* 31(7): 1017-1031.

Ichiki, H., M. Hagiwara and N. Nakagawa. (1991). Self-Organizing Multi-Layer Semantic Maps. *Proceedings of International Conference on Neural Networks*, Seattle, WA. 357-360.

Kangas, J. (1992). Temporal Knowledge in Locations of Activations in a Self-Organizing Map. *Proceedings of the International Conference on Artificial Neural Networks*, Brighton, England. 117-120.

Kohonen, T. (1982). Clustering, Taxonomy, and Topological Maps of Patterns. *Proceedings of the 6th International Conference on Pattern Recognition*, Munich, Germany. 114-128.

Kohonen, T. (1991). The Hypermap Architecture. In T. Kohonen, K. Mäkisara, O. Simula and J. Kangas (eds.) *Artificial Neural Networks*. 1: Helsinki, Elsevier. 1357-1360.

Kohonen, T. (2001). *Self-Organizing Maps*, Springer.

Leischmann, B. and J. Paraschis. (1998). Solving a Large Scale Districting Problem: A Case Report. *Computers and Operations Research* 15(6): 521-533.

Lobo, V., F. Bação and M. Painho. (2004). Regionalization and Homogeneous Region Building Using the Spatial Kangas Map. In F. Toppen and P. Prastacos (eds.) *7th AGILE Conference on Geographic Information Science*, Heraklion, Greece. 301-313.

Macmillan, W. and T. Pierce. (1994). Optimization Modelling in a GIS Framework: The Problem of Political Redistricting. In A. S. Fotheringham and P. Rogerson (eds.) *Spatial Analysis and GIS*. London, Taylor & Francis. 221-246.

Martin, D. (1997). From Enumeration Districts to Output Areas: Experiments in the Automated Design of a Census Output Geography. *Population Trends* 88: 36-42.

Martin, D. (1998). Optimizing Census Geography: The Separation of Collection and Output Geographies. *International Journal of Geographical Information Science* 12(7): 673-685.

Mehrotra, A., E. Johnson and G. Nemhauser. (1998). An Optimization Based Heuristic for Political Districting. *Management Science* 44(8): 1100-1114.

Miller, H. and J. Han. (2001). *Geographic Data Mining and Knowledge Discovery*. London, UK, Taylor & Francis.

Morphet, C. (1993). The Mapping of Small Area Census Data–A Consideration of the Effects of Enumeration District Boundaries. *Environment and Planning A* 25(9): 1267-1277.

Openshaw, S. (1984). The Modifiable Areal Unit problem. CATMOG, Geo-abstracts. Norwich, UK.

Openshaw, S., M. Blake and C. Wymer. (1995). Using Neurocomputing Methods to Classify Britain's Residential Areas. In P. Fisher (ed.) *Innovations in GIS*. Taylor & Francis. 2: 97-111.

Openshaw, S. and C. Wymer. (1995). Classifying and Regionalizing Census Data. In S. Openshaw (ed.) *Census Users Handbook*. Cambridge, UK, GeoInformation International: 239-268.

Painho, M. and F. Bação. (2000). Using Genetic Algorithms in Clustering Problems. *Proceedings of the 5th International Conference on GeoComputation*. University of Greenwich, UK. (CD-ROM)

Sarac, A., R. Batta, J. Bhadury and C. Rump. (1999). Reconfiguring Police Reporting Districts in the City of Buffalo." *OR Insight* 12(3): 16-24.

Skupin, A. (2003). A Novel Map Projection Using an Artificial Neural Network. *21st International Cartographic Conference*, Durban, South Africa. 1165-1172. (CD-ROM)

Tobler, W. (1970). A Computer Model Simulating Urban Growth in the Detroit Region. *Economic Geography* 46(2): 234-240.

Ultsch, A. and H. Siemon. (1990). Kohonen´s Self-Organizing Feature Maps for Exploratory Data Analysis. *Proceedings of the International Neural Network Conference*, Dordrecht, Netherlands. Kluwer. 305-308.

Vesanto, J., J. Himberg, J. Alhoniemi and J. Parhankangas. (2000). *SOM Toolbox for Matlab 5*. Espoo, Helsinki University of Technology: 59.

Wise, S., R. Haining and J. Ma. (1997). Regionalisation Tools for the Exploratory Spatial Analysis of Health Data. In M. Fisher and A. Getis (eds.) *Recent Developments in Spatial Analysis–Spatial Statistics, Behavioural Modelling and Neurocomputing*. Berlin, Springer. 83-100.

Can Relative Adjacency Contribute to Space Syntax in the Search for a Structural Logic of the City?

Roderic Béra and Christophe Claramunt

Naval Academy Research Institute, Lanvéoc-Poulmic,
BP 600, 29240 Brest Naval, France
{bera,claramunt}@ecole-navale.fr

Abstract. Although network geography has long been recognised as a valid method for exploring geographical information systems, there is a renewed interest in applying its principles to the observation and analysis of urban systems. Over the past several years, space syntax has emerged as a new way of analysing the social, economic and environmental functioning of the city based on a graph computational representation. This paper introduces an analysis of the potential of a relative adjacency operator in comparison with current measures of connectivity and distance used in space syntax studies. We analyse how space syntax evaluates complexity and patterns in the city and show that the relative adjacency can provide a valuable complement to those measures. The study is illustrated by an application to the reference case of the village of Gassin in France.

1 Introduction

Geographical Information Science still requires computational models that take into account the complex interactions, relations, and degrees of connectivity between things in space and time. An essential measure of the organisation of a complex geographical system relies on the way its different elements are inter-related and connected. These inter-relationships form networks whose characteristics can be explored to study local properties, clusters and different patterns of spatial organisation. The science of networks, although long established as a fundamental component of geographical analysis [8, 15], has been recently considered as still offering many new opportunities for the modelling of biological, internet and spatial structures [1, 6, 21]. In particular, spatial networks offer alternative modelling representations to many situations where Euclidean spaces are not appropriate. This is one of the modelling principles of space syntax research, where the objective of the study is to treat the built environment and urban networks as systems of space, and to analyse their configurations in order to understand the possible relationships with the structure and the social, economic and environmental functioning of the city [11].

Space syntax treats an urban network by modelling streets as nodes (the usual approach is to consider lines of sight as the elementary modelling units), and intersections between routes as edges of the equivalent graph. Space syntax provides different degrees of integration and connectivity measures that characterise the structural

M.J. Egenhofer, C. Freksa, and H.J. Miller (Eds.): GIScience 2004, LNCS 3234, pp. 38–50, 2004.
© Springer-Verlag Berlin Heidelberg 2004

properties of each network component in the graph [10, 17]. These extrinsic measures attempt to characterise each element in terms of how it relates to the system as a whole from its particular position in the graph, and to analyse the distribution of these relationships throughout the graph. This differs from intrinsic properties and binary relationships, where one object is related to another (e.g., topological, metric relationships). However, and despite its successful application to many urban studies, some interpretation issues are unresolved by space syntax. In particular, there is still a need for some extrinsic measures that come closer to the way we intuit the relative location of a given node within a graph, and for measures of accessibility that better relate to the form and the structure of the city. In particular, accessibility is influenced by heterogeneity in locational and topological distributions as well as truncating edge effects [20]. The objective of this paper is to study to what extent a measure of relative adjacency, introduced in a related work [3], provides a measure of contextual distance in urban networks and complements extrinsic measures used in space syntax studies.

The relative adjacency evaluates how each element of a graph is related to the other while taking into account the topological heterogeneity and structural properties of the represented system. In other words, that measure evaluates the probability of reaching one node from another assuming some random displacements within the graph. This gives an asymmetric concept of proximity, a property that is often considered as relevant for non-Euclidean spaces [22]. The measure is flexible as it takes into account different distance decay factors within the graph by maximising and minimising the proximity of close and remote nodes.

The reminder of the paper is organised as follows. Section 2 describes space syntax principles, local, global and contextual measures of distances in a graph. Section 3 introduces the relative adjacency. Section 4 applies these measures to the reference case of the village of Gassin, and illustrates the respective advantages and limitations of space syntax measures compared to the relative adjacency. Finally, Section 5 provides some concluding comments.

2 Space Syntax Principles

Space syntax can be considered as a form of study of geometrical accessibility where a functional relationship is made, or attempted, between the structure of the city and the social, economical and environmental dimensions. A lot of space syntax empirical studies have been devoted to the exploration of the complexity and configuration of urban space by searching for a computational representation of a form of geometric order. For instance, it has been found that highly integrated streets usually attract more people than segregated ones [10]. The relationship between the structure of the city and criminality [14], way-finding [12, 18], and pollution [16] are other examples of empirical studies where space syntax has been successfully applied.

Space syntax is intimately linked to graph theory. Taking a broader view, a graph can represent a system of relationships where the nodes are the elements, and the edges the relationships between those elements. In space syntax, the elements of interest are the streets of a given urban network, and the edges the intersections between those streets. Streets present the particularity to be defined in terms of units of uninterrupted movement and visibility. As a given street may be connected to more than two other streets, the resulting graph is not planar. Variables currently used in space syntax studies range from local, shortest path to contextual measures.

A graph is denoted as a pair $G(V,E)$ where V is a finite set of nodes (*i.e.*, the streets) and E a finite set of edges (*i.e.*, connections between those streets) defined as a subset of the Cartesian product $V \times V$. Connectivity and clustering coefficient parameters evaluate the local properties of a given urban network in terms of integration or segregation. An immediate measure of connectivity is defined as

$$r_{ij} = \begin{cases} 1 & \text{if } (i,j) \in E \\ 0 & \text{otherwise} \end{cases}$$

Another local measure evaluates the extent to which the neighbours of a node i are also linked one to another. The *clustering coefficient* of a given node i is defined by

$$C(i) = \frac{2 \cdot l_i}{m_i(m_i - 1)} \tag{1}$$

where m_i denotes the number of immediate neighbours of the node i (*i.e.*, the degree of i) and l_i the number of edges among the immediate neighbours of the node i. $C(i)$ is drawn from the unit interval $[0,1]$. The closer to 1 its *clustering coefficient* is, the more clustered is the node.

Several shortest path-based measures have been applied to space syntax studies. They are based on some variations of the average of shortest paths to all other nodes in the graph. Amongst many variants, the *average path length*, *total depth* and *closeness centrality* are some examples of comparable measures often practiced. Given two nodes $i, j \in V$, let $d_{\min}(i,j)$ be the shortest path between these two nodes. The *average path length* of a node i is given by

$$L(i) = \frac{1}{n} \sum_{j=1}^{n} d_{\min}(i,j) \tag{2}$$

where n is the total number of nodes of the graph G. The *total depth* of a node i is given by

$$MD_i = \frac{\sum_{j=1}^{n} d_{\min}(i,j)}{n-1} \tag{3}$$

A related measure of *closeness centrality* $Cc(i)$ [19] is given by

$$C_c(i) = \frac{1}{\sum_{j=1}^{n} d_{\min}(i,j)} \tag{4}$$

Average path length, *total depth* and *closeness centrality* are based on some aggregation of distance measures. They give a sense of integration versus segregation of a given node in the graph, that is, a topological evaluation of its accessibility within the graph. Measures of shortest path provide an approximation of distances although they do not take completely into account the context and structure of the graph. Considering the structure of the graph should lead not only to consider shortest paths, but also the role played by a given node in those shortest paths. An example of such a measure is given by the *betweenness centrality* which evaluates the ratio of shortest paths on which a node lies [5]. The *betweenness centrality* $C_b(i)$ of a given node i is defined as [7]

$$C_b(i) = \sum_{k \neq j \neq i} \frac{S_{jk}(i)}{S_{jk}} \qquad (5)$$

where S_{jk} denotes the number of shortest paths from j to k and $S_{jk}(i)$ the number of shortest paths from j to k on which i lies. A contextual measure of distance should also consider indirect paths between the elements considered. This is the main principle behind the contextual distance recently introduced [2]. The *contextual distance* is given by the number of paths between two nodes i and j

$$\tilde{d}(\ell)_{ij} = \sum_{s} \lambda^s \ell_{ij}^s \qquad (6)$$

where ℓ_{ij}^s is the number of paths of length s between the nodes i and j. The coefficient λ weights this measure in such a way that successively longer step lengths get successively lesser weighting with $0 < \lambda < 1$. A usual value suggested for λ is in the order of 0.05 but there are still some difficulties in interpreting that value. Small variations of λ have an important impact on the resulting values of the *contextual distance*. Also the contextual distance, as for the previous measures of distance, is also asymmetric. A weighted measure of *contextual accessibility*, derived from the *contextual distance*, and computed for a node i, is given as

$$\tilde{d}(\ell)_i = \sum_{j} \tilde{d}(\ell)_{ij} \qquad (7)$$

All together, those space syntax measures provide complementary computational representations of the structure of the city. They evaluate different properties of the underlying graph representation of an urban network. They participate in the study of the relationship between the form and the function of the city. While distance measures based on shortest paths are immediate (and those can be combined with Euclidean distances), identifying contextual measures of distance is still a non straightforward task. To which extent and how the context should be considered are still the open questions to address.

3 Relative Adjacency

In a related work we introduced a relative adjacency operator that encompasses several properties required for a measure of contextual proximity within a graph [3]. This operator gives a form of topology-based proximity measure. The relative adjacency evaluates to which degree regions in a given spatial system are mutually distant in the dual graph derived from adjacency relationships. The operator is flexible enough to evaluate relative adjacencies at different levels of magnitude, that is, by minimising or maximising the impact of outlying regions. More formally, the relative adjacency between two nodes i and j is given by

$$\widetilde{rd}(\ell)_{ij} = \lambda \ell_{ij} + (1-\lambda) \sum_{k \in \mathrm{Adj}(i)} \frac{\widetilde{rd}(\ell)_{kj}}{\#\mathrm{Adj}(i)} \tag{8}$$

where $\mathrm{Adj}(i)$ is the set made of the edges of the graph incident with i, $\#\mathrm{Adj}(i)$ the cardinality of $\mathrm{Adj}(i)$ (equivalent to the degree of i), λ is a coefficient that counterbalances the relative importance of close and remote nodes with $0 \le \lambda \le 1$. Higher values of λ lead to a smaller account of outlying edges, whereas smaller ones minimise the importance of closer edges (note that when $\lambda=1$ then $\widetilde{rd}(\ell)_{ij} = \ell_{ij}$, where $\ell_{ij} = \ell_{ij}^1$).

The resolution of $\widetilde{rd}(\ell)_{ij}$ for a graph with n lines is equivalent in complexity to the resolution of a system of n linear equations with n unknowns. This results from the fact that from a given edge, all edges of the graph are recursively used at least once in the expression. The higher the number of possible paths between two given nodes in the dual graph, the higher their relative adjacency value. The relative adjacency's asymmetry reflects the fact that the likelihood of reaching a node j from a node i is generally different from the reverse. Mathematically speaking this operator is not an inverse distance as some of the properties a distance should possess are not verified. Values are drawn by the unit interval as it is made of a sum of adjacency values (*i.e.*, 0s and 1s) weighted by the cardinality of the adjacency set and the multiplicative coefficient λ. Additional properties of the relative adjacency are as follows:

- Asymmetry: to reflect the fact that distant things are not always symmetrically related.
- Closeness and accessibility: the operator not only measures the closeness between two given nodes in a graph but also evaluates the range of accessibilities between the two (accessibility taken in a purely topological sense).
- Flexibility: different impacts of neighbourhoods and outlying nodes can be considered within the graph.

The relative adjacency is derived from a network of relationships, that is, the adjacency in the context of spatial partitions. In a related work, the relative adjacency has been applied to study proximity relationships, clusters and the structure of spatial partitions [4]. In the context of space syntax, the relationships to consider are those that form the underlying graph that models a given urban network.

4 Benchmark Example

We illustrate the potential of the relative adjacency in comparison with current space syntax measures using the original example of the French village of Gassin, first introduced by [9], and since considered as a reference in further studies [2, 13]. Gassin is a typical Provençal village that lies on top of a hill not far from the bay of Saint Tropez (Fig. 1). It is made of a network of intricate pedestrian streets bounded by a sort of circular street that constitutes a natural boundary. A functional graph representation of the village of Gassin is derived (Fig. 1). It represents a network of 41 lines of sight labelled from a to z and aa to ao (Figures 2-8 where we use indifferently the terms street and line of sight although their meaning is slightly different).

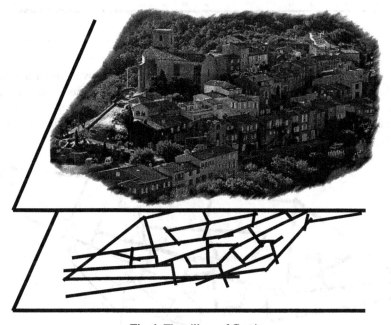

Fig. 1. The village of Gassin

A first analysis of local properties is derived by a computation of the *clustering coefficient* $C(i)$ (Eqn. 1). This gives an evaluation of the local integration of each street in the graph and the derivation of local clusters. Fig. 2 illustrates the results where a line's thickness is proportional to the value of the *clustering coefficient*.

The fact that the *clustering coefficient* $C(i)$ is very local generates some unsatisfactory results when analysing the whole graph. Some of the patterns revealed by the *clustering coefficient* seem to be overestimated. First, some outlying streets that appear highly clustered are relatively isolated and not very well integrated within the network (*e.g.*, ae and ah) whereas some relatively central streets appear as the contrary (*e.g.*, t, l and u). Second, the patterns that appear in Fig. 2 do not exactly fit our intuition of the clusters that form this graph.

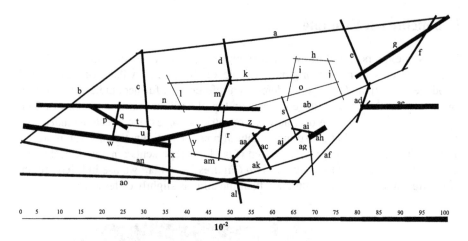

Fig. 2. Clustering coefficient C(i)

Global measures tend to produce an evaluation of how a given street lies in the network. The *shortest path length* estimates the number of transitions that are necessary to reach one street from another. It gives an evaluation of higher order neighbours. Fig. 3 gives an example of such path lengths from the line of sight labelled *n*.

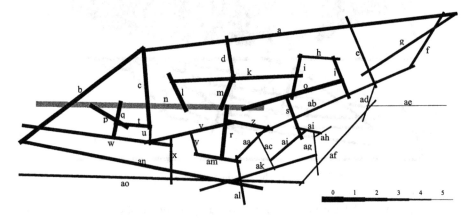

Fig. 3. Shortest path lengths from the line of sight "n"

Fig. 3 shows how the shortest paths give a sense of the global integration of a street by calculating the graph distance of a given line of sight to all others. The application of shortest paths to a reference street generates a sort of potential map. Streets can be regrouped into different classes according to their distances to the reference street. However, this measure does not provide a good discrimination between the different streets. Also there is no direct information about how often the different streets belong to the shortest paths.

A second example of global measure is the *closeness centrality* $C_c(i)$ (Eqn. 4, Fig. 4). The *closeness centrality* reveals some structural properties of the village of Gassin (a sort of skeleton). Central streets *n* and *ab* are revealed, so are the inner streets *b* and *a*. Those lines of sight have in common the facts that they are well connected and can be reached with only few transitions from most of the network. To a lesser extent, this is also true for the line of sight *z*, which participates in the junction of the two central lines *n* and *ab*. However, some relatively central lines of sight are not particularly rewarded by the closeness centrality (*e.g.*, *t, u, k* and *v*) as they do not participate very much in the calculation of the shortest paths. So although closeness centrality detects the general structure and the central streets of the network, it does not reveal the exact role of the other streets.

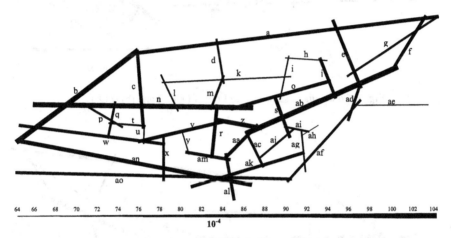

Fig. 4. Closeness centrality

The *contextual distance* (Eqn. 6) evaluates the number of opportunities between two given streets by calculating the number of paths between them. Although there is a clear integration of the context, the influence of longer paths decreases significantly due to the role of the coefficient λ arbitrary fixed to 0.05. This leads to a situation where there is a difficulty in making the difference between the different streets with respect to *n*, and also in interpreting the results due to the nature of the valuation of the coefficient λ (Fig. 5a). A logarithmic visualisation is an alternative but the output is too obvious to be successfully analysed (Fig. 5b).

A generalization of the contextual distance of the whole network is given in Eqn. 7, where average contextual distances are given for each street to give a measure of *accessibility* for each street (Fig. 6). This output is somehow similar to the result given by Fig. 4 and the *closeness centrality*, where the main structure of the network is revealed. The two main lines of sight *n* and *ab* are positively discriminated. And aside the peripheral lines *a, b, an* and *ao*, two sets of secondary lines emerge: the secondary path *v-w* (parallel to *n*), and *aa-al* continuing *ab* to the peripheral nexus *ao-an-ak*.

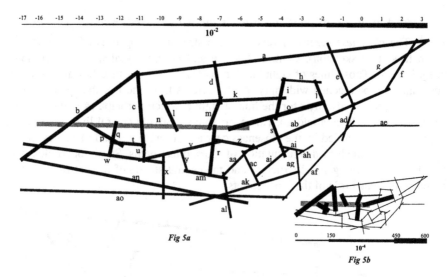

Fig. 5. Contextual distance as (a) Log($\widetilde{d}(\ell)_{ij}$) and (b) $\widetilde{d}(\ell)_{ij}$

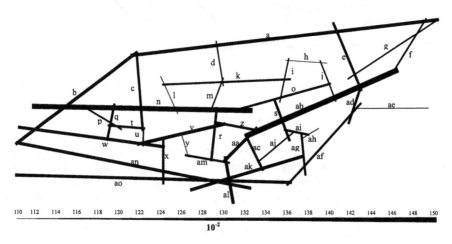

Fig. 6. Accessibility measure $\widetilde{d}(\ell)_i$

The next step is the application of the relative adjacency (Eqn. 8) to the street labelled n, with first a relatively high importance given to higher order neighbourhood in the calculation (λ =0.1, Fig. 7). Despite some similarities with the contextual distance(both measures consider power law of path lengths); the application of the relative adjacency to the village of Gassin shows several differences. First the valuation of the coefficient λ in the case of the relative adjacency marks the relative importance of closer (values of λ that tend to 1) versus outlying streets (values of λ that tend to 0). The relative adjacency is normalized, thus producing a result that is directly interpretable on a decimal scale. Secondly, discrimination between the different streets is satisfactory. Fig. 7 denotes several patterns that can be interpreted visually. Several local clusters and preferential attachments tend to emerge (*e.g.*, the one formed by the streets labelled *p*, *q*, *t* and *c* or the one formed by the streets *p*, *q*, *t*, *u*, *w*, *b* and *c* depending of the level of abstraction chosen by the observer).

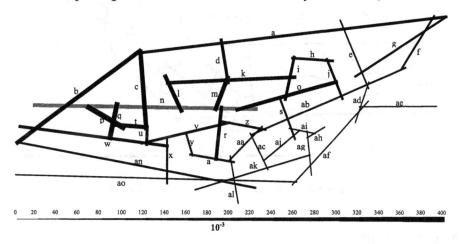

Fig. 7. Relative adjacency with respect to "n", $\lambda = 0.1$

A second example illustrates the relative adjacency with respect to the street labelled *ab* and a value of λ equal to 0.5 (Fig. 8). The structure of the network appears differently due to the change of reference location. With respect to the trends illustrated in Fig. 7. The importance of closer streets is increased due to the lower value of λ. It is also worth noting the emergence of local clusters composed of streets with similar relative adjacencies to the street labelled *ab* (*e.g.*, *ad*, *ae*, *e*, *g*, *f* at the North East). These clusters present the particularity of being derived relatively from a given reference location in the graph and not when considering the whole graph globally as for the clustering coefficient.

This case study illustrates how an application of space syntax measures to the village of Gassin can partially reveal the spatial structure and some sort of order from the analysis of the underlying graph representation. Local integration values, local clusters and the most central streets are estimated. The relative adjacency complements these measures with a finer consideration of the context. The relative adjacency reveals a topological form of distance that approximates structural relationships between lines of sight in the network. The study of the distribution of relative adjacency values in the network also reveals additional patterns, such as emergent clusters relative to a given street and the main structure of the network. The village of Gassin is relatively symmetric so the application of the relative adjacency to urban networks with higher degrees of asymmetry is likely to reveal more interesting patterns. As space syntax studies are often applied to the analysis of the structure of larger cities, the interest of the relative adjacency should be explored in connection with those studies. This should allow for further analysis of the relationships between the form of the city, the emergent structure and order, and the functional aspects of the city.

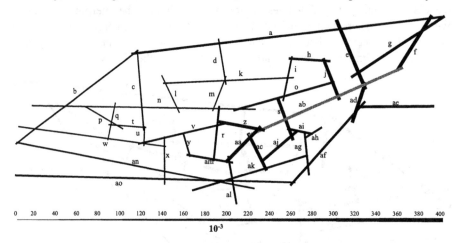

Fig. 8. Relative adjacency with respect to "n", $\lambda = 0.5$

5   Conclusions

Space syntax is one of the emerging computational and experimental domains where there is a clear attempt to identify some relationships between spatial complexity, emerging structures and functioning of the city. Despite real successes in its diffusion and application to the study of many city configurations around the world, space syntax still requires new theoretical developments and representations to measure and interpret configurations and patterns in the city.

This paper shows how a relative adjacency operator can contribute to the space syntax theory. While current space syntax measures evaluate different forms of integration and distances within the dual graph that denotes the structure of the city, the relative adjacency provides a flexible form of relative distance that takes into account the context in its calculation. The relative adjacency possesses an asymmetrical property, which is a valid factor in the analysis of distances within an urban network. It also reveals a range of opportunities from a given street, thus making a step forward in the analysis of relative street functions and accessibilities. The operator is flexible enough to reveal different patterns (relationships with urban studies). Following a suggestion made by Batty [2] we plan to study other measures that incorporate higher-order adjacencies and to investigate the potential of the relative adjacency for clustering analysis.

Space syntax can be also considered as one of the application domains of network geography where structure and patterns can be revealed from the application of graph analysis to the spatial domain. Despite the fact that network geography has been long used in geographic science, recent developments of graph analysis and computational models provide new avenues of opportunity of research for GIS scientists. In particular, we believe that some of the theoretical principles identified and emerging from space syntax research can also be beneficial to the analysis of spatial structures and spatio-temporal networks.

Acknowledgments

We are grateful to Bin Jiang for introducing us to the world of space syntax studies, and to Mike Batty whose recent work on distances in graph and space syntax have inspired the material and the ideas presented in this paper. We also thank the anonymous referees for their constructive comments and suggestions that helped us to improve the quality of this paper.

References

1. Batty, M.: *A New Theory of Space Syntax*. Working Paper 75, Centre for Advanced Spatial Analysis, UCL, London (2004)
2. Batty, M.: *Distance in Graphs*. Working Paper 80, Centre for Advanced Spatial Analysis, UCL, London (2004)
3. Béra, R. and Claramunt, C.: Relative Adjacencies in Spatial Pseudo-Partitions. In W. Kuhn, M. Worboys, and S. Timpf (eds.), *Spatial Information Theory (COSIT '03)*, Lecture Notes in Computer Science 2825, Springer, Ittingen, Switzerland (2003) pp. 218-234
4. Béra, R. and Claramunt, C.: Topology-Based Proximities in Geographical Spaces. *Journal of Geographical Systems* 5(4): 353-379 (2003)
5. Brandes, U.: A Faster Algorithm for Betweenness Centrality. *Journal of Mathematical Sociology* 25(2): 163-177 (2001)
6. Dorogovtsev, S. and Mendes, J.: *Evolution of Networks: From Biological Nets to the Internet and WWW*, Oxford University Press, Oxford, UK (2003)
7. Freeman, L.: A set of Measures of Centrality Based on Betweenness. *Sociometry*, 40: 35-41 (1977)
8. Haggett, P., and Chorley, R.: *Network Analysis in Geography*, Edward Arnold, London (1969)
9. Hillier, B. and Hanson, J.: *The Social Logic of Space*, Cambridge University Press, Cambridge, UK (1984)
10. Hillier, B., Penn, A., Hanson, J., Grajewski, T. and Xu, J.: Natural Movement or Configuration and Attraction in Urban Pedestrian Movement. *Environment and Planning B*, 20: 29-66 (1993)
11. Hillier, B.: *Space is the Machine: A Configurational Theory of Architecture*, Cambridge University Press, Cambridge, UK (1996)
12. Jiang, B.: Multi-Agent Simulation for Pedestrian Crowds. In Bargiela A. and Kerckhoffs E. (eds.), *Simulation Technology: Science and Art, 10th European Simulation Symposium and Exhibition*, Nottingham, UK (1998) pp. 383-387
13. Jiang, B., Claramunt, C. and Batty, M.: Geometric Accessibility and Geographic Information: Extending Desktop GIS to Space Syntax. *Computers, Environment and Urban Systems* 23 (1999) 127-146

14. Jones, M. and Fanek, M.: Crime in the Urban Environment. In *Proceedings of First International Symposium on Space Syntax*, University College London, UK (1997) pp. 25.1-25.11

15. Kansky, K.: *Structure of Transportation Networks: Relationships between Network Geometry and Regional Characteristics*. Research Paper 84, Department of Geography, University of Chicago, IL (1963)

16. Penn, A. and Croxford, B.: Fingerprinting Urban Kerbside Carbon Monoxide Concentrations: The Interaction between the Street Grid Configuration, Vehicle Flows and Local Wind Effects. *International Journal of Vehicle Design* 20 (1-4): 60-70 (1998)

17. Penn, A., Hillier, B., Banister, D. and Xu, J.: Configurational Modelling of Urban Movement Networks. *Environment and Planning B*, 25: 59-84 (1998)

18. Peponis, J., Zimring, C. and Choi, Y.: Finding the Building in Wayfinding, *Environment and Behavior* 22: 555-590 (1990)

19. Sabidussi, G.: The Centrality Index of a Graph. *Psychometrika* 31: 581-603 (1966)

20. Tiefelsdorf, M.: Misspecifications in Interaction Model Distance Decay Relations: A Spatial Structure Effect. *Journal of Geographical Systems* 5(1): 25-50 (2003)

21. Watts, D.: *Six Degrees: The Science of a Connected Age*, W. W. Norton and Company, New York (2003)

22. Worboys, M.: Metrics and Topologies for Geographic Space. In: Kraak M.-J. and Molenaar M. (eds.) *Advances in GIS Research II*, Taylor & Francis, London (1996) pp. 365-375

Semi-automatic Ontology Alignment
for Geospatial Data Integration*

Isabel F. Cruz, William Sunna, and Anjli Chaudhry

Department of Computer Science
University of Illinois at Chicago
851 S. Morgan St. (M/C 152), Chicago, IL 60607, USA
{ifc,wsunna,achaudhr}@cs.uic.edu

Abstract. In geospatial applications with heterogeneous databases, an ontology-driven approach to data integration relies on the alignment of the concepts of a global ontology that describe the domain, with the concepts of the ontologies that describe the data in the local databases. Once the alignment between the global ontology and each local ontology is established, users can potentially query hundreds of databases using a single query that hides the underlying heterogeneities. Using our approach, querying can be easily extended to a new data source by aligning a local ontology with the global one. For this purpose, we have designed and implemented a tool to align ontologies. The output of this tool is a set of mappings between concepts, which will be used to produce the queries to the local databases once a query is formulated on the global ontology. To facilitate the user's task, we propose semi-automatic methods for propagating such mappings along the ontologies. In this paper, we present the principles behind our propagation method, the implementation of the tool, and we conclude with a discussion of interesting cases and proposed solutions.

1 Introduction

In an ontology-driven approach, an application that needs to use the integrated data from a domain expresses its information requests in terms of the concepts in the global ontology, thus giving users the appearance of an homogeneous view over heterogeneous data sources.

An ontology ranges from a simple taxonomy to an axiomatized set of concepts and relationship types [9]. In our approach, we focus on taxonomies for land use coding in the state of Wisconsin. Our work is in collaboration with the Wisconsin Land Information System (WLIS), which is a distributed web-based system with heterogeneous data residing on local and state servers.

In order to develop WLIS, it is necessary to overcome data heterogeneity that originates from having different state and federal agencies involved in acquiring and storing geospatial data. On one hand, proposing a standard ontology to be adopted by all agencies would lead to loss in the resolution of the collected

---

* This research was supported in part by the National Science Foundation under Awards EIA-0091489 and ITR IIS-0326284.

M.J. Egenhofer, C. Freksa, and H.J. Miller (Eds.): GIScience 2004, LNCS 3234, pp. 51–66, 2004.
© Springer-Verlag Berlin Heidelberg 2004

data. A more feasible approach is to allow for the different agencies to maintain their own ontologies locally while specifying how concepts in their own ontology correspond to concepts in a global ontology. On the other hand, we note that even when agencies use a particular standard, their exact implementation of that standard will differ. In the presence of such heterogeneities, however small they may be, queries to one ontology will likely not work on another ontology.

We focus primarily on *ontology alignment*, that is, on establishing mappings between related concepts in the global ontology and a local ontology without combining the ontologies. We also discuss *ontology merging* when the two ontologies are not only aligned but also included in a single, coherent ontology.

In order to integrate the constituent data models, the mappings between concepts in the global ontology and those in the local data sources, e.g., as described by a local ontology, have to be determined. In our approach, the mappings are determined semi-automatically, that is, partly established manually by the user and partly deduced using an automatic process.

The paper is organized as follows. In Section 2, we give an overview of related work in the area. We present the land use codes and how we encode them using an ontology in Section 3. In Section 4, we present the mappings we consider in performing the alignment between two ontologies, we describe our approach to automatically propagate mappings along the two ontologies involved in the mapping, state the assumptions we make that are necessary for the automatic propagation to work, and give an example on how alignments can be used for merging ontologies. In Section 5, we describe the implementation of the mapping tool that supports the semi-automatic deduction process. With the ultimate goal of extending our semi-automatic alignment method to any ontologies, in Section 6 we give examples of ontologies that do not satisfy our previously made assumptions and propose ways of extending our semi-automatic propagation to handle those new cases. Finally, in Section 7, we draw conclusions and outline future work.

2   Related Work

Hovy [4] proposes a semi-automatic ontology alignment approach for combining and standardizing large ontologies based on their lexical similarity. Multiple languages are also supported by consulting dictionaries. Different kinds of matches, including text and hierarchy matches are used in the heuristics for the alignment process. Initially the unaligned ontologies are loaded and brought into a partially aligned state. For all unaligned concepts, the heuristics mentioned earlier are used to create a set of cross-ontology match scores. Then, a new set of alignment suggestions are created by a function that combines the match scores. The user can then check the suggestions and retain the best matches.

Chimaera [5] is a software tool developed by the KSL group at Stanford, also used for merging and testing large ontologies. It provides tools for merging different ontologies created by different authors and for diagnosing individual or multiple ontologies. Similarly to Hovy's approach, the tool allows users to check the automated merging procedure. Classes that need the user's attention are highlighted. PROMPT [7], which also uses a frame-based model for the ontology, supports merges at the slot level, in addition to merges at the class (frame) level.

In the ontologies that we have considered so far for land use management, an ontology is a hierarchy where concepts refer to the codes (the vertices in the hierarchy) and relationships are established between a parent code and a child (the edges between the corresponding vertices). Such relationships represent generalization/specialization between the codes. In our case, there are no explicitly represented properties or attributes associated with the codes. We have a simpler structure than that found in other systems [5, 7], and therefore the decision of whether two concepts match has to be solely based on the codes.

Another system for the alignment of ontologies proposes a comprehensive approach encompassing a rich set of mappings types [8]. The authors stress the importance of declaratively specified mappings (our mappings are declarative), the necessity for automation, and of addressing the problem of defining a measure for the expressiveness of the mappings supported. However, several of these issues are left as future work.

The MOMIS approach [1, 2] has several similarities with our approach. It include the semi-automatic processing of ontology alignment, but it has several different characteristics. It uses subsumption and lexical comparison, but in our case we have strings instead of frames, therefore subsumption is not a possibility, and lexical comparison, as that provided by WordNet [6] would only play a limited role in the types of ontologies we currently consider.

3   Land Use Data

Our examples focus on the Wisconsin Land Information System (WLIS) and on Land Use Data. A land use database system stores information about land parcels in XML format. Our land parcel data contains an identification number for the parcel (represented by the tag *lid*), the category of land usage under which it is classified (*lucode*), the file containing the pertinent shape information (*shape_file*), and information about the owner of the parcel (*owner_id*).

Land use categories include *agriculture, commerce, industry, institutions*, and *residences*. Storing the land use codes of land parcels helps in better planning for township development, transportation, taxation, and so on. A typical query such as *"Where are all the crop and pasture lands in Dane County?"* would be relatively straightforward when using one data set but more difficult when posed over a larger geographic area. Table 1 illustrates the heterogeneity of attribute names and values that would satisfy the criteria of the query over selected multiple data sets.

1. *Lucode, Tag, Lu1* and *Lu_4_4* must be resolved as synonyms for the attribute that represents the land use code in the ontology.
2. The descriptions are not exact matches. For example where one code is used for the remaining classifications, Racine County uses two codes.

To represent the ontologies we use our own XML DTD [3].

Table 1. Heterogeneity of attribute names and values

Planning Authority	Attribute	Code	Description
Dane County RPC	Lucode	91	Cropland Pasture
Racine County (SEWRPC)	Tag	811	Cropland
		815	Pasture and Other Agriculture
Eau Claire County	Lu1	AA	General Agriculture
City of Madison	Lu_4_4	8110	Farms

4   Ontology Alignment

An important step in the data integration process is ontology alignment—the identification of semantically related entities in different ontologies. While establishing a semantic relationship between concepts in the global ontology and concepts in the local ontologies can be challenging, the thorough identification of such relationships is essential for the development of accurate machine-based techniques to handle them. Aligning very heterogenous ontologies can be a difficult process. In our approach, we expect ontologies to be close to each other in a given domain.

In the examples, we represent the ontologies as trees. The vertices of the trees correspond either to existing entities in the ontology (*real vertices*) or to entities created with the end of logically grouping entities (*virtual vertices*). In the figures, the left tree represents the global ontology and the right tree represents the local ontology.

For example, in Figure 1, the codes *Agriculture - Woodlands - Forests* and *Agriculture - Woodlands - Non-forests* in the ontology are mapped to the land use codes *Forestry (91)* and *Non-forest woodlands (92)* in the local ontology (used by Dane County in Wisconsin). There is no local land use code corresponding to *Agriculture - Woodlands*. The land use codes of the land parcels in the database are stored only as *91* or *92*, corresponding to *Forestry* and *Non-forest woodlands*. To better align the local ontology with the ontology, a virtual vertex was introduced corresponding to *Agriculture - Woodlands*.

4.1   Mapping Types

In Figure 1, *Agriculture Woodlands - Forests*, *Agriculture Woodlands - Non-forests*, *Forestry* and *Non-forest woodlands* are semantically at the same level of

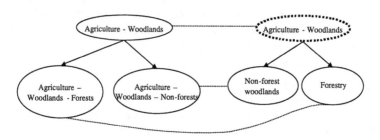

Fig. 1. Real and virtual vertices

detail in the two ontologies. Similarly, the two vertices corresponding to *Agriculture - Woodlands* also at the same level. We say that such concepts are *aligned.* Initially, the information as to which entities in the different ontologies are aligned must be provided by the user, who is the geospatial expert for the local database. Once two entities are known to be aligned, the nature of the relation between them can be characterized using the following mapping types: *Exact,* the connected vertices are semantically equivalent, *Approximate,* the connected vertices are semantically approximate, *Null,* the vertex in the ontology does not have a semantically related vertex in the local ontology, *Superset,* the vertex in the ontology is semantically a superset of the vertex in the local ontology, and *Subset,* the vertex in the ontology is semantically a subset of the vertex in the local ontology.

A mapping can establish the connection between vertices in their entirety or only to parts of a vertex, based on the semantics. Even though the global ontology is usually developed by a team of domain experts, who take utmost care in making sure that every semantically unique entity is represented by exactly one vertex, the local ontology might consist of entities organized or grouped using different criteria. As a result, the semantic equivalent of an entity in the global ontology could be distributed over several vertices or parts of a vertex in the local ontology and vice versa. A county in which agriculture is the main occupation will have more categories of agricultural land usage than the ontology drawn up for the state. Such differences in the resolution of the data can also lead to complications in ontology alignment.

Figure 2 illustrates several mappings between vertices in two ontologies for land use patterns in a centralized integrated system. We show the global ontology subtree on the left side of figure and the local ontology subtree on the right. The vertices corresponding to *Industry, Mining,* and *Manufacturing* in the global

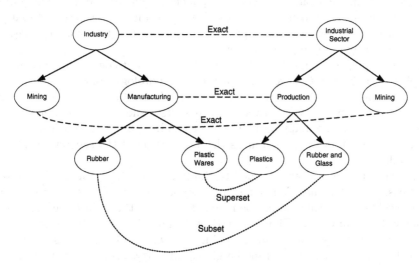

Fig. 2. Mapping types

ontology can be mapped to those corresponding to *Industrial Sector, Mining* and *Production* in the local ontology. In the global ontology, the vertex *Plastic wares* denotes entities that are made of plastic or glass. However, in the local ontology, there is a vertex *Plastics* and another vertex *Rubber and Glass*, which denotes manufactured objects made of rubber or glass.

The *Manufacturing* and the *Production* vertices are aligned. Similarly, the two *Mining* vertices are also aligned. *Manufacturing* is semantically equivalent to *Production*, as both denote a collection of industries producing plastics, glass, and rubber products. Hence, this mapping is of type *Exact* as denoted in the mapping from the *Manufacturing* vertex to the *Production* vertex. *Plastic wares* is semantically a *superset* of the *Plastics* vertex and *Rubber* is semantically a *subset* of the *Rubber and Glass* vertex.

4.2   Semi-automatic Alignment

To allow for the semi-automatic processing of the ontology alignment, we propose a framework that defines the values associated with the vertices of the ontology in two possible ways: as functions of the values of the children vertices or of the user input.

We need to establish two assumptions to guarantee the correctness of the deduction process. The first one is that the specialization of a vertex in the ontology must be total, that is, each lower-level concept must belong to a higher-level concept in the hierarchy. The second one is that "bowties" [4], which are inversions in the order of the two ontologies that are being aligned, do not occur.

The mapping techniques described in Section 4.1 can be integrated in a semi-automatic alignment methodology to simplify the task of aligning large ontologies [4]. The user initially identifies the hierarchy levels in the two ontologies that are aligned. Then the alignment component propagates. When ambiguities or inconsistencies are encountered, or the the algorithm can not propagate values any further, those vertices are singled out. The user can then manually assist the algorithm by mapping concepts by hand.

For example, in Figure 3, vertex b in the global ontology is mapped using mapping type *Superset* to vertex e in the local ontology, and vertex c in the global ontology is mapped using mapping type *Exact* to vertex f in the local ontology. The mapping type between their parents a and d can be deduced to be *Superset* based on the mapping between the children, because we consider that the semantic content of the parent is the aggregation of the semantic contents of its children.

All the children of d are mapped to children of a. This is the *Fully Mapped (FM)* case. The *Partially Mapped (PM)* case occurs if there are some children in the local ontology that cannot be mapped to any of the children in the global ontology. For example, in Figure 4 vertices b and c in the global ontology are mapped to vertices e and f using mapping type *Exact*. But vertex g cannot be mapped to any of the vertices in the global ontology. As a result, vertex a is mapped using mapping type *Subset*.

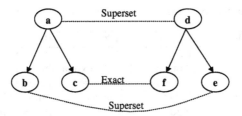

Fig. 3. Fully mapped deduction operation

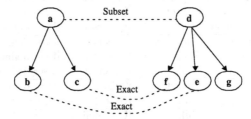

Fig. 4. Partially mapped deduction operation

Table 2 lists the different possible combinations of vertex mappings and the resulting mappings for their parents. The table assumes that a vertex in the first ontology has two children that are mapped to the children of a vertex in the second ontology. Column 1 in the table shows the mapping type of the first child and column 2 shows the mapping type of the second child. Column 3 shows the deduced mapping type between the two vertices in the *Fully Mapped (FM)* scenario and column 4 shows the deduced mapping type between the two parents in the *Partially Mapped (PM)* scenario. A *User-defined* entry in the table indicates that the parent's mapping type cannot be automatically deduced and the user has to provide the appropriate mapping type by hand. These deduction operations can easily scale up to include the cases where a vertex has more than two children. They will be performed recursively, starting from the vertices that are aligned and traveling up the global ontological tree, to deduce the mapping types between the ontology and the local ontologies. As previously mentioned, all combination results can be overridden by the user.

4.3   Ontology Merging

Each local ontology might have a different organization of the entities based on the primary function of the agency maintaining it. For example, a county in which agriculture is the main occupation will have more categories of agricultural land usage than the global ontology drawn up for the state. When such a local ontology is aligned to the global ontology, there might be several places where the mapping type is *Null* or the user has to provide mappings because the automatic alignment process fails. This can indicate that a particular criterion of classification is missing in the global ontology and leads to loss in the resolution of the data when local ontologies using that classification technique are aligned.

Table 2. Automatic mapping deduction operations

Child 1	Child 2	FM	PM
Exact	Exact	Exact	Subset
Exact	Approximate	Approximate	Subset
Exact	Superset	Superset	User-defined
Exact	Subset	Subset	Subset
Exact	Null	Superset	User-defined
Approximate	Approximate	Approximate	Subset
Approximate	Superset	Superset	User-defined
Approximate	Subset	Subset	Subset
Approximate	Null	Superset	User-defined
Superset	Superset	Superset	User-defined
Superset	Subset	User-defined	User-defined
Superset	Null	Superset	User-defined
Subset	Subset	Subset	User-defined
Subset	Null	User-defined	User-defined
Null	Null	User-defined	User-defined

In such cases, the expert in charge of maintaining the global ontology could add the missing classification, that is, could *merge* concepts from a local ontology into the global ontology. This can be viewed as merging concepts from local ontologies into the global ontology.

In the global ontology of Figure 5, commercial land usage is classified as *Commercial Sales* and *Commercial Service* (based on the primary function of the commercial establishment). In the local ontology, commercial land usage is sub-classified as *Commercial Intensive* and *Commercial Non-intensive* (based on the size of the operations). The two parent vertices are considered aligned, because their level of detail is similar. As shown in Figure 5, vertices *Commercial Sales* and *Commercial Service* cannot be mapped to any of the vertices in the local ontology and hence have their mapping type as *Null*. Therefore, the mapping type between *Commerce* and *Commercial Sector* cannot be automatically deduced and is specified as *Approximate* by the user. This mapping type denotes that the mapping between two vertices seems right, but the subclassification is not along the same characteristics.

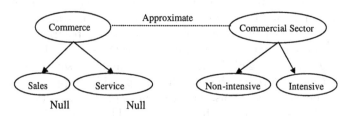

Fig. 5. Ontology alignment before the deduction process

Classification of commercial land usage, based on the scale of operations, is missing from the ontology and could be introduced to better align local ontologies using that classification scheme. The alignment of the ontologies after the additional level of classification was introduced is shown in Figure 6.

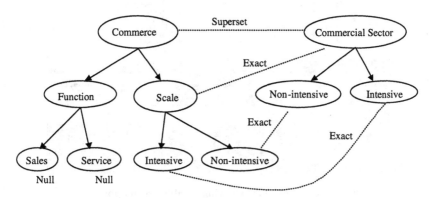

Fig. 6. Ontology alignment after deduction

5   Agreement Maker

The *Agreement Maker* is a software tool that is used to create the mappings between the global ontology and a local ontology and generate an agreement document, which is used by the query processor. The query processor maps a query expressed in the terms used in the global ontology to the local ontologies [3].

The local expert maps the ontology of the local database to the global ontology with a user interface, which shows the two hierarchies that represent the ontologies in two separate panes, allowing the expert to browse through the contents of each of the ontologies and establish the mappings between an entity (or entities) in the global ontology and an entity (or entities) in the local ontology. Figure 7 shows that interface.

5.1   Agreement Document

The local expert selects semantically related entities in both ontologies and chooses one of the mapping types from Section 4.2, which would best explain the relation between the concepts. There is a button labeled 'Update Mapping' that saves the mapping into a table. Each row in the table shows the names of the two vertices that are mapped (one from the global ontology and the other from the local ontology), and the type of mapping between them. As soon as the mapping is saved, all the vertices that have been mapped get clearly marked (with '⟶', as shown in Figure 7). In the backend, the agreement document also gets updated.

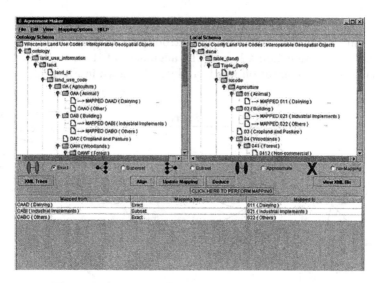

Fig. 7. User interface showing established mappings

5.2   Deduction Module

The semi-automatic alignment methodology from Section 4.2 is integrated into the Agreement Maker system to simplify the task of aligning large ontologies. Once the children of two vertices (one belonging to the global ontology and the other one to the local ontology) are mapped by the user, the mapping between the parents of those vertices can be inferred by the automatic mapping deduction operations of Table 2. These results are then presented to the user, who may choose to integrate them in the agreement document or discard them.

However, not only mappings that result from the deduction operations are added to the agreement document. Mappings can also result from the comparison of the labels of vertices, which have either the same name (e.g., person_name) and are therefore mapped using *Exact*, or because by using a vocabulary it is found that two terms are synonyms, or because the comparison of the respective strings (e.g., using an edit distance algorithm) produces a high score. Such mappings also reduce the load on the user. A variety of such algorithms for comparison of terms is available. Of course, the user's decision overrides any mappings by the automatic deduction mechanism. Also, if the user has already mapped a vertex, no deductions will be considered for that vertex. The results obtained by using the automatic deduction module are displayed in a separate pop-up window, which is shown in Figure 8.

The tabular results shown in Figure 8 indicate the vertices of the global ontology that are mapped to vertices of local ontology, and the respective mapping type. It also indicates whether the mapping was performed by the user, by the deduction mechanism, or by the tool box (which we named "Magic Box") for the comparison of the labels of the vertices. The user can either commit the

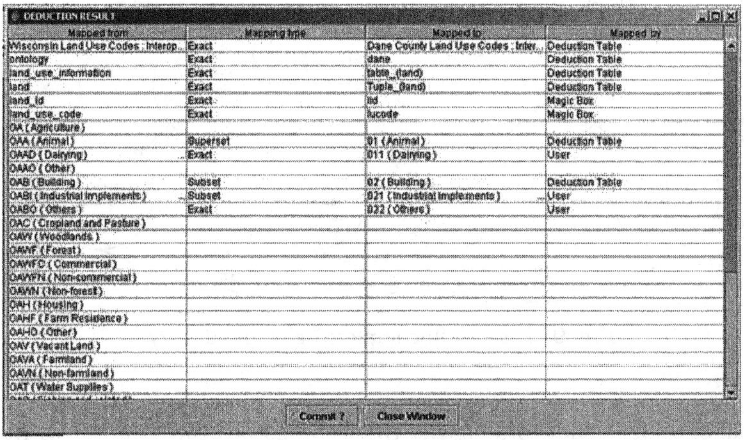

Fig. 8. The deduction result window

changes to the Agreement File if the results seem appropriate or discard them. If the changes are committed, they will be displayed in the main user interface.

6 Discussion

In Section 4 we present the assumptions we have made for the semi-automatic alignment to work. One of them is that the specialization must be total, otherwise the automatic deduction operation may produce semantically incorrect results. Next we consider two cases: in the first case, the specialization of the global ontology is not total and in the second case, the specialization of the local ontology is not total. For the sake of our discussion, we consider a simple example based on residential parcels where we assume that the only possible specializations of *Residential Buildings* are *One-family residence* and *Two-family residence*.

Case 1. In the first case we consider two subtrees of the local and of the global ontologies, where all the children of *Residential Buildings* and all the children of *Apartment Buildings* are represented in Figure 9. In Figure 9 the user initially maps *Two-family residence* to *Multiple-family residence* using *Subset* as the mapping type and maps *One-family residence* in the global ontology subtree to *One-family residence* in the local ontology subtree using *Exact* as the mapping type. Running the deduction process of Table 2, the parent vertex of the global ontology subtree *Residential Buildings* is mapped as *Subset* of the parent of the local ontology subtree *Apartment Buildings*. This result is semantically incorrect: *Apartment Buildings* should be a subset of *Residential Buildings*, not the opposite.

The system may notice a problem when the user overrides the automatic alignment module to make *Residential Buildings* a *Superset* of *Apartment Buildings*. It is clear in this case that the specialization of *Residential Buildings* in

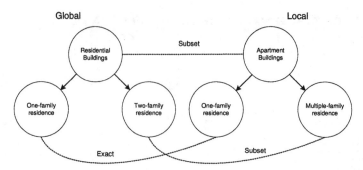

Fig. 9. Case 1: "Incomplete" global ontology

the global ontology is not total: either there should be other children of *Residential Buildings* that are siblings of *One-family residence* and *Two-family residence* (e.g., *Three-family residence*, *Multiple-family residence*) or there should be a vertex that generalizes *One-family residence* and *Two-family residence*, is a child of *Residential Buildings*, and is a subset of *Apartment Buildings* in the local ontology. Note that ideally one would want the global ontology to be as "complete" as possible, but, as mentioned before, sometimes the local domain might contain more detail on certain concepts, since it specializes in a certain domain.

Case 2. In this case, we interchange the ontologies of the previous case as shown in Figure 10. This problem illustrates the case where the local ontology is less "complete" than the global ontology. This may be a rather common case, where the local ontology models the reality of a situation that is more limited. The local ontology may represent a town district where a *Residential Building* can only be a *One-family residence* or a *Two-family residence* and therefore these are all the representatives of *Residential Buildings*. Again, the deduction mechanism captures the structural aspect but not the semantic one, and erroneously concludes that *Apartment Buildings* is a superset of *Residential Buildings*.

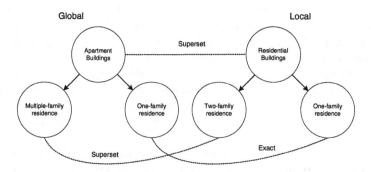

Fig. 10. Case 2: "Incomplete" local ontology

Proposed Solution. To analyze the situation we consider Figure 11, where we represent "monotonic" mappings between two different trees. For example, vertex B maps as *Exact* to vertex G. Since vertex G represents a *Superset* of vertex H and a *Subset* of vertex F, then vertex B is a *Superset* of vertex H and a *Subset* of vertex F. In order to guarantee monotonicity, the situation where we have a "bowtie" cannot exist (recall that it was our assumption in Section 4 that "bowties" were not allowed and we will not be relaxing that assumption now).

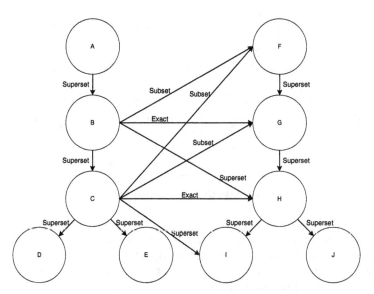

Fig. 11. Monotonic mappings

A possible solution for the first case is to augment the global ontology so as to encompass the local ontology. When the user maps *Residential Buildings* as a *Superset* of *Apartment Buildings* the system should catch such inconsistent mapping with the one produced by the automatic deduction mechanism and suggest the introduction of a virtual vertex (see Figure 12). A virtual intermediate vertex *Apartment Buildings* is inserted in the global ontology that carries the same name of its equivalent in the local ontology. The mapping remains *Subset* because structurally what *Apartment Buildings* contains in the global ontology is a *Subset* of what *Apartment Buildings* contains in the local one. Because a set is both a subset and a superset of itself the deduction mechanism is correct (but not as tight as it could be; in this case, the string comparison mechanism that we have mentioned in Section 5 would produce the correct *Exact* mapping). As for *Residential Buildings* in the global ontology, it can possibly be mapped to the immediate parent of *Apartment Buildings* in the local ontology or an exact match could be found further up in the local ontology. The introduction of *Apartment Buildings* in the global ontology will improve the granularity of the queries on the local ontology, as the user can now specifically query for "Apartment Buildings" and obtain a precise response from the local ontology. Likewise, more children can be added to the vertex *Residential Buildings* to encompass other local ontologies.

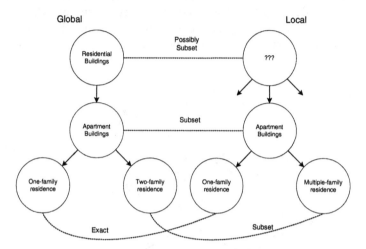

Fig. 12. Insertion of a virtual vertex

As for the second case, and taking into consideration the monotonic characteristic of the mappings, an ancestor of *Apartment Buildings* in the global ontology will be more likely to yield an exact match to the vertex *Residential Buildings* in the local ontology (Figure 13).

Scoring functions [4] can also capture the essence of the monotonicity of the trees. For example the value of a scoring function would improve until we find

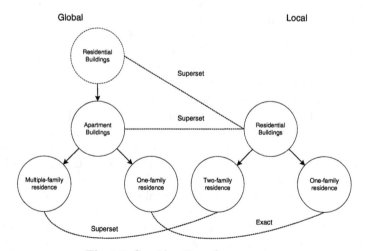

Fig. 13. Consideration of an ancestor

a more suitable vertex that is a parent of *Residential Buildings*. After an exact match is found, the scoring function would deteriorate for the parents of the vertex that yielded an exact match.

7   Conclusions

In this paper, we describe an ontology-driven approach to data integration and how we use such an approach in the geospatial domain. Data is integrated based on a global ontology of land usage patterns. The end user poses queries in terms of entities in the global ontology, which are then executed against the local ontologies. To enable query processing, the different ontologies have to be aligned. The relationships between aligned entities in the ontology and in each data source are expressed using different types of mappings, which we have defined so as to capture the range of possible semantics in our domain.

To allow for the semi-automatic processing of the ontology alignment, we have proposed a framework that defines the values associated with the vertices of the global ontology in two possible ways: as functions of the values of the chidren vertices or of the user input. We discussed some of the problems that we would encounter by relaxing one of our simplifying assumptions and possible solutions to those problems. We are currently conducting experiments on aligning ontologies to test the effectiveness of our semi-automatic processing.

There are several related issues that still need to be investigated, pertaining to the alignment process. In particular, more applications need to be investigated to possibly find other mapping types and to consider different kinds of ontologies. While our mapping types are "adequate", to express the mappings in our current application, we would like to characterize the notion of *adequacy* for such mappings, especially as integrated with the alignment process. Another issue we are trying to resolve is that of vertices that intersect in meaning.

We have made some simplifying assumptions concerning the alignment process, which can be rather complex. Our simplifying assumptions may or may not hold in other applications or for other types of ontologies. Therefore, we plan to continue the work we started in Section 6.

Acknowledgments

We would like to thank Nancy Wiegand and Steve Ventura, from the Land Information & Computer Graphics Facility at the University of Wisconsin-Madison, for the discussions on land use problems. We are also indebted to Naveen Ashish for pointing to us some interesting problems and to Afsheen Rajendran for the initial implementation of the prototype.

References

1. H. Benetti, D. Beneventano, S. Bergamaschi, F. Guerra, and M. Vincini. An Information Integration Framework for E-Commerce. *IEEE Intelligent Systems*, 17:18–25, 2002.

2. S. Bergamaschi, F. Guerra, and M. Vincini. A Data Integration Framework for E-Commerce Product Classification. In *1st International Semantic Web Conference (ISWC)*, pages 379–393, 2002.
3. I. F. Cruz, A. Rajendran, W. Sunna, and N. Wiegand. Handling Semantic Heterogeneities using Declarative Agreements. In *International ACM GIS Symposium*, pages 168–174, 2002.
4. E. Hovy. Combining and Standardizing Large-Scale, Practical Ontologies for Machine Translation and Other Uses. In A. Rubio, N. Gallardo, R. Castro, and A. Tejada, editors, *First International Conference on Languages Resources and Evaluation (LREC), Granada, Spain*, pages 535–542. 1998.
5. D. L. McGuinness, R. Fikes, J. Rice, and S. Wilder. An Environment for Merging and Testing Large Ontologies. In *Seventeenth International Conference on Principles of Knowledge Representation and Reasoning (KR-2000)*, pages 483–493, 2000.
6. G. A. Miller. WordNet: An Online Lexical Database. Technical report, Princeton University, 1990.
7. N. F. Noy and M. A. Musen. PROMPT: Algorithm and Tool for Automated Ontology Merging and Alignment. In *The Sixteenth National Conference on Artificial Intelligence (AAAI)*, pages 450–455, 2000.
8. J. Park, J. Gennari, and M. Musen. Mappings for Reuse in Knowledge-based Systems. In *11th Workshop on Knowledge Acquisition, Modelling and Management, KAW 98*, 1998. http://ksi.cpsc.ucalgary.ca/KAW/KAW98/park/.
9. J. Sowa. Building, Sharing, and Merging Ontologies, 2001.
   http://www.jfsowa.com/ontology/ontoshar.htm.

Modeling Surface Hydrology Concepts with Endurance and Perdurance

Chen-Chieh Feng[1], Thomas Bittner[2], and Douglas M. Flewelling[1]

[1] Department of Geography, University at Buffalo
105 Wilkeson Quad, Amherst, NY 14261, USA
{cfeng,dougf}@geog.buffalo.edu
[2] Institute for Formal Ontology and Medical Information Sciences, University of Leipzig
Härtelstraße 16-18, 04107 Leipzig, Germany
thomas.bittner@ifomis.uni-leipzig.de

Abstract. Integration of GIS and hydrologic models has been a common approach for monitoring our ever-changing hydrologic system. One important issue in adapting such an approach is to ensure the right correspondence of data across databases. To reach this goal, it is necessary to develop a description of the surface hydrology concepts that is internally consistent and semantically rich. In this paper, we apply the notions of endurance and perdurance to model the semantics of hydrologic processes in surface hydrology. Three hydrologic models were examined to identify concepts used in surface hydrology. The paper demonstrates the usefulness of applying the notions of endurance and perdurance to surface hydrology. The result is a set of primitive entities, aggregate entities, and relations between these entities that are necessary to cover surface hydrology concepts.

1 Introduction

Modeling the semantics of spatial dynamics has always been a key issue in advancing geographic information science. By incorporating the temporal dimension and the concept of change into a geographic information database, reality can be portrayed more faithfully than through the lenses of cartographic constructs [7]. Capturing such semantics is especially desirable in an interoperable environment, because the identification of suitable geographic information usually cannot be done with the help of human experts. While methods for the identification of such geographic information are available [5, 13], a complete description of such geographic information is still needed for those methods to be useful.

The objective of this research is to improve the usability of geographic information for modeling spatial dynamics in an interoperable environment. As an example of the fields that requires such information, this work focuses only on the surface hydrology domain. Over the past decade, there have been great demands in incorporating geographic information for modeling hydrologic systems [18]. This task typically requires the establishment of the right correspondence of data across databases, which is a problematic issue [2]. Solving this problem requires developing a data model that captures the semantics of hydrologic concepts used in disparate databases so that the

M.J. Egenhofer, C. Freksa, and H.J. Miller (Eds.): GIScience 2004, LNCS 3234, pp. 67–80, 2004.
© Springer-Verlag Berlin Heidelberg 2004

right correspondence between them can be made. This paper seeks to address this issue by examining how the notions of endurance and perdurance [6, 9] can be used to develop a data model for such a purpose.

Data models have previously been developed for the hydrology domain, among which the Arc Hydro data model [11] comes closest to this research. The Arc Hydro data model focuses on the description of surface hydrology and hydrography, and was built on the Environmental Systems Research Institute ArcObject data model. In this work, we are not interested in a particular GIS data model, but in a more generalized and robust mechanism to organize information about surface hydrology concepts. We apply the Basic Formal Ontology (BFO) proposed by Grenon and Smith [8], which is based on the notions of endurance and perdurance and on mereology [15], as a tool to build a data model that centers around hydrologic processes in the surface hydrology domain. The data model developed is rich in surface hydrology semantics and, therefore, offers better potential for facilitating the modeling of surface hydrology phenomena.

The remainder of this paper is structured as follows: Section 2 discusses the method used to extract concepts from the surface hydrology domain. Section 3 introduces BFO as a way to encode surface hydrology concepts. Sections 4 and 5 present the results of organizing surface hydrology concepts with the notions of perdurance and endurance, respectively. Section 6 concludes the paper.

2 Methodology

The development of a data model is an iterative process that typically involves the following steps: determining the purpose of developing a data model, extracting concepts, defining relations between concepts, and evaluating the results [12]. In this research, a similar methodology was adapted. Specifically, the following three steps were adapted:

1. Extract knowledge and identify the ontological commitments in a hydrologic model. The results of this step are informal.
2. Assign each entity, identified in the first step, to a category defined in BFO. We first identify perdurants, then endurants following the relations between these two entity types (Section 3.2). This sequence is important in this research, because it allows us to identify entities that are absolutely essential for describing hydrologic processes.
3. Cross-check the meanings of these entities identified in the first step against the definitions of categories and the relations between categories in BFO. The cross-checks minimize any mistake in assigning a concept to a wrong category.

The knowledge and ontological commitments about the surface hydrology domain were extracted from sources including books, journal articles, model programs, and source code of particular hydrologic models. These sources complement each other in a way that the model programs and source code provide "proofs" or verification for the conceptual models of the hydrologic systems laid out in books and journal articles, because those programs can be tested by comparing their estimates with the actual values obtained from a real hydrologic system.

Three hydrologic models were selected with all aforementioned sources in presence, including the HEC-HMS [17], TOPMODEL [1], and BLTM [10]. HEC-HMS is a hydrologic model that can be used to predict flood stage in response to a precipitation event. TOPMODEL is a hillslope scale hydrologic model that predicts the amount of runoff of a catchment, based on the idea that the area contributing to the runoff changes with the degree of soil saturation. BLTM is a river routing model that describes the speed of a contaminant traveling to a downstream location. These models have been chosen for two reasons. First, they have been developed over an extended period of time and have undergone several refinements. The validity of those hydrologic models is thus higher than those which have not gone through such revisions. Second, they were developed using different modeling approaches for the transportation of water on the land surface. This allows us to examine various conceptualizations of the same hydrologic process.

3   Encoding Surface Hydrology Concepts

Surface hydrology is concerned with the flow of water and its constituents over the land surface [4]. In order to describe such phenomena, various entities are involved, such as surface runoff or a watershed. Surface runoff is a process in which water is moved from one place to another. It is an entity that unfolds through time. A watershed is typically an entity that certain hydrologic processes act upon. The hydrologic processes may change a watershed's shape or boundary. However, the watershed does not unfold through time, but remains its identity over a prolonged period of time (i.e., the watershed you see today is the same one that you see tomorrow). An entity of the first type is called an *endurant*, while an entity of the second type is called a *perdurant*.

Differentiating these two distinctive types on surface hydrologic concepts is important in the development of a data model for the following reasons. First, the entities encoded should have a one-to-one correspondence to concepts comprehended by domain experts. This avoids any ambiguity of an entity's meaning. An entity that unfolds through time should be recognized as such. The same principle applies to an entity that retains its identity through time. Second, relations between entities can be identified more easily if the entities' types (*perdurant* or *endurant*) are known. This leads to fewer errors in the development a data model. This section describes a framework that recognizes both types of entities, that is, the Basic Formal Ontology (BFO) as presented by Grenon and Smith [8], to encode concepts in surface hydrology.

3.1   Endurant and Perdurant

The most basic categorial distinction between entities relates to different modes of persistence through time. Two categories of persistent entities can be distinguished:

endurants and perdurants. Endurants are wholly present (i.e., all their proper parts are present) at any time at which they exist. For example, you (an endurant) are wholly present in the moment you are reading this. No part of you is missing.

Perdurants, on the other hand, are extended in time in virtue of possessing different temporal parts at different times. As opposed to endurants, they are only partially present at any time at which they exist—they evolve over time. For example, at this moment only a (tiny) part of your life (a perdurant) is present. Larger parts of your life—such as your childhood—are not present at this moment.

Since those distinctions are so fundamental, it is critical to distinguish between perdurants and endurants at the very top of the classification hierarchy. In BFO, perdurants and endurants form disjoint category trees. Endurants are divided into two major categories: independent endurants, such as catchments and channels, and dependent endurants, such as qualities like catchment area or channel sinuosity. BFO distinguishes the following kinds of independent endurants: substances, fiat parts of substances, aggregates of substances, boundaries of substances, and cavities (Figure 1).

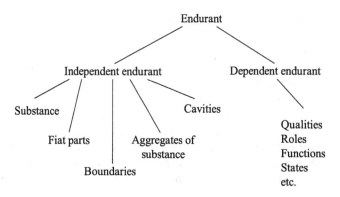

Fig. 1. Category tree of endurant (adapted from [8])

Substances are maximally connected entities. They have connected bona fide boundaries that correspond to discontinuities in the underlying reality. For example, neither a channel section nor a tributary are substances. Both are fiat parts of a channel. Fiat parts are demarked from a substance by boundaries that are not bona bide in nature but rather the result of human demarcation. Aggregates of substances are not substances either. Examples of aggregates are aggregates of several sub-catchments, channel and hillslope, etc.

Dependent endurants are entities that cannot exist without some other entity or entities upon which they depend. Dependent endurants can include, among others, qualities, roles (a canal can play the role of a channel), functions (one function of a riverbank is to accommodate excess water), and states (a catchment is in the state with full water storage).

Perdurant, the second top-category in BFO, distinguishes processes, their fiat parts, aggregates, and boundaries (Figure 2). Precipitation is a process, as is the graduation of a channel [16]. The development of a flood plane by a channel is a fiat part of the perdurant channel gradation. The interception of precipitation can be an aggregate of canopy interception and surface interception.

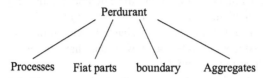

Fig. 2. Category tree of perdurant (adapted from [8])

3.2 Cross-Categorial Relations

Cross-categorial relations are relations that hold between the categorical distinct perdurants and endurants. The most important cross-categorial relations are *participation, realization, involvement*, and *affecting* [8]. In this paper, we mainly focus on *involvement*. The *participation* relation in our case is equivalent to the *involvement* relation in a reverse direction (i.e., one is the converse of the other). Involvement is a relation that holds between a perdurant and a substance endurant. Every perdurant acts upon at least one such endurant. Examples for *involvement* relations are a channel flow process that involves a flow plane and water, evapotranspiration that involves soil matrix and vegetation, etc. These examples can be easily translated into the *participation* relations: flow plane and water participates in the channel flow process.

3.3 A Case Study – Catchment, Watershed, and Basin

In surface hydrology the terms catchment, watershed, and basin are often used interchangeably to refer to the same entity. Results from the first step of our methodology showed that in HEC-HMS watershed and catchment are used to refer to identical entities and in TOPMODEL the same situation applies to basin and catchment. In different occasions, they may or may not carry the same meaning depending on how water flow is described. Figure 3 shows some possible meanings identified from the three hydrologic models.

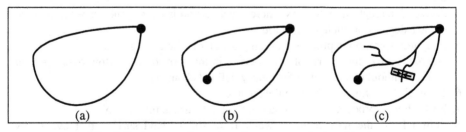

Fig. 3. Various meanings of catchment, watershed, and basin

The case (a) indicates that the catchment is represented by the outlet indicated by the black dot. The spatial variation within the catchment is not very important. In case (b) the catchment is similar to that of case (a), but now a channel (represented by the curved line in the catchment) is added into the catchment, indicating the consideration of water velocity in modeling the water flow. Case (c) is a more complex version of the case (b) in the sense that it considers a channel network instead of one channel. In addition, it also considers the hillslope flow to the channel network, as indicated by the strips on both side of the channels.

The discussion above demonstrates the need to develop a consistent data model for surface hydrology, if one needs to integrate data from disparate sources. Without such a data model it is difficult to distinguish whether the notion of catchment employed in model A is equivalent or similar to the notion of watershed (or basin) employed in model B. The rest of this section discusses how the notions of endurant and perdurant can provide the basis for developing a data model in surface hydrology. We first apply the notion of perdurant to classify concepts of surface hydrology. This reflects the priority of process in surface hydrology. Through the involvement (or participation) relation between perdurants and endurants, we eventually define the concept of a catchment and other concepts needed to support such a definition.

4   Perdurants in the Surface Hydrology Domain

In this section, we introduce entities that fall under the perdurant category. The discussion starts with processes and how they are determined. It continues with entities that are aggregates and fiat parts in the perdurant categorical tree. It ends with discussion on how involvement (or participation) relation leads to the identification of endurants.

4.1   Processes

Following the methodology specified in Section 2, eleven kinds of process have been identified under the scope of this research:

1. Precipitation: a process that moves water from the atmosphere to a land surface.
2. Canopy interception: a process in which vegetation catches and stores precipitation.
3. Surface interception: a process where precipitation is caught and stored in a small surface depression such as a puddle.
4. Evapotranspiration: a process that returns water to the atmosphere.
5. Overland flow: surface runoff that occurs in the form of sheet flow over the land surface without concentrating in clearly defined channels.
6. Channel flow: water flowing in the channel.
7. Infiltration: a process of water entry at the land surface into the soil.
8. Return flow: the movement of water from soil to land surface that eventually reaches to a channel.

9. Through flow: the movement of water within unsaturated or saturated soil layers located immediately beneath the surface.

10. Percolation: the movement of water through the interstices of rock or soil.

11. Base flow: the portion of channel flow originating from underground aquifer.

These processes were obtained first by determining if they qualify as perdurants. By accepting them as processes, we are able to reason about their temporal parts. For example, it is possible to obtain water transported by evapotranspiration process of yesterday, or to describe the behavior of channel flow (e.g., how fast water travels to a down stream location). The concept of perdurance is especially helpful in cleaning up ambiguity and duplications in the following cases:

1. Endurant or Perdurant?

In the first step of our methodology, precipitation volume was identified as an input to all three hydrologic models. Precipitation volume is a quality and, according to the BFO classification, a dependent endurant. It is dependent on another endurant, such as a catchment in HEC-HMS or a hillslope in TOPMODEL. While the *precipitation volume* serves as an important factor in describing hydrologic behavior, we did not find any description about the *precipitation process* in the hydrologic models examined. None of the models addresses the important question of how the precipitation process is formed or how its intensity or location changes through time. In this research, we represent precipitation as a process because talking only about qualities of processes does not allow us to reason about when and where a process occurs. What we really need is the process itself.

2. Naming Heterogeneity Between Processes

A case of naming heterogeneity was identified in the three hydrologic models. Naming heterogeneity refers to the circumstances where the semantically-alike entities are named differently [2]. Such cases can be identified through comparing meaning carried by entities of the same type, which in this case are those of process type. The case of semantic heterogeneity was found for depression storage and surface storage. Depression storage is a perdurant indicating water volume held in land surface depressions such as small ponds. Surface storage for runoff generation is a perdurant indicating water volume held in local depressions in the ground surface [17]. Given that depression storage and surface storage both perform the same function and occur on the same endurants, they are identified as a single process.

3. Process Specialization

Besides naming heterogeneity, we also found examples were one hydrologic process is a special case of another hydrologic process. Consider the processes lateral flow and source flow. Lateral inflow moves water from a flow plane into a channel. The flow plane can be on the left side or right side of a channel. On the other hand, source flow moves water from a flow plane to a source of a channel. Both types of flow move water in the form of sheet flow. The meanings of lateral inflow and source flow are essentially overland flow, except that the flow destination is explicitly defined. Both are therefore considered as two specific cases of the more general concept overland flow.

4.2 Aggregates Perdurants

Some processes identified from the three hydrologic models are composed of the eleven processes listed above. Examples of such aggregates are listed in Table 1.

Table 1. The list of aggregate perdurants identified from the three hydrologic models

Aggregate Perdurant	Processes Included in an Aggregate Perdurant
Loss	evapotranspiration, interception, infiltration
Precipitation excess	precipitation, loss
Interception	canopy interception, surface interception
Runoff generation	precipitation, loss, overland flow, channel flow
Direct runoff	precipitation, overland flow, through flow, infiltration, return flow

To see the usefulness of the notion of aggregates of processes consider the notion of direct runoff. It can be understood as the composition of overland runoff and through flow [17] of a catchment and thus is an aggregate formed by multiple perdurants. According to the analysis in Section 4.1, overland flow takes place on the land surface and through flow occurs in the soil layer immediately beneath the land surface. Due to the aggregation of these two processes, the resulting aggregate process, direct runoff, includes the amount of water flowing on the land surface and immediately below the land surface. Aggregation of those two hydrologic processes causes us to consider two other hydrologic processes occur in the same spatial region, namely infiltration and return flow, to be included for a complete description of direct runoff. The notion of aggregate endurant thus helped us to find and revise an incomplete definition of a hydrologic process in the models we analyzed.

Aggregates can also help identify semantically similar concepts. In a sub-model (i.e., Kinematic Wave Model) of HEC-HMS that describes overland flow, lateral flow is an aggregate of precipitation and loss. It is therefore identical to precipitation excess as defined in Table 1. The concept of lateral inflow in this sense was dropped.

4.3 Fiat Parts

The processes identified in Section 4.1 can have arbitrary temporal parts or parts that are based on the endurant parts. The first three hours of a precipitation event is an example of arbitrary temporal parts. From the three hydrologic models examined in this research, tributary flow was identified as a fiat part that is based on the endurant parts. Tributary flow is the movement of water in a tributary. The tributary can be defined arbitrarily depending on the modeling needs, but its existence depends on a channel. If a tributary (an endurant) is part of a channel (defined in 5.5), and if tributary flow is governed by the same physical laws as that of channel flow, then tributary flow can be considered a fiat part of the channel flow. Tributary flow in the three hydrologic models that we examined always exhibited these characteristics. It was, therefore considered a fiat part of a channel flow, and was merged with the channel flow concept.

All perdurant entities identified above are shown in Figure 4.

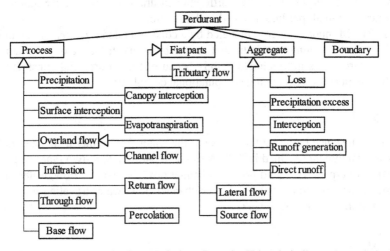

Fig. 4. Perdurants in the Surface Hydrology Domain (Triangle indicates is-a relation)

4.4 Involvement Between Perdurants and Endurants

Perdurants can have certain relations with endurants. For example, in order to describe the evapotranspiration process in TOPMODEL, three variables are needed: root zone storage, maximum root zone storage, and potential evapotranspiration rate. All three variables are *qualities* of a catchment (defined in 5.7). In addition to the catchment, sufficient amount of water is needed so that evapotranspiration can take place. The evapotranspiration process thus involves a catchment and water (defined in 5.1).

By tracking all variables needed to describe the hydrologic processes mentioned in this Section, additional independent endurants that possess these qualities were identified and are listed in Section 5. Dependent endurants are not included in this discussion for the purpose of clarity.

5 Endurants in the Surface Hydrology Domain

In this section, we will discuss eight independent endurants (substances) that were identified following the methodology set forth in Section 2. These independent endurants are entities that are wholly present at any time at which they exist. For each kind of endurant, the definition and its participation relation to the kinds of perdurants that were identified in Section 4 will be discussed.

5.1 Water

Water is an entity that is moved between all other endurants mentioned in this section. It may lose parts (water on the land surface may evaporate), or its acidity may change. Nonetheless the water still remains the same entity. It is therefore an endurant. Water can be moved around by different perdurants mentioned in Section 4, and thus participates in all perdurants listed in that section.

In the BLTM model, water is modeled by a concept called a *parcel*. A parcel has a pre-defined water volume and is treated as an object that travels along a channel. In this work, parcel was not included mainly because it is an artificial construct introduced for the purpose of reducing computational complexity.

5.2 Sink

A sink is a substance endurant that exhausts the water from either a channel (defined in 5.5) or a catchment (defined in 5.7). A sink can participate in two types of process: source flow (a type of overland flow) or channel flow. For source flow, water running on a flow plane converges at the sink. For channel flow, water traveling in a channel drains out of the channel at its sink.

5.3 Source

A source contributes water to a receiving endurant, which is normally a channel (defined in 5.5) or a catchment (defined in 5.7). It can participate in channel flow, where source provides input water volume to the channel. It can also participate in overland flow (in the form of a spring) to provide water input in a flow plane.

5.4 Junction

A junction connects at least two channel parts, such as a tributary and a channel section. It participates in channel flow in the sense that water flow from two channel sections is merged or flow from one channel section underwent bifurcation.

5.5 Channel

A channel is a linear feature connecting at least one source to a sink. A channel participates in channel flows, return flow, and base flow. With these flows, the direction of a channel can be defined as the flow of water from the source to the sink. If the channel has branches, the channel will also have junctions at the location where the channel and the branches meet. A channel is associated with several flow planes (defined in 5.6). Water on those flow planes flows into the channel.

A channel can have parts, which are themselves channels. Examples of those channel parts include sub-channels, tributaries, branches, or any channels of a lower order. The channel order designates the relative position of a sub-channel in a channel. A sub-channel without a tributary is a first-order channel. The convergence of two first-order sub-channels produces a second-order channel; the convergence of two second-order sub-channels produces a third-order channel. These channel parts participate in channel flow in the sense that they participate in channel flow parts, such as tributary participates in tributary flow indicated in 4.3.

The conditions for a channel y to be a sub-channel of another channel x are as follows:

- Channel y is part of channel x.
- The sink of channel y is a junction or a sink of channel x.
- The water flows from channel y to channel x.

If the sub-channel is a channel section, an addition rule applies:

- It has only one source and one sink.

5.6 Flow Plane

A flow plane is an area that contributes water to a channel. It participates in processes including precipitation, surface interception, overland flow (or lateral inflow), infiltration, and aggregate perdurants including precipitation excess, direct runoff, and loss. The flow plane itself does not have any channel, given that only sheet flow exists on a flow plane.

Depending on the specific relationship with a channel, four types of flow planes can be identified: left plane, right plane, source area, and inter-catchment area. The left flow plane contributes flow to a channel from its left side. The side of the channel is defined in terms of the channel flow direction. The right flow plane contributes flow to a channel from its right side. The source area contributes flow to the source of a channel. The last type of flow plane, the inter-catchment area [14], contributes flow to particular channels. Two conditions must hold for the inter-catchment area:

1. Those channels have to be at least of channel order 2. This means that there exist some channels of order 1 merging at junctions.
2. The inter-catchment area is within a catchment that encompasses those channels.

5.7 Catchment

A catchment is an areal feature that is defined by an outlet, the channel that drains to the outlet, and the area that drains to the channel. It participates in precipitation, surface interception, evapotranspiration, overland flow, channel flow, baseflow, infiltration, as well as in aggregate perdurants including direct runoff and interception.

Similar to the channel, a catchment can also be designated a catchment order. The order is defined as the largest channel order contained in the catchment. A catchment's boundary is defined as the area upstream of the outlet that, with the presence of water, drains to the outlet. Within this area, a catchment has flow planes and a channel. A catchment of order 1 has only a left flow plane, a right flow plane, and its source area. A catchment of order higher than 1 has all four types of flow planes. A catchment also includes earthy material, such as soil and rock, as well as land cover, such as tree and grass. The catchment may contain or carry fluid (e.g., water), but the fluid is not a part of it.

A catchment can have sub-catchments, as long as each sub-catchment is homogeneous in certain catchment-dependent entities, such as the intensity of the precipitation or the soil's infiltration rate. A sub-catchment has a sink and it has its own channel. In this respect, a sub-catchment is a catchment if the catchment met the characteristic of homogeneous property. The conditions for a catchment *y* to be a sub-catchment of another catchment *x* are as follows:

- The earthy material of the catchment *y* is part of that of the catchment *x*.
- The channel of the catchment *y* is part of that of the catchment *x*.
- The outlet of the catchment *y* is a junction of the channel of catchment *x*.

5.8 Atmosphere

Atmosphere is the gaseous envelope of a celestial body. It participates in precipitation and evapotranspiration.

All endurant entities identified above are shown in Figure 5.

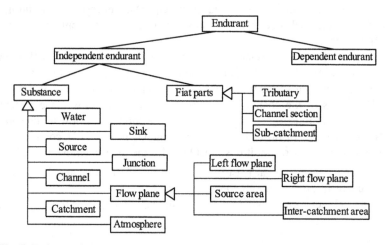

Fig. 5. Endurants in the Surface Hydrology Domain (Triangle indicates is-a relation)

6 Discussion and Conclusions

In this paper, we demonstrated how the notions of endurance and perdurance can help us to develop a data model focusing on processes in surface hydrology. We incorporated the following notions in our data model: substance endurants, fiat parts, aggregates, and dependent endurants to identify endurants such as catchment and channel, as well as process, fiat parts, and aggregates to identify perdurants such as precipitation and overland flow. In Section 4 and 5 we identified entities that are indispensable to understanding the movement of water on the surface of the earth. We showed how BFO can be used not only as a tool to assist in capturing concepts for surface hydrology, but also in identifying missing, duplication, or specialized concepts. The use of the endurance and perdurance notions thus contributes to a better understanding of hydrologic concepts, in the sense that they are now more structured and less ambiguous. The data model proposed in this paper is more consistent than those obtained in step 1 listed in Section 2. It is more consistent because redundant entities were either removed (e.g., a river or a reach as to a channel) or merged through parthood relations (e.g., tributary, sub-reach as part of a channel), artificial entities introduced for the purpose of computational simplicity were removed, and several missing entities were identified. It is possible to know what a process is by examining its variables encoded as dependent endurants. Through the participate relation between endurants and perdurants, it is possible to identify the functions of an endurant in monitoring water flow on the land surface.

One interesting issue that is not handled in this work is the issue of granularity [3]. Some dependent endurants identified for catchment, such as the two routing coefficients in a sub-model of HEC-HMS, indicate the intention of describing overland flow and channel flow behavior in the model. However, the substance endurants involved do not include a channel or a flow plane of any sort. This may suggest that it is necessary to introduce the concept of granularity in the development of a data model so that it more completely describes surface hydrology.

Future works will need to provide a formalization of the presented data model using formal tools like predicate logic, and to verify the completeness of the data model. The completeness of the data model can be verified by using Haskell, as suggested by Winter and Nittel [19]. We also wish to broaden the scope of this research to the ground water domain.

Acknowledgment

Support for the second author from the Wolfgang Paul Program of the Alexander von Humboldt Foundation and from the National Science Foundation Research Grant BCS-9975557: Geographic Categories: An Ontological Investigation, is gratefully acknowledged. Partial Funding for the third author was provided by Grant # R82-7961 from the U.S. Environmental Protection Agency's Science to Achieve Results (STAR) program. This paper has not been subjected to any EPA review and therefore does not necessarily reflect the views of the Agency, and no official endorsement should be inferred. The authors also acknowledge the helpful suggestions of the anonymous reviewers of this paper.

References

1. Beven, K.J. and Kirkby, M.J.: A Physically Based, Variable Contributing Area Model of Basin Hydrology. *Hydrological Sciences.* 24 (1979) 43-68
2. Bishr, Y.: Overcoming the Semantic and Other Barriers to GIS Interoperability. *International Journal of Geographical Information Science.* 12(4) (1998) 299-314
3. Bittner, T. and Smith, B.: A Theory of Granular Partitions. In Duckham, M., Goodchild, M., and Worboy, M. (eds.): *Foundations of Geographic Information Science.* Taylor & Francis, London. (2003) 117-151

4. Chow, V., Maidment, D., and Mays, L.: *Applied Hydrology*. McGraw-Hills Inc, New York (1988)
5. Feng, C.-C. and Flewelling, D.: Assessment of Semantic Similarity between Land Use/ Land Cover Classification Systems. *Computers, Environment, and Urban Systems*. 28(3) (2004) 229-246
6. Gangemi, A., Guarino, N., Masolo, C., Oltramari, A., and Schneider, L.: Sweetening Ontologies with DOLCE. *AI Magazine*. 24(3) (2003) 13-24
7. Goodchild, M.: Communicating Geographic information in Digital Age. *Annals of the Association of American Geographers*. 90(2) (2000) 344-355
8. Grenon, P. and Smith, B.: SNAP and SPAN: Towards Dynamic Spatial Ontology. *Spatial Cognition and Computation*. 4(1) (2004) 69-104
9. Hawley, K.: *How Things Persist*. Clarendon Press, Oxford (2001)
10. Jobson, H. and Schoellhamer, D.: *Users Manual for A Branched Lagrangian Transport Model*. United States Geological Survey: Reston, Virginia. (1987)
11. Maidment, D.: *Arc Hydro: GIS for Water Resources*. ESRI Press, Redlands (2002)
12. Noy, N. and McGuinness, D.: *Ontology Development 101: A Guide to Creating Your First Ontology*.
    http://protege.stanford.edu/publications/ ontology_development/ontology101.pdf, (2001)
13. Rodríguez, M.A. and Egenhofer, M.: Determining Semantic Similarity among Entity Classes from Different Ontologies. *IEEE Transactions on Knowledge and Data Engineering*. 15(2) (2003) 442-456
14. Schumm, S.: *Evolution of Drainage Systems and Slopes in Badlands at Perth Amboy*, New Jersey. Department of Geology, Columbia University: New York, NY. (1954)
15. Simons, P.: *Parts, A Study in Ontology*. Oxford University Press, Oxford (2000)
16. Strahler, A. and Strahler, A.: *Introducing Physical Geography*. John Wiley and Sons, Inc., New York (2003)
17. USACE: *Hydrologic Modeling System HEC-HMS*, Technical reference manual. US Army Corps of Engineers, Hydrologic modeling center, HEC: Davis, CA. (2000)
18. Westervelt, J. and Shapiro, M.: Combining Scientific Models into Management Models, in *4th International Conference on Integrating GIS and Environmental Modeling (GIS/ EM4)*: Banff, Canada. (2000)
19. Winter, S. and Nittel, S.: Formal Information Modelling for Standardization in the Spatial Domain. *International Journal of Geographical Information Science*. 17(8) (2003) 721-741

Procedure to Select the Best Dataset for a Task

Andrew U. Frank[1], Eva Grum[1], and Bérengère Vasseur[2]

[1] Institute for Geoinformation and Cartography, Technical University of Vienna,
Gusshausstrasse 27-29, 1040 Vienna, Austria
{frank,grum}@geoinfo.tuwien.ac.at
[2] Université de Provence, Laboratoire des Sciences de l'Information et des Systems LSIS,
Rue Joliot-Curie, 13453 Marseille cedex 13, France
berengere.vasseur@lsis.org

Abstract. This paper models the decision process when selecting among different datasets the one most suitable for a task. It shows how metadata describing the quality of the dataset and descriptions of the task are used to make this decision. A simple comparison of task requirements and available data quality is supplemented with general, common-sense knowledge about effects of errors, lack of precision in the data and the dilution of quality over time. It consists of two steps: first, compute the data quality considering the time elapsed since the data collection; and second, assess the utility of the available data for the decision. A practical example of an assessment of the suitability of two datasets for two different tasks is computed and leads to the intuitively expected result.

1 Introduction

The selection of a dataset or a map from a choice of several potentially useful ones seems to be nothing else than a numerical comparison between the requirements of the task with the available data quality listed in the metadata description. It is often done intuitively, but a computational model for this seemingly simple task is missing. A computational model is necessary for search engines to use the metadata for automatic ranking of available sources with respect to an intended use. Unless search engines use metadata, especially data about geographic data quality, the investment in metadata production does not yield the expected benefits.

The collection of metadata is often advocated, for example in the EU project INSPIRE, but metadata is not always available and sometimes in a grotesquely unsuitable form (Hunter and Goodchild 1995). One may conclude that they are not often used (UNE 2001). Is the lack of practical interest in metadata due to the absence of use? Do we really know how to use data quality descriptions? In this paper, we give a computational model for the use of metadata describing the quality of geographic data to assess the usability or "fitness for use" (Chrisman 1984). We assume a situation where more than one dataset from different sources could be useful for a task. The task may be navigation in town, where the decisions are to turn left or right at a street corner or it may be the decision about the location of a facility. A dataset is contributing to a task if it improves the decision. Valuable data lead to an execution that uses fewer resources or produces a better result (Krek 2002).

The rational, programmable method proposed here leads to the selection of the datasets that provide most improvements in the decision for the task. The approach is novel in several aspects:

M.J. Egenhofer, C. Freksa, and H.J. Miller (Eds.): GIScience 2004, LNCS 3234, pp. 81–93, 2004.
© Springer-Verlag Berlin Heidelberg 2004

- it is a quantitative assessment,
- it breaks user groups into individual tasks and decisions,
- it is based on qualitative models of processes (Kuipers 1994),
- it separates temporal currency from other descriptions of data quality, and
- it leads to a programmable selection using existing metadata.

We describe here a rational method for comparing the utility of different datasets with respect to an intended use. It is based on a detailed assessment of the data quality for each theme available in the data sets following the usual model of data quality description for geographic data (Moellering 1987; Chrisman 1988; Morrison 1988; Goodchild and Gopal 1990). A decision process connects the data with the task. In this process the quality of the data is transformed into the quality of the task. A model of this decision process contains all the information necessary to assess the utility of a dataset to improve the task. We use this "common sense" knowledge when we make an intuitive decision about which data source to use for a decision.

The paper demonstrates the method using a simplified example: two data sets are available and assessed for use by a tourist and firefighters. The data sets are patterned after the multi-purpose digital map of Vienna and a digital tourist map (Wilmersdorf 1992; Wien 2004). The resulting recommendation agrees with our intuitive choice and are formalized, ready to be integrated into a search engine for geographic data.

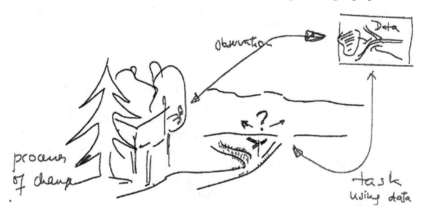

Fig. 1. Closed loop semantics (Frank 2003): the observation process is linked to the task using the data (section 3). Changes occur in the world between data collection and execution of task (section 4) and task using data for decisions (section 5)

The paper is structured as follows (fig. 1): The second section of the paper describes the terminology used and how we model the question in the abstract. It covers the user task and its requirements; what is data quality and how to describe the usability of a map. It introduces the processes involved, namely data collection, changes in the world, and decision. The third section assesses the available data. We assume that geometric precision, attribute accuracy and temporal currency are the most dominant components of data quality. The data quality description therefore concentrates on these aspects. It is an abstract description of the observation and data collection processes. Section 4 then describes how the data quality assessment is updated, considering the usual changes in the world. Section 5 then assesses the usability of a single data set for a single decision. Section 6 summarizes the result and gives a comprehensive description of the process that leads to the selection of a dataset for a decision and applies it quantitatively to an example. The seventh section introduces the notion of a user group and shows why a decision among several datasets is usually not made for individual tasks, but for user groups. Section 8 concentrates on the issue of clutter: a data set that contains irrelevant data is less suitable than one that has exactly what is required. The concluding section summarizes the result and points out that the same method can be used as a guideline to design new usable geographic information products.

2 Processes

In this section we explain how we conceptualize the situation and introduce the terms we use subsequently.

A user needs to make a decision in a task and considers the acquisition of a dataset (or a conventional map). The example task in this paper is navigation in a city, that is, the decision about turning left or right at a street corner, extensively studied by Krek (2002). Other spatial decisions have different requirements for the data necessary, but they can be dealt with the same logic.

The user has the choice of several data sets that are potentially beneficial for her task. They have *utility* for a task if they lead to making better decisions and a better execution of the task. The data sets contain data describing different aspects of reality, which we will call *themes*. Examples are: street name, the position of an ATM or the number of the bus line serving a bus stop. We use the term data *quality* to describe the correspondence between an object in reality and its representation in the data set. Data quality will be differentiated for several *aspects*, such as precision and completeness.

The description of the quality of the data available to the user is a description of the processes that were used for the collection; the quality of the data is directly linked to these processes (Timpf, et al. 1996). This description of data quality is independent of the task; the provider of the data can only describe the data from this perspective.

Originally, the data collection process is linked to the decision process for which the data are collected, which is abstracted in the description of the transformation of data quality to utility of the decision. In this *closed loop* from the real world to data collection, to data use for a decision, to a task executed in the real world (fig. 1), semantics of data quality measures can be defined (Frank 2003). Data are usually collected long before they are used in a decision; reality is changing in the intervening interval. These changes must be modeled as a decrease of the data quality (section 4).

The models of processes used here are second order; the first order process is the decision process itself. Therefore, only rough estimates of how these (second order) processes of data quality transformation work are sufficient. The model of the influence of data quality on decision quality leads to a rational assessment of usability of a dataset, even in the absence of detailed knowledge.

3 Description of the Data Quality of a Data Set

This section describes how we assess the quality of the data sets when it is collected. It follows the literature on metadata and data quality descriptions (Moellering 1987;

Chrisman 1988; Morrison 1988; Goodchild and Gopal 1990).The quality of a dataset results from the processes used for data collection. Data collected with the same processes have similar quality; it is therefore reasonable to give data quality indications for whole data collections or large parts that were collected in the same way.

Table 1. Example of data quality description (fictitious values)

Theme	Multi purpose map			City map		
	when collected	complete	precision (m)	when collected	complete	precision (m)
Bus line				2001	90%	0,5
Church				1998	99%	0,1
Fire hydrant	2001	99%	0,05			
Building number	2003	99%		2000	90%	
Points of interest				1999	90%	0,5
Stations for public transport				2002	95%	1
Street name	2001	95%		1999	95%	
Street network	2002	98%	0,1	2003	95%	0,2
Tram line				2001	90%	0,2
Walls between and inside buildings	2003	99%	0,05			

The measurement and observation procedures determine the *precision* of the data, which are typically described by the standard deviation of random errors – assuming no gross errors and no systematic bias. The data collection methods used determine the *completeness* of a theme. For brevity, completeness is used as a summary assessment for omissions and commissions; precision of data not measured on a continuous scale is lumped together with completeness. Last but not least, the *date* of the data collection indicates when the correspondence between reality and the data were assessed by precision and completeness.

The two datasets used here as examples for typical content in publicly available maps useful for in-city navigation are the multi-purpose map of the city of Vienna and the city map of Vienna (Wien 2004). Table 1 lists some of the themes in these datasets and fictitious quality assessments for them. The data quality assessment values are for demonstration purposes only and do not reflect the data quality found in the products distributed by the Magistrate Vienna.

4   Dilution of Precision and Completeness over Time

The metadata that come with a dataset describe the data quality at the time of collection or last major update. With time the world changes and the quality is reduced (figure 2). For a decision about the utility of a dataset, the quality today is of interest.

This section shows how to update the given precision and completeness with estimated rates of dilution in quality. It addresses the paradoxes that the data do not change with time, but nevertheless their quality dilutes (data do not age with time like good wine!).

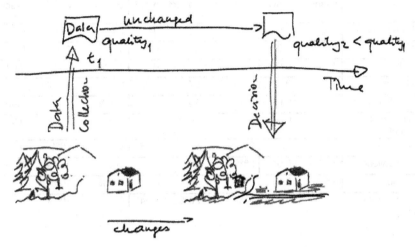

Fig. 2. The world changes, but data remains fixed

Figure 2 gives a more understandable explanation of the previously published statement that precision, completeness and up-to-datedness are *not* orthogonal (Frank 1987; Frank 1998). The utility of a dataset of high precision, but observed some time ago, and a not-so-precise dataset observed more recently is perhaps the same.

4.1 Precision

The precision of the location of objects is diluted by random movements. Such movements are more important for natural themes, like forest boundary, streams etc., than for the manmade objects in a city (Burrough 1996). The precision is reduced proportionally to the time intervened:

$$p_i = p_0 - r_p * dt_i$$

where $dt_i = t_i - t_o$
t_0 the time of data collection
p_0 the precision at that time
r_p the rate of change in the precision

Similar to location, the precision of other values can be transformed to a current date considering the rate of dilution in the precision of this data element. Some descriptive values, for example prices, change gradually over time by the inflation rate, which is a systematic, first-order effect and can be taken into account when transforming the data. The dilution in precision is due to the differential change in prices, which reduces the precision with which we can know the current price given the price several years ago, second order effect. Estimation of quantitative values is not crucial; we leave it for future work to study the influence of errors on these estimates.

Table 2. Updated (diluted) data quality

Data quality description for a data set	r_c	r_p	Multi purpose map			City map of Vienna		
	change	move	complete	correct	precision	complete	correct	precision
Building number	2%		97%	97%	0	98%	77%	0
Bus line	5%	0.01	0%	100%	0	105%	100%	0.47
Church	1%	0.01	0%	100%	0	105%	100%	0.04
Fire hydrant	4%	0.01	87%	87%	0.02	0%	100%	0
Points of interest	4%	0.01	0%	100%	0	110%	100%	0.45
Stops of public transport	5%	0.01	0%	100%	0	105%	100%	0.98
Street name	1%		92%	92%	0.08	95%	90%	0
Street network	5%	0.01	88%	88%	0	100%	100%	0.19
Tram line	4%	0.01	0%	100%	0	102%	100%	0.17
Walls between buildings	3%	0.01	96%	96%	0.04	0%	100%	0

4.2 Completeness

The completeness is diluted by the rate of new objects appearing and old ones disappearing (a different factor could be used to transform omissions and commissions separately).

$$c_i = c_0 - r_c * dt_i$$ where c_0 the completeness at the time of data collection

Table 2 gives the assumed values for rate of change of precision and completeness.

5 Description of User Task

One cannot determine the suitability of data for a task just by studying the data quality description alone, but has to consider the task and the decision that should be made with the information from the dataset and how it is influenced by the quality of the data. A compact description of the user's decision situation and how it is influenced by the available quality of the data is necessary. The description of the task given here is reduced to the aspects relevant for the selection of a data set considering data quality alone and is only one component of a comprehensive description of a task. This is the goal of this section.

Krek has shown how the quality of spatial information affects the quality of a decision (Krek and Frank 2000; Krek 2002); for a given decision, it is possible to identify which themes influence the decision. Only the data quality of the themes that influence the decision affects the quality of the decision.

Krek computed the value of a dataset as the economic contribution it makes to improve the decision and thus the outcome of the task, comparing with the decision made without information available. Here, we want to assess the utility of a dataset and compare it with the perfect dataset, that is, the dataset that contains no error, which has by definition utility 1. Any other dataset has a lesser utility, expressed as a percentage. The unusable dataset has utility 0.

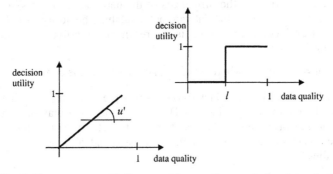

Fig. 3. The two types of influence of data quality: gradual and threshold

5.1 Influence of Single Decision: Precision and Correctness of One Theme

The influence of the quality of the dataset on the decision depends on the dataset and the decision. Two typical cases can be differentiated (figure 3):

- **gradual influence:** "the higher the quality, the better the decision;" it is described with the rate u' with which the utility increases with an increase in the data quality.
- **threshold:** if the quality is above a certain level, the decision is the same and the utility 1; if the quality is lower, then the decision is the same as if there were no data, that is, the utility is 0; it is characterized by the threshold value l.

These two influences are sometimes combined: up to a threshold the quality does not influence the decision, and from then it increases with the quality proportionally (Fig. 4). Note that precision is usually expressed such that lower values mean higher quality!

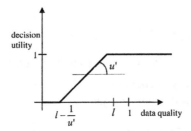

Fig. 4. The influence model with a threshold value and an influence rate

$$u_{pi} = \begin{cases} 0 & for\ p_i < l - \dfrac{1}{u'} \\ 1 & for\ p_i > l \\ 1 - (l - p_i) \cdot u' & otherwise \end{cases}$$

where u_{pi} is the resulting utility from precision, p_i the available data precision, l the threshold and u' the rate of increase.

5.2 Influence on Many Decisions: Completeness of One Theme

The formula above computes the influences of the data quality for a single decision where the data are available. If the data are not available, the utility is 0. The average utility given for an incomplete dataset is the result of multiplying u with the completeness factor.

$$u = u_p * c \qquad \text{where } u \text{ is average utility, } c \text{ is the completeness}$$

There may be cases where the lack of data is interpreted as negative fact, invoking the closed world assumption (Reiter 1984). Then the decision may be worse than in the absence of any information – a negative utility! This is not further considered here, because it is related to the discussion of omission and commissions, which is left for future studies.

5.3 Multiple Themes Used in One Decision

Decisions may require several themes as inputs. Two different situations can be differentiated:

Single decision: Each theme contributes individually and independently to the decision. The utility of the decision is the weighted average of the contributions of the themes. The weighted average is used because the utility is normalized to the interval [0..1].

Related decisions: One or several themes are enabling. If the data are not present, data from other themes cannot be used. This is in decision making sometimes called a K.O. criterion and related to Maslow's pyramid, where requirements on a higher level are only considered if the lower level requirements are fulfilled (Jahn 2004). If a theme is crucial and the utility of the theme for the decision is less than a low threshold (say 20%), then the total utility of the dataset for this decision is 0.

5.4 Summary and Example

Every decision related to a task is characterized by four parameters:

- Threshold l for data quality to count as available.
- Rate of increase u' in utility with data quality.
- Is it a K.O. criterion?
- Relative weight compared to other themes.

For three example decisions, the themes used and the parameters are shown in table 3.

6 Calculate Utility for a Specific Decision

A specific decision requires only few inputs themes. For these the utility is computed and summed up. Table 4 shows the computation for the task *navigation to a street*. For the two themes *street name* and *street network* the contribution of the theme to this decision is computed; it uses the updated data quality from table 2 and the task description from table 3 and computes from right to left the usability due to precision and correctness and then multiplies it with the completeness value. The contributions for the two themes are then averaged and yield the total utility.

Table 3. Characterization of tasks

Task		Threshold l	u'	K.O.	Weight
Navigation to a street	Street name			1	1
	Street network	5	0.1	1	1
Find building on street	Street name				1
	Building number	2	0.2	1	1
Public transportation use	Stop of public transportation	10	0.1	1	1
	Bus lines	20	0.1		1
	Tram lines	20	0.1	1	1

Table 4. Utility of themes for a task

Navigation to a street	Task description			Multi-purpose map		
Theme	Threshold l	u'	K.O.	complete	correctness	precision
Street name			1	92%	95%	0
contributions				87%	95%	1
Street network	5	0.1	1	88%	100%	0.08
contributions				88%	100%	1
Utility total				**88%**		

With the same formulae, the utility for other tasks and datasets can be computed as in table 5.

Table 5. Utility for different tasks

Task	Multi-purpose map	City map of Vienna
navigation to a street	88%	83%
find building on street	90%	70%
public transportation use	0%	79%
fire response planning	92%	0%
find interesting places	30%	58%

The multi-purpose map of Vienna is assumed to be more up-to-date, which results in a higher utility for a task in navigation to a street and finding a building on a street; the higher geometric precision does not influence the result at all.

7 User Groups

In the previous sections, decisions for the suitability of a dataset are made for a single task and a single decision. This is, for example, the case when a dataset is acquired for a single administrative decision.

In practice a decision to acquire a dataset or a map is often made in order to cope with a complex mixture of future decisions including contingencies, like where is the next pharmacy, or where is the next police station. The cost of acquiring a dataset and the cost of learning how to read and use it is considerable. A multi-purpose set, which one can use in many situations and for many tasks – even if not completely optimized for my present task – is preferable than to learning each time how to use the specially constructed data set for this task.

Traditionally, maps where produced for special user groups. There are city maps for tourists and for locals; special maps are prepared for emergency services. Topographic maps are produced to serve a very wide audience. Recently some more specialized maps have appeared, such as maps with bicycle paths, or maps with bigger letters for the elderly.

A user group is characterized by a weighted set of task and the related decisions. The utility of a dataset is assessed as the weighted mean of the utility of the dataset for each of the decisions. For example, the user group "tourists" relates to the tasks:

- identify interesting objects, and find locations where they are clustered,
- finding an object to visit on the map, and
- use the map for navigation and identification of the object.

Firefighters have different needs: they are interested in buildings and space to drive to a building as close as possible, independent of traffic regulations, the location of fire hydrants and fire resistant walls inside buildings. Some of the tasks in a user group can be declared as K.O. criteria and the utility of a dataset is zero if this criterion is not satisfied to some reasonable degree.

Table 6 gives the utility of the two maps for different tasks of importance to the user groups and gives the weights for each task and user group. With these weights, the multi-purpose map is most suitable for the fire-fighters and the city map suits the tourist best (table 7). If K.O. criteria were assigned, the same result would be obtained.

Table 6. Tasks for two user groups, the utility of two datasets for these tasks and weights two user groups assign to these tasks

	Utility		Weight	
	Multi purpose map	City map of Vienna	Tourist	Firefighter
Navigation to a street	88%	83%	3	4
Find building on street	90%	70%	2	4
Public transportation use	0%	79%	3	0
Fire response planning	92%	0%	0	3
Find interesting places	30%	58%	4	0

Table 7. The overall utility of two datasets and two user groups

	tourist	firefighter
Multi purpose map	47%	**89%**
City map of Vienna	**72%**	56%

8   Clutter Reduces Usability

If two datasets contain all the data required, but one offers additional data that are not required for any decision of this user group, the two datasets are numerically assessed the same utility, but intuitively they are not equally usable. We prefer the dataset with less clutter. The additional effort to filter out the unwanted data is a cost for the user and may lead to errors in decisions.

To reduce usability due to clutter we can reduce the usability by a factor proportional to the amount of data required, compared with the amount of data presented. The clutter factor is less than 1 and is multiplied with the usability measure introduced above. To calculate clutter ratio, the number of data elements per theme must be available in the metadata.

$$clutter\ ratio = \frac{data\ used}{data\ presented}$$

For the example data sets and user groups, we find that the clutter factor for the tourist using the multi-purpose map is much lower (worse) than for the city map (table 8). This reflects the intuitively higher usability of the city map for tourists.

Table 8. Usability reduced for clutter

Map elements		Multi-purpose map	clutter factor	City map	clutter factor
offered		1919.84		739.33	
required	firefighter	1796.74	**0.93**	724.65	**0.98**
	tourist	801.32	**0.42**	733.83	**0.99**

9   Conclusions

This method to calculate the utility of a data set for a task emerged from a method to describe suitability of a dataset for a task in a matrix, where on one axis the available data themes were listed and on the other axis the requirements for the task. It was refined by separating the assessment of the dataset into one table and the required data quality into another table (Grum and Vasseur 2004). This reduced the effort to prepare data quality and task descriptions from *(n x m)* to *(n + m)*, where *n* is the number of task and *m* is the number of themes.

The new method is based on a model of the processes involved and a very generalized qualitative assessment of these processes (Kuipers 1994). Only rough estimates of the quantities involved are sufficient to the metadata into an assessment of the utility of the dataset for a task or user group.

The method can be used – as described – to select the most useful dataset for a specific task. It can be used equally well when considering the design of an information product (Krek and Frank 1999). It leads to a multi-step procedure for a design of an information product:

- Identify a user group.
- List the tasks the users have to solve and identify the decisions included.
- For each decision, consider information inputs that can improve the decision.

The research reported here creates the framework for precise and operational definitions for data quality measures. Open questions relate to the treatment of omissions and commissions, with respect to objects and attributes recorded for these objects. Another long standing question is how to describe the quality of data not measured on a continuous scale

Understanding the decisions related to a task is quite limited. Two Ph.D. candidates are working on a detailed analysis – at the level of detail demonstrated by Hutchins (Hutchins 1995) – of two tasks, namely in town movement, using public transportation, and orienteering in open landscape.

The method is based on quantitative estimates of the influence of data quality on decisions; it should be investigated how sensitive the decision between two datasets is to variations in these estimates.

Acknowledgments

Many of my colleagues and students have helped me advance to this point. Of importance where the many discussion in the REVIGIS project, led by Robert Jeansoulin. Particular thanks go to Werner Kuhn and Max Egenhofer for valuable discussion over the years. The comments from three reviewers were very helpful to improve this final version of the paper.

This work was funded as part of the REVIGIS project #IST-1999-14189. We also thank the magistrate Wien for providing us with datasets and descriptions and apologiz for the liberty we have taken with assessing the quality in this demonstration.

References

1. Burrough, P. (1996). Natural Objects with Indeterminate Boundaries. *Geographic Objects with Indeterminate Boundaries*. P. Burrough and A. Frank (eds.). London, Taylor & Francis: 3-28.
2. Chrisman, N. (1984). The Role of Quality Information in the Long-Term Functioning of a Geographic Information System. *Cartographica* 21(3 and 4): 79-87.
3. Frank, A. (1987). Overlay Processing in Spatial Information Systems. *Eighth International Symposium on Computer-Assisted Cartography (AUTO-CARTO 8)*, Baltimore, MD: 16-31.
4. Frank, A. (1998). Metamodels for Data Quality Description. *Data Quality in Geographic Information-From Error to Uncertainty*. R. Jeansoulin and M. Goodchild (eds.). Paris, Editions Hermès: 15-29.

5. Frank, A. (2003). Ontology for Spatio-Temporal Databases. In M. Koubarakis, T. Sellis, A. Frank, S. Grumbach, R. Güting, C. Jensen, N. Lorentzos, Y. Manolopoulos, E. Nardelli, B. Pernici, H.-J. Schek, M. Scholl, B. Theodoulidis, and N. Tryfona (Eds.) *Spatiotemporal Databases: The Chorochronos Approach.* Berlin, Springer, Lecture Notes in Computer Science 2520: 9-78.

6. Goodchild, M. F. and S. Gopal (1990). *The Accuracy of Spatial Databases.* London, Taylor & Francis.

7. Grum, E. and B. Vasseur (2004). How to Select the Best Dataset for a Task? In A. Frank. (ed.) *International Symposium on Spatial Data Quality 04.* Vienna, Austria, pp. 197-206.

8. Hunter, G. and M. Goodchild (1995). Dealing with Error in a Spatial Database: A Simple Case Study. *Photogrammetric Engineering & Remote Sensing* 61(5): 529-537.

9. Hutchins, E. (1995). *Cognition in the Wild.* Cambridge, MA, The MIT Press.

10. INSPIRE *INfrastructure for SPatial InfoRmation in Europe* [http]. 2003 [cited 14.07. 2004]. Available from http://www.ec-gis.org/inspire/.

11. Jahn, M. (2004). User Needs in a Maslow Schemata. in: A. Frank (ed.) *International Symposium on Spatial Data Quality 04.* Vienna, Austria, pp. 169-182.

12. Krek, A. (2002). *An Agent-Based Model for Quantifying the Economic Value of Geographic Information.* Ph.D. Department of Geoinformation. Technical University Vienna, Vienna, Austria..

13. Krek, A. and A. Frank (1999). Optimization of Quality of Geoinformation Products. in P. Whigham (ed.), *Proceedings of the Eleventh Annual Colloquium of the Spatial Information Research Centre,* Dunedin, New Zealand, pp. 151-159.

14. Krek, A. and A. Frank (2000). The Economic Value of Geo Information. *Geo-Informations-Systeme* 13(3): 10-12.

15. Kuipers, B. (1994). *Qualitative Reasoning: Modeling and Simulation with Incomplete Knowledge.* Cambridge, MA, The MIT Press.

16. Moellering, H. (1987). *A Draft Proposed Standard for Digital Cartographic Data,* National Committee for Digital Cartographic Standards.

17. Morrison, J. (1988). The Proposed Standard for Digital Cartographic Data. *American Cartographer* 15(1): 9-140.

18. Reiter, R. (1984). Towards a Logical Reconstruction of Relational Database Theory. In M. Brodie, M. Mylopolous and L. Schmidt (eds.). *On Conceptual Modelling.* New York, Springer Verlag, pp. 191-233.

19. Timpf, S., M. Raubal, and W. Kuhn (1996). Experiences with Metadata. in: *Proceedings of the 7th International Symposium on Spatial Data Handling, SDH'96,* Delft, The Netherlands, IGU: 12B.31-43.

20. UNE. *The Use of Metadata in University Libraries and Campuses* [http]. 2001 [cited 28.06.04 2004]. Available from http://www.caul.edu.au/surveys/Metadata.doc.

21. Wien, Magistrat (2004). Maps of the City of Vienna.

22. Wilmersdorf, E. (1992). The Implementation of A Multidisciplinary GIS-Network for An Urban Geocommunication System. *Proceedings of EGIS '92,* Munich, Germany, EGIS Foundation, pp. 1408-1416.

Floating-Point Filter
for the Line Intersection Algorithm*

Andras Frankel, Doron Nussbaum, and Jörg-Rudiger Sack

School of Computer Science, Carleton University,
1125 Colonel By Drive, Ottawa, Ontario, Canada, K1S 5B6
Andras.Frankel@sympatico.ca
{nussbaum,sack}@scs.carleton.ca

Abstract. The use of floating-point arithmetic in geometric computation represents a formidable challenge for development and implementation of geometric algorithms. On one hand, one thrives to develop algorithms that are robust and produce accurate results, while on the other hand, one attempts to achieve rapid execution time. In particular for GIS applications, where large problem sizes are frequently encountered, efficiency considerations are important. In this paper, we present a floating-point filter written specifically for the important line intersection operation that is robust, outperforms existing general purpose filters and results in an accurate discovery and representation of topology from the geometric information.

1 Motivation

Geometric algorithms are often designed upon the assumption that numeric computations are accurate (namely, computation is performed by using real numbers $\Re$). However, computer hardware does not support exact arithmetic of real numbers. Instead computers employ finite precision floating-point data representation and arithmetic (6-15 decimal points using 32-64 bits representation). This results in a loss of accuracy, since not all real numbers can be stored exactly as floating-point numbers and hence have to be rounded [9]. Floating point operations may produce results which are fundamentally geometrically wrong. For example, Ramshaw [18] showed how two line segments may intersect more than once due to floating point errors.

Geometric computation errors may be compounded during the execution of many geometric/GIS algorithms, because the algorithms rely on previously computed data values. For example, intersection points between line segments may become vertices of new objects. These vertices are processed subsequently in other operations. It has been observed [7, 10] that for graphical output this may be visible as glitches and GIS application programs may

---

* This R&D project has been supported in part by Sun Microsystems, the Natural Sciences and Engineering Research Council of Canada and Nortel Networks.

M.J. Egenhofer, C. Freksa, and H.J. Miller (Eds.): GIScience 2004, LNCS 3234, pp. 94–105, 2004.
© Springer-Verlag Berlin Heidelberg 2004

- crash: The reason for crashes is that programs enter into states that are illegal, e.g., into states that a program should have never reached if the topology was geometrically correct. One or more incorrect decisions may result in incorrect and unanticipated branching in the program's flow.
- produce incorrect results: Wrong decisions are made which produce an output, but the output is incorrect. For instance, Shewchuk [20] observed that precision error problems may produce a "Delaunay triangulation," which has intersecting edges (Figures 1 and 2.) (For further details on this example see [20].) Sometimes an incorrect output is easily seen, while other times the output "looks right," but is in fact not correct. This motivates also the study of efficient algorithms which verify the correctness of an output independent on the construction methodology used to create it.
- go into infinite loop: Due to contradicting decisions algorithms may jump between two or more states indefinitely.

Geometric inaccuracy not only impacts the correctness of the coordinates of the spatial objects, but also the topological relationship between objects and other topological information, which is an integral part of spatial and geographic information systems. In many cases topological information is extremely important. For example, in a GIS for electrical utility, it is not only important to know where a man-hole is but also to know on what side of the road it is located. Incorrect geometrical computations can lead to incorrect topological information in the databases and thus impact future topological queries (e.g., containment, intersect).

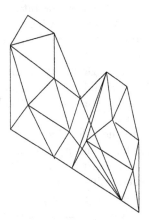

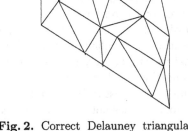

Fig. 1. Incorrect Delauney triangulation due to numerical instability. *Reproduced from [20]*

Fig. 2. Correct Delauney triangulation. *Reproduced from [20]*

Most computational geometry algorithms are not tolerant to such numerical problems [10]. Let us take the line intersection paradigm as an example. Even for the standard Bentley-Ottmann's plane sweep algorithm [16], the robustness

cannot be guaranteed. It is because the algorithm relies on the correct maintenance of topological relationships to find the intersection points, which cannot
be computed accurately with floating-point arithmetic.

Numerical analysis techniques do not seem to be applicable to geometric calculations. There are several methods that handle geometric numerical problems.
These methods are classified into three broad categories based on the required
quality of the output: program robustness only, topological correctness and geometric and topological correctness. All the techniques have to at least guarantee
the robustness of the program. The first category includes methods that guarantee only the robustness of the program but not the correctness of the result
whether it is geometric or topological. Epsilon tweaking, which is defined in Section 2, is an example of a method that falls into this category. The second category encompasses techniques that guarantee correctness of the resulting topology, but not the numeric accuracy of the result. There are numerous methods
that lie in this category. Examples are floating-point filter algorithms, structural
filters and multi-precision floating-point arithmetic with interval technique. The
third category contains techniques that guarantee both the correctness of the
topology and exactness of the numeric results. Exact arithmetic fits into this
last category.

2   Overview of Previous Techniques

One of the earliest related work from the areas of GIS is the band set method
by Peucker [22]. It computes the intersection point of two curves (chains of line
segments) by drawing a bounding box, called band, around the curve. The intersection point of two curves lies in the parallelogram formed by the intersection
of the two bands. This area can be refined further by drawing a band around
the portions of the two curves that lie within the original parallelogram, and
constructing a smaller parallelogram with the intersection of the new bands.
Peucker's method works assuming two curves have a single intersection point.
Moreover, the reduction speeds up only the search for candidate segments within
the curves where an intersection may occur. The final stage of this method still
requires correct geometrical intersection computation.

An often used and simple technique to alleviate the side effect of floating-
point arithmetic is called *epsilon tweaking*. Two points, p_1 and p_2, are considered
identical as long as they are less than a distance of ϵ apart ($\|(p_2 - p_1)\| < \epsilon$).
The value of ϵ is either set to some arbitrarily predetermined value or found by
trial-and-error: substitute an arbitrary small value into ϵ, and run the program.
If the program does not crash, then accept the error level of ϵ as it is. Otherwise,
repeat the execution of the program with a slightly larger value of ϵ [19]. Epsilon
tweaking does not guarantee the accuracy of the result nor the accuracy of the
underlying topology. Moreover, it is not well suited to large data sets, where
re-computing results with different values of ϵ is not only very time consuming,
but can also produce different output depending on the value of ϵ. Setting ϵ to
some arbitrary small value independent of the magnitude of the operands or of
the type of operation performed is another common problem.

A fail-proof method to deal with rational numerical problems is called *exact arithmetic*. Every number is stored as a rational number, which is made of a pair of integers: the numerator and the denominator. Operations on these numbers, called rational arithmetic, support addition, subtraction, multiplication and division. Two rational numbers can be compared exactly. Rational arithmetic in the line intersection algorithm guarantees the accuracy of the intersection points and hence that of topology. The drawback of this approach is that it is very slow in practice. After each operation, the greatest common denominator has to be computed, or the numerator and denominator risk overflowing. Greatest common denominator is implemented in software, and it is CPU intensive to compute [19].

A floating point filter can be described by the saying "Why compute something that is never used?" [19]. Translated into computational geometry terms, approximative computation can be used to compute a result, but comparison between two values has to be exact. Therefore, every result is computed using fast, floating-point arithmetic. An error bound is computed dynamically with every computation. Two values can be compared using their error bounds. If the two values lie beyond their error bounds, then the comparison was definitive. Otherwise, the filter is not applicable, and a more definitive technique (such as exact arithmetic) is required to compute the result. Hence, for every computation, a list of pointers to its operands has to be maintained in order to be able to recompute its value would the filter be inapplicable [7]. Structural filters have also been studied in the context of geometric sorting algorithms [6].

Many geometric algorithms use explicitly or implicitly the dot products. Substantial work has been carried out to make this operation more numerically stable. For instance, [1] propose a method to evaluate signs of 2x2 and 3x3 determinants with b-bit integer entries using b and $(b+2)$-bit arithmetic, respectively. (See also the earlier work of [15, 16].)

Another technique that is being investigated is the use of interval arithmetic (e.g., [3, 2] or surveys such as [11, 13]), possibly in combination with multiple-precision floating-point arithmetic. For interesting alternate approaches, including the Core Library, see also [21, 4].

Achieving robustness may be expensive, cases have been reported where robustness costs a factor of 100 in speed (e.g., [14]) or even an initial slow-factor of 10,000 when floating-point arithmetic is replaced by exact rational arithmetic [12]. Efficiency considerations are key in particular for GIS applications where input sizes are huge and thus large constants in complexity functions must be avoided. Even a small improvement in constants has a huge effect when dealing with such large data sizes.

3   Description of the Line Intersection Filter

Our approach to numerical issues is inspired by the floating-point filter methodology. We take advantage of the fact that we design a filter for a specific algorithm, namely the line intersection algorithm. Since this operation is of fundamental importance, it merits attention. Knowledge about the application allows us to construct for specific instances arising during the execution of the algorithm, instead of filtering for every comparison performed by the program.

In general-purpose filtering techniques, every computation is performed with floating-point arithmetic and an error bound is computed dynamically. A list is maintained with the operands that were used to compute the value. Every time two values are compared, error bounds are used to determine the ordering of the two values.

In the line intersection algorithm, which is presented here, we identify the computationally hard cases. These cases correspond to degenerate cases of the line segments intersection algorithm. The degenerate cases are: null line segment, overlapping line segments, intersection near a vertex, and, overlapping intersection points [17]. In "near" and "close range" we mean that the intersection is inside the point's error bound. Just like in the general filter method, every intersection is computed using floating-point arithmetic, along with an error bound on the intersection point. Right after every computed intersection point, we test whether it corresponds to one of the above degenerate cases. If it does, we use exact arithmetic to recompute the intersection.

Our filtering is applied in the context of a plane sweep algorithm, the Bentley-Ottmann algorithm [16]. In this plane sweep the sweep-line status changes only at endpoints and intersection points of segments. These points are referred to as event points and they are held in a structure called the event queue. At every event point a maximum of two new intersection points can be found. This is the part of the algorithm where filtering is applied.

The following is the list of degenerative cases and how to test for them.

1. Null line segment: the endpoints of the line segment are within their error bounds (Figure 3).
2. Overlapping intersection points: If the intersection point is within an error bound of another intersection or endpoint (Figure 4). This case is identified before a new intersection point is inserted to the event queue. If both of the following two conditions are true, use rational arithmetic to recompute the intersection:

   - There is at least one intersection point P_e within the error bound of the new intersection point P_n.
   - At least one of the segments passing through P_e was not used to compute P_n. This is to insure that there are at least three segments that could intersect at the same point.

3. Overlapping line segments: if all of the following three cases are true (Figure 5):

   - The two segments have overlapping X-ranges.
   - The slopes of the line segments are within their respective error bounds.
   - The two lines' Y-intercepts are within their respective error bounds.

4. Intersection near vertex: the distance between a vertex and the intersection point is within the error bounds (Figure 6).

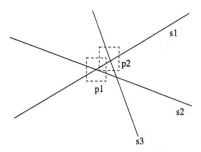

overlapping error
bounds

Fig. 3. Segment of null size

Fig. 4. Intersection points are close to each other

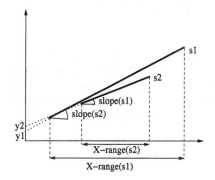

Fig. 5. Segments are potentially overlapping

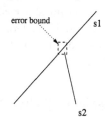

Fig. 6. Uncertain intersection point

Although this filter technique requires a large number of comparisons it does not impose additional computation. The slopes and the Y-intercepts have already been computed to determine the intersection point regardless of the filter. Moreover, the slope and the Y-intercept are computed only once per segment and applied for every intersection point involving this segment.

Figure 7 gives an example where the filter is non-applicable, and rational arithmetic is needed to determine a degenerative case. Take line segment S_1: (0, 1)-(2, 1.0002) and S_2: (0, 0)-(1.0008, 1.0001). Using floating-point arithmetic, the algorithm determines that S_1 and S_2 intersect at (1.0008, 1.0001). Using rational arithmetic, it determines that S_1 and S_2 do not intersect, as point (1.0008, 1.0001) is to the right of S_1.

Figure 8 gives an example where two segments S_3: (0,1)-(2,3) and S_4: (0.5, 1.25)-(1,1.5) are tested for overlap. Their X-ranges are overlapping. Using floating-point arithmetic, their slopes are both computed to .5 and their Y-intercepts are both 1. Using rational arithmetic, slopes and Y-intercepts are recomputed and they are indeed the same. Since S_3 completely covers S_4, S_4 is removed from further computation.

Note that we have used the slope-intercept form of the line equation ($y = mx + b$) to compute the intersection point. We could have used Cramer's Rule [5] instead, which is based on the general equation of the line ($Ax + By = C$).

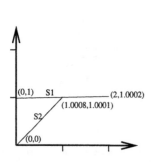

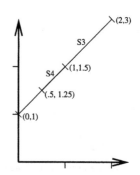

Fig. 7. Example of a degenerate case: do S_1 and S_2 intersect?

Fig. 8. Example of a degenerate case: do S_3 and S_4 overlap?

The advantage of our method over the general purpose filter is that it does not have to maintain a list of operands at every computation. Moreover, the performance of the filter algorithm depends on the number of occurrences of the degenerate cases in the input and not on the number of computations, which is related to the precision of the floating-point filter representation.

4 Discussion of the Results

We implemented the line intersection filtering algorithm, and tested it experimentally against LEDA's implementation of the general purpose floating point filter [7, 8]. LEDA was selected because it is considered an industry and academic state of the art library for geometric computation.

We evaluated the merits of using our filtering technique as follows. We compared two commonly used representations of coordinate systems integer and float with high precision (64 bits floating point representation, which achieves 15 decimal points accuracy). We compared integer geometry and floating point geometry. For testing we used four different data sets. Three of the data sets are sampled from Digital Elevation Models (DEM): africa, madagascar and australia [23]. The fourth data set was created by randomly generating a point set distributed uniformly across the domain and triangulating the vertices. The data sets contain 50,000, 20,000, 30,000 and 50,000 line segments, respectively. The vector data sets, used in our tests, contain topological information. This implies that geometric redundancies do not occur.

The intersection experiments were conducted as follows. Each data set (e.g., africa) was duplicated (e.g., dup(africa)). The line segments in each duplicated data set (e.g., dup(africa)) were translated by ∂x and ∂y, where ∂x and ∂y are arbitrary integers. The purpose of this translation is to generate a large number of intersections of the same magnitude as the number of line segments. Then each pair of data sets were tested for intersection (e.g., africa and dup(africa)). The algorithm was executed several times for each pair for each of the data types (integer or float) and an averaging of the timing results was computed. The test computer was a SUN SPARC single 440 MHz CPU with 512 MB RAM running Solaris. All floating-point arithmetic used 64 bits (double precision) operands.

Table 1. Running time of our filtering algorithm versus LEDA's general purpose filtering algorithm when using integer based coordinates

Input file	size	Running time our filter (s)	Running time LEDA (s)	Improvement of filter over LEDA (%)
random	50000	29	116	75
africa	20000	7.7	20	62
madagascar	30000	32	50	36
australia	50000	53	89	40

Table 2. Running time of our filtering algorithm versus LEDA's general purpose filtering a algorithm when using float based coordinates

Input file	size	Running time our filter (s)	Running time LEDA (s)	Improvement of filter over LEDA (%)
random	50000	20	382	95
africa	20000	5.1	49	90
madagascar	30000	12	173	93
australia	50000	19	290	93

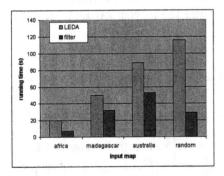

Fig. 9. Running time LEDA versus our filter algorithm with integer coordinates

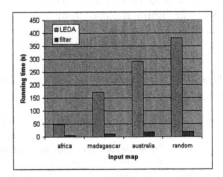

Fig. 10. Running time LEDA versus our filter algorithm with floating-point coordinates

Figures 9 and 10 show the performance of our filtering algorithms versus the general purpose filtering algorithm implemented in LEDA. It shows that our algorithm improves over the performance of LEDA by a factor of two to four if the input is composed of integer coordinates, and by a factor of 10 to 20 in case of fractional coordinates (Tables 1 and 2).

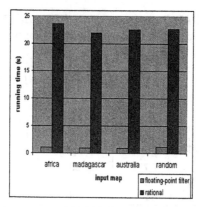

Fig. 11. Time it takes to execute 100,000 floating-point filter and 100,000 rational operations

We further investigated the cause of such a significant improvement in performance. First the "cost" of a filtering operation versus exact arithmetic using rational operation was evaluated. In our algorithm, a rational operation denotes the occurrence of a degenerate case. We observed that our filtering operation is much faster than a rational operation of LEDA by a factor of 20 (Figure 11 and Table 3). Table 3 shows for each data set test the number of times that filtering operations were called and the time to execute the filtering operations. The table also shows the number of times that the filter invoked rational arithmetic and the amount of time it took to execute these invocation. Last, the table shows the time to execute 100,000 operations for each and the percent saving from using the filter.

Table 3. The running time of filter and rational operations of overlaying maps with integer coordinates

Input file	Number of filter operations	Time for filter operations (s)	Number of rational operations	Time for rational operations (s)	Time per 100,000 filter ops (s)	Time per 100,000 rational ops (s)	Filter appli-cability rate (%)
random	349000	4.1	33000	7.5	1.17	22.73	90.5
africa	71000	0.7	14000	3.3	0.99	23.57	80.3
madagascar	158000	1.5	82000	18	0.95	21.95	48.1
australia	281000	2.7	137000	31	0.96	22.63	51.3
average	-	-	-	-	1.02	22.72	-

The goal of the filter operations is to reduce the number of times that a rational arithmetic operation is invoked. The filter operation invokes a rational arithmetic operation only when the filter detects that output falls into one of

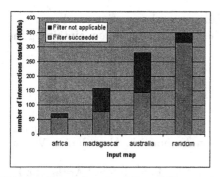

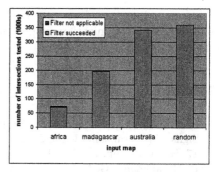

Fig. 12. Rate of applicability of filter with input of integer coordinates

Fig. 13. Rate of applicability of filter with input of fractional coordinates. Note that in the cases of madagascar, australia and random maps, the number of non-applicable filter operations is negligible

the four criteria (presented in Section 3) in order to accurately compute the intersection point.

Our filtering operation is much faster in comparison to a rational arithmetic operation of LEDA and achieves a performance ratio of 1:20. Thus, if one considers counting only float and rational operations and ignores other parts of the intersection algorithm, then by reducing the number of rational operations such that the ratio between rational operations and filter operations is better than 19:20 (19 rational operations for 20 filter operations), our algorithm would achieve superior performance to that of LEDA. In reality, one cannot ignore other parts of the intersection algorithm (e.g., the maintenance of the event queue), and thus our filtering algorithm must achieve a better ratio than 19:20.

In all the cases that we tested (integer coordinates and float coordinates) the "savings" from using the filter is much better than 19:20 (Figures 12 and 13). In the case of an integer coordinate system the ratio was between 1:2 and 1:10, which implies that the filter achieve between 50-90% efficiency. In the case of float coordinates (fractional coordinates), the filter achieves efficiency rate around 100%.

5   Conclusion

In this paper, we discussed several aspects of numerical issues related to geometrical algorithm design and implementations that are often the core of any spatial information system or geographic information system. We proposed a floating-point filter algorithm that provides fast, robust and numerically and topologically correct solutions to the fundamental operation of line segments intersection. Experimental results show that it outperforms existing general-purpose filter algorithms, such as the one implemented in the LEDA library [7]. We foresee that our special purpose filtering technique can be used, potentially, with some minor modifications, in other fundamental operations such as triangulations and polygonization to speed up the performance while maintaining the correctness of the results.

References

1. F. Avnaim, J.-D. Boissonnat, O. Devillers, F. P. Preparata, M. Yvinec, "Evaluating Signs of Determinants Using Single-Precision Arithmetic", Algorithmica, Volume 17, Number 2, pp. 111-132, 1997.
2. H. Brönnimann, G. Melquiond, and S. Pion. "The Boost Interval Arithmetic Library", In Proceedings of the 5th Conference on Real Numbers and Computers, pp. 65-80, 2003.
3. H. Brönnimann, C. Burnikel, and S. Pion, "Interval arithmetic yields efficient dynamic filters for computational geometry", Discrete Applied Mathematics, 109:25-47, 2001.
4. The Core Library, see http://www.cs.nyu.edu/exact/core/.
5. S. Lang, Linear Algebra, Springer-Verlag, 3rd Edition, 1987.
6. S. Funke, S. Näher, and K. Mehlhorn "Structural Filtering – A Paradigm for Efficient and Exact Geometric Programs", Proc. (electronic) 11th Canadian Conference on Computational Geometry (CCCG), Vancouver, Canada, 1999.
7. K. Mehlhorn and S. Näher, "The Implementation of Geometric Algorithms" 13th World Computer Congress IFIP 94, Elsevier Science B.V., Vol. 1, pp. 223-231, 1994.
8. K. Mehlhorn and S. Näher, "The LEDA Platform of Combinatorial and Geometric Computing", Cambridge University Press, Cambridge, United Kingdom, 1999.
9. D. Goldberg "What Every Computer Scientist should know about Floating-Point Arithmetic", ACM Computing Surveys, v. 23 no 1, pp. 5-48, March 1991.
10. D. P. Dobkin, D. Silver "Recipes for geometry and numerical analysis - Part I: an empirical study", Proceedings of the Fourth annual Symposium on Computational Geometry, pp. 93-105, 1988.
11. B. Hayes, "A Lucid Interval", American Scientist, Vol. 91, No. 6, pp. 484-488, 2003.
12. M. Karasick, D. Lieber, and L. R. Nackman. "Efficient Delaunay Triangulations Using Rational Arithmetic", ACM Transactions on Graphics, 10(1), pp. 71-91, 1991.
13. R.B. Kearfott, "Interval Computations: Introduction, Uses and Resources", http://www.cs.utep.edu/interval-comp/, 2004.
14. G. Knott and E. Jou, "A Program to Determine Whether Two Line Segments Intersect", Technical Report CAR-TR-306, CS-TR- 1884, DCR-86-05557, Computer Science Deptartment, University of Maryland, College Park, 1987.
15. U.W. Kulisch and W.L. Miranker, "Computer Arithmetic in Theory and Practice", Academic Press, New York, NY, 1981.
16. J. L. Bentley, T. A. Ottmann "Algorithms for reporting and counting geometric intersections" IEEE Transactions on Computers, C-28:643-647, 1979.
17. E. Rafalin, D. Souvaine, I. Streinu "Topological Sweep in Degenerate cases", Proceedings of the 4th International Workshop on Algorithm Engineering and Experiments, Springer-Verlag, Berlin, Germany, pp. 155-165, 2002.
18. L. Ramshaw "The Braiding of Floating Point Lines", CSL Notebook Entry, Xerox Parc, 1982.

19. S. Schirra "Robustness and Precision Issues in Geometric Computation", in J.-R. Sack, J. Urrutia (eds.) "Handbook of Computational Geometry", pp. 597-632, 1999.

20. J.R. Shewchuk "Adaptive Precision Floating-point Arithmetic and Fast Robust Geometric Predicates", Technical Report CMU-CS-96-140, School of Computer Science, Carnegie Mellon University, 1996.

21. C. Yap "On Guaranteed Accuracy Computation", in: "Geometric Computation" (eds. Falai Chen and Dongming Wang) World Scientific Publishing Co., 2004 (to appear).

22. T.K. Peucker, "A Theory of Cartographic Line", Proceedings, Second International Symposium on Computer-Assisted Cartography, AUTO-CARTO II, pp. 508-518. 1975.

23. M.A. Lanthier, Shortest Path Problems on Polyhedral Surfaces, Ph. D. thesis, School of Computer Science, Carleton University, Ottawa, Ontario, Canada, 1999.

Project Lachesis:
Parsing and Modeling Location Histories

Ramaswamy Hariharan[1] and Kentaro Toyama[2]

[1] School of Information and Computer Science,
University of California, Irvine,
Irvine, CA 92697, USA
rharihar@ics.uci.edu

[2] Microsoft Research, One Microsoft Way,
Redmond, WA 98052, USA
kentoy@microsoft.com

Abstract. A datatype with increasing importance in GIS is what we call the *location history*–a record of an entity's location in geographical space over an interval of time. This paper proposes a number of rigorously defined data structures and algorithms for analyzing and generating location histories. *Stays* are instances where a subject has spent some time at a single location, and *destinations* are clusters of stays. Using stays and destinations, we then propose two methods for modeling location histories probabilistically. Experiments show the value of these data structures, as well as the possible applications of probabilistic models of location histories.

1 Introduction

A datatype with increasing importance in GIS is what we call the *location history*–a record of an entity's location in geographical space over an interval of time. In the past, location histories have been reconstructed by archaeologists and historians looking at migrating populations or census takers tracking demographics, at temporal resolutions of decades or centuries and spatial resolutions of tens or hundreds of kilometers. Recent advances in location-aware technology, however, allow us to record location histories at a dramatically increased resolution. Through technologies such as GPS, radio triangulation, and localization through mobile phones, 802.11 wireless systems, and RFID tags, it becomes feasible to track individual objects at resolutions of meters in space and seconds in time–in some cases, even greater resolution is possible.

Although this increase in resolution is merely quantitative, the sheer volume and granularity of data opens up possibilities for intricate analysis and data mining of a qualitatively different nature. In this paper, we propose generic data structures and algorithms for extracting interesting information in high-resolution location histories, develop probabilistic models for location histories, and some present applications of these analytical tools.

In the geographic sciences, Hägerstrand is credited with introducing the first rigorous tools for the analysis of human migratory patterns. His *space-time prism* provided a useful visualization of movement, both of groups and individuals [3]. The space-

M.J. Egenhofer, C. Freksa, and H.J. Miller (Eds.): GIScience 2004, LNCS 3234, pp. 106–124, 2004.

time prism plotted time on the independent axis, and an interval, rather than a point, on the dependent axis to indicate the extent or uncertainty in location of a population. He used data from government censuses and manually collected logs to study migration. Geographers have built on this work considerably since its introduction, but have so far restricted attention to coarse location histories [12].

More recently, a body of work has focused on modeling location histories using "bead and necklace" representations, which capture the uncertainty of an object's location given point samples; the beads fatten as they move away from known samples at a rate proportional to bounds on object speed [4, 6, 9]. One version of this representation allows a scaled inspection of the data dependent on choice of data granularity [6]. This work has also been applied to track health-related information over many individuals [9].

Another application of location histories is in optimization of mobile phone networks, some of which allow consumers to keep track of "buddies." Mobile phones operate by switching their connectivity from tower to tower as the phone moves between cells. By predicting the movement of mobile devices, the number of location updates with each phone in service can be minimized [2, 7, 11]. Work in this area is often targeted to the task of optimizing mobile-phone operations, but at core, there are similar data structures and algorithms for handling location histories. Most often, geography is represented as a partition into cells, and movement is modeled as transition probabilities between cells [2, 7]. Others propose a more continuous approach where traditional filtering and smoothing techniques are used to estimate future state [11].

Consumer-oriented applications use similar predictive algorithms to help form personal to-do lists [10] or to give trusted friends and co-workers a better sense of one's current location [8]. A number of single-user and multiple-user applications that are made possible by using location-aware wearable computers [1]. If a wearable computer includes a GPS, clusters of logged GPS coordinates can be used to determine destinations of interest, and transitions between clusters can provide training data for developing a probabilistic model of personal movement [1].

There has also been some work in efficient updating of location histories in databases [14, 15]. Since moving objects continuously change positions, algorithms for avoiding overly frequent updates are desirable. Solutions here propose representations of movement as function of time and other parameters to predict future movements. Hence an update to the database is made only when motion parameters change.

Thus, most work to date with location histories has focused on specific applications or on particular methods for logging location, with the processing of location histories tailored to the task at hand. In this paper, we attempt to define general data structures that are independent of both application and method of acquisition. Our algorithms are likewise independent of the method or resolution at which location histories are gathered. Applicability, however, is not sacrificed at the expense of generality, and we illustrate the kind of analysis that can be performed with the proposed tools.

After defining some notation in the next section, Section 3 discusses parsing of raw location histories into *stays* and *destinations*, which we take as fundamental data structures in Section 4, for building probabilistic models of location histories. These sections strive for generality with respect to representation of location and resolution of data. Finally, in Section 5, we show the kind of analyses that can be accomplished when our basic data structures and algorithms are applied to data collected by GPS.

2 Notation

We assume the simplest possible representation of raw location data: data consist of a time-stamp and a point location. A body of raw data is, therefore, a set, $R = \{r_i\}$, consisting of pairs, $r_i = (t_i, 1_i)$, each containing a time-stamp and a location. Without loss of generality, we assume the data is labeled such that $1 \le i \le R$ (where $R = |R|$) and sorted in time order: $r_i < r_j$ if $t_i < t_j$, for any i and j.

We define locations in the most general way. They may be any identifier that iden-tifies a single, unique, geographic point location–n-tuples of real values are probably the most typical, but alternate representations, such as a text label, are possible. What is critical, however, is that the locations exist in a metric space. That is, there must be a metric function, $Distance(1_i, 1_j)$, which computes the distance between two loca-tions, and which satisfies all of the criteria of a true mathematical metric, namely that the function is (1) positive definite: $Distance(1_i, 1_j) \ge 0$ for any 1_i, 1_j; (2) Dis-$tance(1_i, 1_j) = 0$, if and only if 1_i and 1_j represent the same location; and (3) the triangle inequality holds: $Distance(1_i, 1_j) + Distance(1_j, 1_k) \ge Distance(1_i, 1_k)$.

We point out that the data structures and algorithms proposed below require only that this metric function exists–they are not dependent on how location *per se* is repre-sented. Fig. 1a shows an example of a location history overlaid on a map.

3 Parsing Location Histories

In order to analyze location histories, we parse raw location data to extract symbols that approximate intuitive semantic notions of location. In particular, we believe the following four concepts are intuitively meaningful (we will use the word *place* to mean a neighborhood around a point location):

- A *stay* is a single instance of an object spending some time in one place.
- A *destination* is any place where one or more objects have experienced a stay.
- A *trip* occurs between two adjacent stays (made by a single object).
- A *path* is a representation of the description of a set of trips between destinations.

For example, four hours spent at the office today could be a single *stay*. The office itself would be a *destination*. The particular timed trajectory going from home to office would be a *trip*. Multiple trips over the same spatial trajectory would form a *path*.

Stays and destinations are identified with places, whereas trips and paths are con-cerned with trajectories between places. Destinations and paths can be thought of as "timeless" generalizations of their time-dependent counterparts, respectively stays and trips. This paper focuses on what can be done with stays and destinations. A future paper will focus on trips and paths, which require their own in-depth treatment.

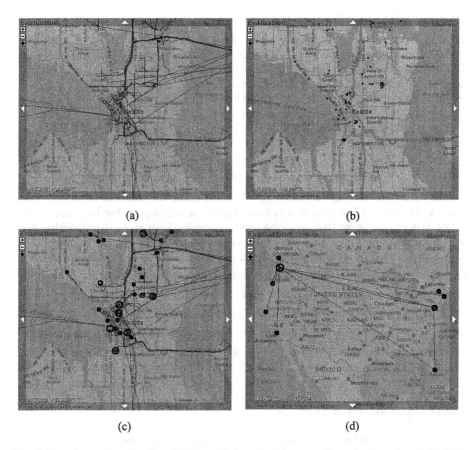

(a)   (b)

(c)   (d)

Fig. 1. Data from a few months of Subject A's location history, collected using a handheld GPS device: (a) line segments connecting adjacent points in the location history; (b) extracted stays marked as dots; (c) destinations marked as circles; and (d) stays and destinations extracted at a much coarser resolution

In the subsections to follow, we present rigorous definitions of stays and destinations, as well as algorithms for extracting them from a location history. Our approach considers a data-driven approach using variations of clustering algorithms; destinations are defined independently of *a priori* information about likely destinations. In particular, we postpone attempts to correlate stays and destinations with geographic entities defined by an existing map or GIS, and focus on destinations that appear naturally in the data themselves. We believe this is a more general approach, as it would be straightforward to associate data-driven destinations *post hoc* with existing geographical entities, if necessary.

3.1 Stays

A stay is characterized by "spending some time in one place." We would like to capture this concept rigorously while maintaining the breadth required to encompass the semantic intuition. In particular, we note that a five-minute visit to the restroom, a half-day lounge at the beach, and a one-week vacation in Hawaii all represent different stays, even though they might all occur within the same two-week time interval. This sort of nested or overlapping structure happens throughout a given object's location history and what creates it is *scale*: stays can occur at various geographic and temporal scales. Stays at some scale might be relevant for some applications, but not for others. A hierarchical nesting of scales might be useful for yet other applications.

In any case, these examples show that the extraction of stays from a location history is dependent on two scale parameters, one each for time and spatial scale. We call these the *roaming distance* and the *stay duration*. The roaming distance, Δl^{roam}, represents the maximum distance that an object can stray from a point location to count as a stay; and a stay duration, Δt^{dur}, is the minimum duration an object must stay within roaming distance of a point to qualify as staying at that location. These parameters can be set according to the needs of the application, or the algorithm can be run multiple times with increasing scale values to create a hierarchy of stays.

A single stay is characterized by a location vector, start time, and end time: $s_i = (\mathbf{l}_i, t_i^{start}, t_i^{end})$. Our algorithm, which recovers a set of stays, $S = \{s_i\}$, from the raw data, is given in Table 1. The functions $Medoid(R, i, j)$ and $Diameter(R, i, j)$ are computed over the set of locations represented in the set of raw data $\{r_k : r_k \in R\}$, for $i \le k < j$. The *Diameter* function computes the greatest distance between any two locations in a set, and the *Medoid* identifies the location in a set that minimizes the maximum distance to every other point in the set (i.e., it is the data point nearest to the "center" of the point set). The algorithm essentially identifies contiguous sequences of raw points, which remain within the roaming distance for at least as long as the stay duration.

Table 1. Algorithm for extracting stays from raw data

Input: raw location history, $R = \{r_i\}$	**Output:** a set of stays, $S = \{s_i\}$

Initialize: $i \leftarrow 1$, $S \leftarrow \varnothing$
while $i < R$
$\quad\quad j^* \leftarrow \min j$ s.t. $r_j \ge r_i + \Delta t_{dur}$;
$\quad\quad$**if** $(Diameter(R, i, j^*) > \Delta l_{roam})$
$\quad\quad\quad\quad i \leftarrow i + 1$;
$\quad\quad$**else**
$\quad\quad$**begin**
$\quad\quad\quad\quad j^* \leftarrow \max j$ s.t. $Diameter(R, i, j) \le \Delta l_{roam}$;
$\quad\quad\quad\quad S \leftarrow S \cup (Medoid(R, i, j^*), t_i, t_{j^*})$;
$\quad\quad\quad\quad i \leftarrow j^* + 1$;
$\quad\quad$**end**
end

In the worst case, the algorithm is an $O(n^2)$ algorithm for n data points, since medoid and diameter computations require distance computations between all pairs in a stay cluster. In practice, however, clusters over which these computations must take place are far smaller than n, and performance is effectively $O(n)$. Many of the problems of clustering unordered points (e.g., as encountered in [15]) are avoided because of the temporally ordered nature of the data.

Examples of stays extracted in this manner are shown in Fig. 1. In Fig. 1b, stays were extracted with a roaming distance of 50 m and stay duration of 10 minutes, whereas Fig. 1d shows the results for a roaming distance of 20 km and a stay duration of 24 hours.

3.2 Destinations

A destination is any place where one or more tracked objects have experienced a stay. Destinations are dependent on geographic scale, but not on temporal scale (i.e., beyond the temporal scales used to identify stays). The scale determines how close two point locations can be and still be considered part of the same destination. As with stays, the scale of a destination is dependent on the intended usage, and so it is a parameter that must be set explicitly. For example, a scale representing ~3 m might be appropriate for extracting destinations corresponding to offices in a building, but a scale of ~100 m would be necessary for identifying whole buildings as destinations.

Given a set of locations, $L = \{l_i\}$, our aim is to extract all the destinations $D = \{d_j\}$ at a particular geographic scale Δl^{dest}. Each destination will be represented by a location and the scale used: $d_j = (l_j, \Delta l_j^{dest})$.

Determining destinations from a set of location vectors is a clustering task. There are many options for clustering points, ranging from k-means clustering to hierarchical clustering techniques. We choose to use a type of agglomerative clustering, because it allows us to specify the spatial scale of the clusters, rather than the number of clusters or the number of points contributing to a cluster, neither of which we know *a priori*.

Let a cluster be characterized by a set of point locations: $c = \{l\}$. The clusters are initialized by assigning each input point location to a cluster, and hence there are as many clusters as location points at the beginning. During each iteration of the algorithm, the two closest clusters are identified. If the cluster resulting from merging the two clusters would be within the specified scale, Δl^{dest}, they are merged. Otherwise, the algorithm stops and outputs all remaining clusters as destinations. This is an $O(m^2)$ algorithm for m stays, because of the need to compute distances between all pairs of stays.

Table 2 shows pseudocode for this algorithm. The function *FindClosestPair* finds the closest two clusters from the cluster set, *Radius* computes the combined radius of the two clusters assuming that they are merged, and *Merge* combines two clusters into one. The *Radius* of a set of locations is the distance from the set's medoid to the location within the set, which maximizes the distance.

It will be useful for later sections to define a function $d(l)$, which returns the nearest destination to location l. This may be further extended to $d(l, \Delta l^{dest})$, which returns a null value if the location is not within Δl^{dest} of any known destination.

Finally, destinations can be further computed hierarchically across scales, by allowing the medoids of each cluster created at one scale, Δl_j^{dest}, to be used as input locations to compute destinations at a greater scale, Δl_{j+1}^{dest}.

Table 2. Algorithm for computing destinations

Input: a set of point locations, $L = \{l_i\}$ **Output:** a set of destinations, $D = \{d_j\}$
Initialize: $c_i \leftarrow l_i$, for $1 \leq i \leq L$, and $C = \{c_i\}$
loop
$(c_i, c_j) \leftarrow FindClosestPair(C)$;
if $Radius(c_i, c_j) \leq \Delta l^{dest}$
$c_i \leftarrow Merge(c_i, c_j)$;
$C \leftarrow C - c_j$;
else
exit
end
foreach $c_i \in C$, create destination $d_i = (Medoid(c_i), \Delta l^{dest})$;

Fig. 1c and Fig. 1d show destinations after clustering stays with this algorithm. The circles indicate both the location and radius of each destination. The destinations in Fig. 1c were clustered with a scale setting of $\Delta l^{dest} = 250\,\text{m}$, in Fig. 1d, $\Delta l^{dest} = 25\,\text{km}$.

Armed with data structures for stays and destinations, we can proceed to construct probabilistic models of location histories.

4   Modeling Location Histories

The goal of our location-history models is to condense, understand, and predict the movements of an object over a period of time. We investigate two probabilistic models for location histories, one with and one without first-order Markovian conditioning of the current location on subsequent location. Our experiments in Section 5 show that both have value, depending on the kind of questions that are asked of the model. The next two subsections define some notation and establish assumptions made by our model. The subsections after that describe our model, together with algorithms for training, estimation, and prediction.

4.1   Notation

The *destination set*, $D = \{d_i\}$, is the set of all destinations (as determined in Section 3.2), where $1 \leq i \leq n$ and $n = |D|$ denotes the total number of destinations.

We need to distinguish between three different units of time. A *time instant*, t, represents an instantaneous moment in time; if time is thought of as a real-valued entity of one dimension, a time instant represents a single point on the real number line. Next, for a given interval unit of time, δt (e.g., an hour), a *time interval*, $\mathbf{t}$, represents a half-open unit interval on the real number line, aligned to the standard calendar and clock. For example, for δt equal to an hour, $\mathbf{t}$ might be a time interval starting at 18:00UTC today and going up to, but not including 19:00UTC. Finally, a *recurring time interval*, τ, is the set of all time intervals that represents a regularly recurring interval of time. Continuing the example, τ might be the set of all times occurring between 18:00 and 19:00, regardless of date. A set of non-intersecting, recurring time intervals that covers all times will be denoted $\mathsf{T} = \{\tau_k\}$, for $1 \le k \le m$, with $m = |\mathsf{T}|$ indicating the number of recurring time intervals required to cover all of time.

The granularity, δt, of a recurring time interval and the period with which it recurs is something that must be decided for a particular model *a priori*. Thus, we might decide for a particular model that δt represents an hour and recurring time intervals cycle each day (in which case, $m = 24$) or that each hour of the week should be different recurring intervals ($m = 168$). If so, then $\mathbf{t}_p \subset \tau_k$ if $\mathbf{t}_p$ represents the particular hour between 18:00 and 19:00 on September 30, 2003, and τ_k represents the recurring time interval 18:00-19:00.

Finally, we define a function, $\tau(t)$, that extracts the recurring time interval that contains a time instance: $\tau(t_p) = \tau_k$, if and only if $t_p \in \tau_k$. With minor abuse of notation, we also let $\tau(\mathbf{t}_p) = \tau_k$, if and only if $\mathbf{t}_p \subset \tau_k$.

4.2 Model Assumptions

Both of the location-history models presented in this paper are based on the following assumptions:

- At the beginning of a given time interval, an object is at exactly one destination.
- During any given time interval, an object makes exactly one transition between destinations. A transition may occur from a destination to itself (a *self-transition*).

These are not ideal assumptions, by any means. For example, the possibility of multiple transitions occurring within a time interval is not explicitly modeled by our current algorithms. We chose these assumptions, however, to strike a compromise between allowing arbitrary transitions and expressive power of the model–a compromise that would not require unreasonable amounts of data to train.

Based on the above assumptions we define the following probability tables, in a manner analogous to Hidden Markov Models [13] . The critical difference from the standard HMM formulation is that we incorporate time-dependence into the model, where transition probabilities are conditioned on recurring time intervals, rather than being fixed regardless of the time. This was a deliberate design decision that allows us to capture cyclical behavior that is, for example, dependent on time-of-day. With this modification, we can model the fact that at 8am, it is far more likely that we travel from home to office than at 4am.

The probability of the object starting time interval τ_k at destination d_i is represented by a matrix of probabilities, $\Pi = \{\pi(d_i, \tau_k)\}$ where

$$\pi(d_i, \tau_k) = \Pr(d = d_i \text{ at the start of } \tau_k) \tag{1}$$

and

$$\pi(d_i, t_p) = \pi(d_i, \tau_k), \text{ for } t_p \in \tau_k. \tag{2}$$

such that, $\sum_{i=1}^{n} \pi(d_i, \tau_k) = 1$.

Next, the probability that the object makes a transition from destination d_i to d_j during interval τ_k is given by a table, $A = \{a(d_i, d_j, \tau_k)\}$,

$$a(d_i, d_j, \tau_k) = \Pr(d = d_j) \text{ where at the start of } \tau_{k+1} \mid d = d_i \text{ at the start of } \tau_k \tag{3}$$

such that, $\sum_{i=1}^{n} a(d_i, d_j, \tau_k) = 1$. Also, $a(d_i, d_j, t_p) = a(d_i, d_j, \tau_k)$ where $t_p = \tau_k$.

To complete the HMM analogy, we include the observation probability. $B = \{b(d_i, d_j)\}$ represents the probability of observing that the object is at destination d_j, given that the object is actually at destination d_i, with

$$b(d_i, d_j) = \Pr(d^{observed} = d_j \mid d^{actual} = d_i) \tag{4}$$

Together as $\lambda = (\Pi, A, B)$, these tables represent a probabilistic generative model of location for the object modeled. Once the parameters are learned, this model can be used to solve problems such as finding the most likely destination occupied at a particular time, determining the relative likelihood of a location history sequence, or stochastically generating a location history sequence.

4.3   Training the Model

We now present algorithms for learning model parameters λ from training data. Our training data consist of a set of stays, $S = \{s_i\}$, as extracted from the raw data in Section 3. Recall that each stay, s, is a 3-tuple containing a start time, an end time, and a destination: $s_i = (d_i, t_i^{start}, t_i^{end})$.

4.3.1   Computing Π
To compute Π, we simply count the number of occurrences in the training data where the object started a recurring time interval in a particular destination and normalize it over all training data for that recurring interval. Table 3 shows pseudocode.

Table 3. Algorithm for computing Π, the prior probabilities of being at a destination at a given recurring time interval

Input: set of stays, $S = \{s_i\}$ $\qquad\qquad$ **Output:** probability table, $\Pi = \{\pi(d_i, \tau_k)\}$
 	
 Initialize: $count(d_i, \tau_k) \leftarrow 0$, for $1 \le i \le n$ and $1 \le k \le m$
 // count
 for each $s_i \in S$
 $\qquad$ **if** $\tau(t_i^{start}) = \tau(t_i^{end})$ **and** $t_i^{end} - t_i^{start} < \delta t$
 $\qquad\qquad$ **continue**
 $\qquad$ **else**
 $\qquad\qquad$ **for** $t \leftarrow Ceiling(t_i^{start}) : t_i^{end} : \delta t$
 $\qquad\qquad\qquad count(d^{(i)}, \tau(t)) \leftarrow count(d^{(i)}, \tau(t)) + 1$;
 $\qquad\qquad$ **end**
 end
 // normalize
 for each i, k in $1 \le i \le n$ and $1 \le k \le m$
 $\qquad \pi(d_i, \tau_k) \leftarrow count(d_i, \tau_k) / \sum_j count(d_i, \tau_k)$;
 end

An example of the result of this algorithm is given in Section 5.4.

4.3.2 Computing A

To compute A, we count the number of occurrences in the training data where the object makes a transition from a particular destination to another destination (or itself) during a recurring time interval and normalize it over all the training data for that recurring interval. This algorithm is shown in Table 4.

4.4 Location History Analysis

We now use the location history model, λ, to estimate the relative likelihood of a new location history, $\tilde{H} = \{d(t_u)\}$, defined over $u \in [start, finish]$. We propose two different processes for doing this.

4.4.1 Non-Markovian Solution

We determine the probability of the location history by computing the joint probability $\pi(d(t_u), t_u)$ and $b(d(t_u), d(t_u))$ from time t_{start} to time t_{finish}, and marginalizing (summing) the joint probabilities over all possible location history sequences. This can be represented by the following equation:

$$\Pr_\pi(\tilde{H} \mid \lambda) = \sum_{H \in \{d_{h_{start}},...,d_{h_{finish}}\}} \prod_{u=start}^{finish} \pi(d_{h_u}, t_u) b(d(t_u), d_{h_u}) \qquad (5)$$

Table 4. Algorithm for computing A, the probability table showing the likelihood of transition between destinations at a given recurring time interval

---

Input: $S = \{s_i\}$ **Output:** probability table, $A = \{a(d_i, d_j, \tau_k)\}$

Initialize: $count(d_i, d_j, \tau_k) \leftarrow 0$, for $1 \le i, j \le n$ and $1 \le k \le m$

for each $s_i \in S$

    // count self-transitions

    if $\tau(t_i^{start}) = \tau(t_i^{end})$ **and** $t_i^{end} - t_i^{start} < \delta t$

        continue

    else

        for $t \leftarrow Ceiling(t_i^{start}) : t_i^{end} : \delta t$

            $count(d^{(i)}, d^{(i)}, \tau(t)) \leftarrow count(d^{(i)}, d^{(i)}, \tau(t)) + 1$;

        end

    end

    // count other transitions

    if $i \ne |S|$ **and** $\tau(t_i^{end}) \ne \tau(t_{i+1}^{start})$

        $t = count(t_{i+1}^{start})$;

        $count(d^{(i)}, d^{(i)}, \tau(t)) \leftarrow count(d^{(i)}, d^{(i)}, \tau(t)) + 1$;

    end

end

// normalize

for each i, j, k in $1 \le i, j \le n$ and $1 \le k \le m$

    $a(d_i, d_j, \tau_k) \leftarrow count / \sum_j count(d_i, d_j, \tau_k)$;

end

If observations are accurate, this reduces to

$$\Pr_\pi(\widetilde{H} \mid \lambda) = \prod_{u=start}^{finish} \pi(d(t_u), t_u) \tag{6}$$

This approach assumes that there is no conditional dependency of state between time intervals.

4.4.2 Markovian Solution

Another method of determining the probability of location history is by computing the joint probability of the observation sequence and the state sequence and marginalizing over all possible location history sequences:

$$\Pr_A(\widetilde{H} \mid \lambda) = \sum_{H \in \{d_{h_{start}}, \ldots, d_{h_{finish}}\}} \pi(d_{h_s}, t_s).b(d(t_u), d_{h_s}) \prod_{u=start}^{finish-1} a(d_{h_{u+1}}, d_{h_u}, t_u).b(d(t_{u+1}), d_{h_{u+1}})$$

$$\tag{7}$$

This reduces to

$$\Pr{}_A(\widetilde{H} \mid \lambda) = \pi(d(\mathbf{t}_s),\mathbf{t}_s) \prod_{u=start}^{finish-1} a(d(\mathbf{t}_{u+1}),d(\mathbf{t}_u),\mathbf{t}_u) \qquad (8)$$

if observations are accurate. This approach uses the transition probabilities A, and assumes that the object's destination at a time interval is conditionally dependent on the destination at the previous time interval. This is equivalent to the standard "forward algorithm" used to evaluate the probability of a sequence of observations in an HMM [13], but with the modification for time-dependent transition probabilities.

Whichever method is used, the output is a true probability in the strict sense, but only given the assumptions of the respective estimates. In reality, probabilities of events over time intervals are ill-defined–for one thing, the probability of a particular event approaches zero as the event is sampled over shorter sub-intervals. Thus, these values are most meaningful when interpreted as relative likelihoods between events observed using the same interval unit. For example, we can compare the relative likelihoods of two location histories of a week's length with δt equal to one hour and judge their relative rarity. We could also set thresholds for a history dependent on the length of the history, to determine whether an input history appears normal or abnormal. Finally, given multiple models, λ_i, we can determine which model best explains a given history by computing $\arg\max_i \Pr(\widetilde{H} \mid \lambda_i)$.

4.5 Stochastic Generation

Using the model parameters, λ, we can stochastically generate a location history $H_{gen} = \{d(\mathbf{t}_u)\}$ for $u \in [start, finish]$, where $d(\mathbf{t}_u)$ is the destination occupied at time interval $\mathbf{t}_u$. We outline two methods for generating location histories.

In the first, we use only the Π parameters, and randomly sample from the set of destinations for each time interval without conditional dependence between time intervals. Destinations are chosen such that

$$\Pr(d(\mathbf{t}_u) = d_i) \propto \sum_j \pi(d_j,\mathbf{t}_u)b(d_j,d_i) \qquad (9)$$

In practice, this can be done by a basic Monte Carlo "coin-tossing" process to generate an "actual" destination, d_j, using π, which is then followed by another coin toss to determine the observed destination, d_i, using B. This simplifies to a single coin toss per time interval in the case where observations of destinations are noiseless.

In the second technique, we utilize the full Markov model and perform a similar Monte Carlo sampling using the transition probabilities, A, in all but the first time interval. Thus,

$$\Pr(d(\mathbf{t}^{start}) = d_i) \propto \sum_j \pi(d_j,\mathbf{t}^{start})b(d_j,d_i) \qquad (10)$$

as before for $\mathbf{t} = \mathbf{t}_{start}$, but

$$\Pr(d(\mathbf{t}_u) = d_i \mid d(\mathbf{t}_{u-1}) = d_k) \propto \sum_j A(d_k, d_j, \mathbf{t}^{start}) b(d_j, d_i) \qquad (11)$$

for the remaining time intervals. Again, this is implemented in practice as a simple series of Monte Carlo coin tosses.

5   Experimental Results

We conducted experiments with the raw location histories of two subjects, who together collected over two years of data using handheld GPS devices carried on their person. We have 346 days of data for Subject A, and 386 days for Subject B.

5.1   Stays

We extracted stays from the raw points using the algorithm described in Table 1. Generating stays at five different temporal scale parameters shows the effect of time scale on number of stays. Time duration, Δt^{dur}, was set to 10, 20, 40, 80, and 160 minutes. In all cases, the roaming distance was set to $\Delta l^{roam} = 30$ m to account for GPS noise. Fig. 2 summarizes these results.

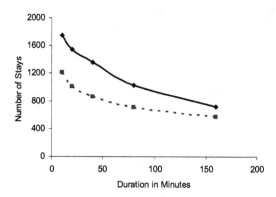

Fig. 2. Plots of the number of stays versus the stay duration parameter used to extract stays, for two subjects

As would be expected, fewer stays are generated if the time threshold for considering a pause a stay is lengthened. While the two subjects show differences in the absolute number of stays, there is an approximately exponential fall-off in the number as stay duration is increased. This confirms the intuition that stays might conform to a power law, where short stays are far more likely than long stays–one is much more likely to make short trips to the bathroom than to take week-long vacations in Hawaii.

5.2   Destinations

Given stays, we then cluster them into unique destinations, using the algorithm described in Table 2. In computing destinations, geographic scale is a key factor. We confirm that scale affects extraction of destinations, with destinations generated at geographic scales of 250, 2,500, and 25,000 m. For both the experiments, stays at $\Delta t^{dur} = 10$ minutes and $\Delta l^{roam} = 30$ m were used. Table 2 summarizes these results. Fig. 1c shows destinations extracted when $\Delta l^{scale} = 250$ m; and Fig. 1d, when $\Delta l^{scale} = 25$ km.

Table 5. Destinations generated for subjects A and B at different geographic scales

Subject	Δl_{scale} (meters)	# Destinations
	250	234
A	2,500	78
	25,000	20
	250	179
B	2,500	72
	25,000	26

5.3 Some Simple Analysis

The power of extracting stays and destinations is illustrated in the following examples, where we compute a variety of statistics about the subject's lives.

Table 6. Top five destinations by number of stays at each, for two subjects

Subject	Destination	# Stays at Destinations	Total Time at Destinations (hours per year)
	work, primary	443	3,365
	home	388	2,190
A	gym	70	135
	mall	40	39
	friend's house	38	131

Subject	Destination	# Stays at Destinations	Total Time at Destinations (hours per year)
	home	411	3,924
	work, primary	180	1,117
B	work, secondary	76	420
	work, other	20	15
	murphy's corner	19	27

In Table 6, we show a sample of the kind of information that can be easily extracted by displaying each of our two subject's top five most frequently visited destinations. In this case, the destination names were provided by the subjects, who viewed the destinations on a map, but even this process could be automated by using a combination of GIS-lookup to map destinations to established place names, and heuristics to learn person-specific destinations (e.g., time of day and amount of time spent at a location will give strong indicators of home and work). One application of this information is for cell-phone location privacy. The location-based services (LBS) industry, for example, is marketing such services as location-sensitive coupons, if users are willing to allow merchants to know their current location. Consumers may be reluctant to give away this information for privacy reasons, if it discloses sensitive destinations, such as when exactly they are at home. By automatically determining where a user's home is, however, the carrier can offer "location dithering" services that would either limit or intentionally coarsen estimates of location when the user is in the neighborhood of a sensitive destination.

We can also analyze subtle patterns of behavior through simple manipulation of the data. For example, in Fig. 3, we show the average number of hours spent at Subject A's primary office location, computed by histogramming stays by day of the week and dividing them by the number of total days of raw data. There is a clear trend where the number of hours spent at work peaks on Tuesdays and gradually trails off toward Friday. Subject A confirms, "I always felt most productive on Tuesdays."

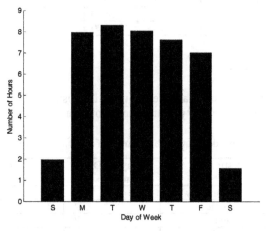

Fig. 3. Histogram of average number of hours spent by Subject A at his primary workplace

In Fig. 4, we show another easily generated plot of the average number of destinations for Subject A, broken down by month. We can instantly see that Subject A has a reasonably steady routine that involves little daily travel, but that there is greater variance in August and December, probably due to vacation activities.

5.4 Experiments with Location History Modeling

We used the low-level processed information, stays and destinations generated from the raw data of user, to train our location-history models. We chose location histories recurring at a period of one week, with recurring time interval, δt, of one hour (m =168 hours per week), for our models. Fig. 5 shows "typical" and "atypical" weeks for Subject A, as picked by the subject. The x-axis plots the day of the week, and the y-axis the index of the destination (an arbitrary number chosen per destination during the clustering process). Indeed, the vast majority of the weeks in this dataset reveal a pattern of activity that qualitatively looks like Fig. 5a, where most of the time is spent either at home or at work.

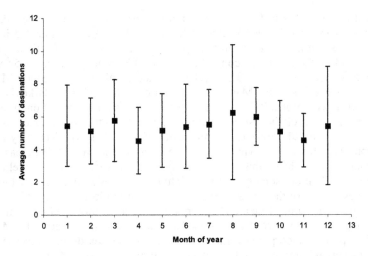

Fig. 4. Average number of destinations visited by Subject A, by month of year

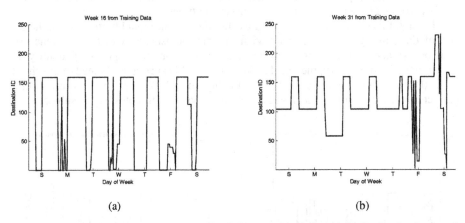

(a)   (b)

Fig. 5. (a) Typical and (b) atypical weeks for Subject A. The index of the destination is plotted against time. (The ordering of the destination indices is entirely arbitrary)

5.5 Evaluation of Location Histories

In this section we present results of the experiments conducted to evaluate the likelihood of a week's location history. This sort of analysis could distinguish between typical and atypical patterns of behavior, and might be used, for example, to provide more sophisticated electronic calendars, which take into account a person's recent location history to predict future location and work cycles.

We computed the model parameters $\lambda = (\Pi, A, B)$ using the algorithms described in Tables 3 and 4. We did not consider stays that were less than an hour while calculating the probability table Π.

As a simple verification of the evaluation process, we computed likelihoods of week-long location histories, given trained models (that did not include the week evaluated). The goal is to find an evaluation process that gives us higher likelihoods for typical weeks and lower likelihoods for atypical weeks. As indicated in Section 4.4, we can use either a Markovian or non-Markovian approach for estimation.

Fig. 6 shows the results of computing the log likelihood of each of 52 weeks in Subject A's location history. The circles indicate results using the non-Markovian evaluation, and crosses, for Markovian evaluation. Although the Markovian evaluation shows lower probabilities overall (due to inclusion of transition probabilities during evaluation), relative estimates are similar between the two instances of data, as expected. Unexpected, however, are the results for week 13 (indicated by arrows in the graph), which was an atypical week according to the subject. Whereas the non-Markovian process shows this to be a highly atypical week with low likelihood, the Markovian evaluation, somewhat counterintuitively, shows an unusually high likelihood. Investigation of the underlying data shows that Subject A engaged in an activity that occurred with almost identical patterns of infrequent movement, exactly once in the training data, and once during week 13 of the test data. Because the Markovian evaluation process incorporates transition matrices between destinations, near-match sequences between training and test data for atypical weeks will come out to be far more likely than typical weeks, which distribute transition probabilities more diffusely across a greater number of destinations and times.

Although this case could be handled by the Markovian model by training on larger sets of data, or by clustering Markovian models themselves with the frequency of their occurrence, our conclusion in this case is that for the purposes of identifying typical patterns of activity, the non-Markovian model is sufficient. Indeed, a threshold of -350 on the log likelihood for this data using non-Markovian analysis results in a perfect identification of atypical weeks, that is in synch with notes by the subject.

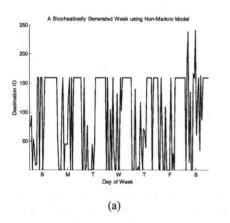
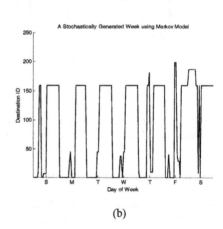

(a)   (b)

Fig. 6. Plots of synthesized weeks, using a model trained on Subject A's data: (a) using the non-Markovian model and (b) with Markovian transitions

It should be noted, however, that even the Markovian generation does not result in histories that match the statistics of true data–careful comparison of Fig. 6b and Fig. 5a reveals that in actual data, the subject spends longer amounts of time at destination 157 (home). This is due to a known flaw of standard Markov chains, in that the length of time spent at a particular destination is necessarily exponential in distribution (whereas real data may contain non-exponential distributions). The problem is actually mitigated in our algorithm, because our transition probabilities are time-dependent, but the effects of considering only first-order effects are still noticeable.

6 Conclusions

This paper proposed rigorous definitions for location histories, as well as algorithms for extracting stays and destinations from location histories in a pure, data-driven manner. Both Markovian and non-Markovian probabilistic models were also developed for modeling a location history. Experiments show that these techniques are effective at extracting useful information about detailed location histories, and that they can be applied to a variety of applications. We find that a non-Markovian approach is better suited for evaluating likelihoods of a location history, while the Markovian approach is superior for purposes of stochastically generating a history.

We believe analysis of location histories to be a rich area of research, with many technical approaches and interesting applications. In future work, we expect to extend the analysis to trips and paths (what happens between stays and destinations), as well as to develop more accurate location-history models.

Acknowledgments

We would like to thank the following people for early brainstorming on this project: Ross Cutler, John Douceur, Nuria Oliver, Eric Ringger, Dan Robbins, Andreas Soupliotis, and Matt Uyttendaele. Thanks also to Chris Meek for discussions on probabilistic modeling.

References

1. Ashbrook, D., Starner, T. Learning significant locations and predciting user movement with GPS. In: Billinghurst, M., eds. *6th International Symposium on Wearable Computers (ISWC)*, 2002, pp. 101-108, Seattle, WA, IEEE Computer Society.
2. Bhattacharya, A., Das, S.K. LeZi-update: an information-theoretic approach to track mobile users in PCS networks. In: Imielinski, T., and Steenstrup, M., eds. *5th Annual ACM/IEEE International Conference on Mobile Computing and Networking*, 1999, pp. 1-12, Seattle, WA.
3. Hägerstrand, T. What about people in regional science? Papers of the Regional Science Association, 1970, vol. 24, pp. 7-21.
4. Hariharan, R. Modeling Intersections of Geospatial Lifelines. M.S. thesis, Department of Spatial Information Science and Engineering, University of Maine, 2001.
5. Hershberger, J. Smooth kinetic maintenance of clusters. In *Proceedings of the Nineteenth Annual Symposium on Computational Geometry*, 2003, pp. 48-57, San Diego, CA.
6. Hornsby, K., Egenhofer, M.J. Modeling Moving Objects over Multiple Granularities. Annals of Mathematics and Artificial Intelligence, 2002, vol. 36 (1-2), pp. 177-194.
7. Lei, Z., Rose, C. Wireless subscriber mobility management using adaptive individual location areas for pcs systems. In *IEEE International Conference on Communications (ICC)*, 1998, vol. 3, pp. 1390-1394, Atlanta, GA.

8. Mantoro, T., Johnson, C. Location history in low-cost context awareness environment. In: Johnson, C., Montague, P., and Steketee, C., eds. *Proceedings of the Australasian Information Security Workshop Conference on ACSW Frontiers*, 2003, vol. 21, pp. 153-158, Adelaide, Australia.

9. Mark, D., Egenhofer, M.J., Bian, L., Hornsby, K., Rogerson, P., Vena, J. Spatio-temporal GIS analysis for environmental health using geospatial lifelines [abstract]. In: Flahault, A., Toubiana, L., and Valleron, A., eds. *2nd International Workshop on Geography and Medicine*, 1999, GEOMED'99, pp. 52, Paris, France.

10. Marmasse, N., Schmandt, C. Location-aware information delivery with *ComMotion*. In: Thomas, P.J., and Gellersen, H., eds. *Handheld and Ubiquitous Computing, Second International Symposium (HUC)*, 2000, pp. 157-171, Bristol, UK, Springer.

11. Pathirana, P.N., Savkin, A.V., and Jha, S.K. Mobility modeling and trajectory prediction for cellular networks with mobile base stations. In 4th *International Symposium on Mobile Ad Hoc Networking and Computing (MobiHOC)*, 2003, pp. 213-221, Annapolis, MD.

12. Pred, A. Space and time in geography: essays dedicated to Torsten Hägerstrand. Lund Studies in Geography, 1981.

13. Rabiner, L.R. A tutorial on Hidden Markov models and selected applications in speech recognition. In *Proceedings of the IEEE*, 1989, vol. 77(2), pp. 257-285.

14. Revesz, P., Chen, R., Kanjamala, P., Li, Y., Liu, Y., Wang, Y. The MLPQ/GIS constraint database system. In: Chen, W., Naughton, J.F., Bernstein, P.A., eds. *SIGMOD 2000*, pp. 601, Dallas, TX.

15. Wolfson, O., Sistla, P.A., Chamberlain, S., and Yesha, Y. Updating and Querying databases that track mobile units. Distributed and Parallel Databases, 1999, vol. 7(3), pp. 257-287.

The SPIRIT Spatial Search Engine: Architecture, Ontologies and Spatial Indexing

Christopher B. Jones, Alia I. Abdelmoty, David Finch,
Gaihua Fu, and Subodh Vaid

School of Computer Science, Cardiff University, Cardiff, Wales, UK
{c.b.jones,a.i.abdelmoty,d.finch,gaihua.fu,s.vaid}@cs.cf.ac.uk

Abstract. The SPIRIT search engine provides a test bed for the development of web search technology that is specialised for access to geographical information. Major components include the user interface, a geographical ontology, maintenance and retrieval functions for a test collection of web documents, textual and spatial indexes, relevance ranking and metadata extraction. Here we summarise the functionality and interaction between these components before focusing on the design of the geo-ontology and the development of spatio-textual indexing methods. The geo-ontology supports functionality for disambiguation, query expansion, relevance ranking and metadata extraction. Geographical place names are accompanied by multiple geometric footprints and qualitative spatial relationships. Spatial indexing of documents has been integrated with text indexing through the use of spatio-textual keys in which terms are concatenated with spatial cells to which they relate. Preliminary experiments demonstrate considerable performance benefits when compared with pure text indexing and with text indexing followed by a spatial filtering stage.

1   Introduction

All aspects of human activity are rooted in geographic space in some respect. As a consequence many types of documents include references to geographical context, typically by means of place names. This common occurrence of geographical references is reflected in documents stored and retrieved on the world-wide web. If users of a web search engine wishes to find resources in which the subject matter is related to a particular place, then they can include the name of the place of interest in the search engine query. Conventional search engines treat the query place name in the same way as any other keyword and will retrieve documents that include the specified name. For some purposes this may be adequate, but there are many situations in which the user is interested in documents that relate to the same region of space as that specified by the place name, but which might not actually include the place name. This could occur if there were documents that used alternative names, or referred to places that were in or nearby the specified place. The process of exact match will also inevitably result in the retrieval of irrelevant documents due to multiple use of certain names to refer to

M.J. Egenhofer, C. Freksa, and H.J. Miller (Eds.): GIScience 2004, LNCS 3234, pp. 125–139, 2004.
© Springer-Verlag Berlin Heidelberg 2004

different places, and the fact that place names often occur in the names of products or organisations that are not associated with the named place. There is a need therefore for spatially-aware search engines that can interpret the presence of a place in a query in an intelligent manner that results in improved quality of information retrieval.

The introduction of spatial awareness in a search engine poses several significant challenges. At the user interface it should be possible to recognize the presence of place names in a query expression. This implies the existence of a directory of place names, such as is found in gazetteers [Ale]. If the search engine is to be entirely general purpose it would need to maintain knowledge of the names of every place on Earth. Users may wish to qualify the spatial aspect of their search using spatial relationships, in which case it becomes necessary to interpret such relationships and transform them to the representation of an appropriate geographical region, or *"query footprint"* that can be used for search purposes. Here we refer to the knowledgebase required for these purposes as a geographical ontology, or geo-ontology. Retrieval of relevant documents requires a process of geographical metadata extraction whereby the geographical context of a document is determined by some form of analysis of the text. Techniques are needed for disambiguation of place names and for establishing the likelihood that the document's information content is actually related to particular names that are present. The geographical coverage of a document is referred to here as the *"document footprint"*, which could consist of multiple parts. Once document footprints have been established, the possibility then arises for spatial indexing of the documents to facilitate fast access to documents pertaining to a given query footprint. Retrieval of documents must be followed by relevance ranking with respect both to their closeness to the query footprint and to thematic, possibly non-spatial, terms that the user has employed in the original query.

Recently a few search engines that are specialised with respect to geographic space have appeared (e.g. the vicinity products in the Mapblast[Vic] and Northern Lights web sites [Nor]; Mirago.com [Mir]; the experimental Google locational search engine [Goo]). Relatively little has been published on the technology that underlies these search engines, though there have been some published accounts of research efforts relating to particular aspects of the development of a geographical search engine functionality [DGS00,BCGM+99,McC01]. In this paper we describe the architecture of the SPIRIT prototype geographical search engine that is currently under development and which is intended to address the challenges to spatial search that are listed above. The main components include a user interface, a test collection of web documents with associated search engine maintenance and retrieval functionality, a geographical ontology, relevance ranking procedures, and document metadata creation and enrichment procedures. The role of each of these components and the interactions between them are introduced, before focusing in more detail on two specific aspects of the search engine, concerning the geographical ontology, and integration of spatial indexing with text indexing. Future articles will focus on other components of the search engine.

In what follows we summarise related work on geographical search engines, before describing the overall architecture of the SPIRIT search engine in section 3. The design of the geographical ontology is then presented in section 4 along with a summary of the access methods that have been implemented for it, in order to support processes of document annotation and metadata extraction, user interaction and relevance ranking. In section 5 we describe a novel approach to combining text indexing with spatial indexing, using spatio-textual keys, and provide a preliminary experimental evaluation of the technique. The paper concludes by highlighting current research issues and summarising future work.

2   Related Work

Geographical web search facilities developed by the company Vicinity [Vic] are present on the Mapblast [Map] web site and the Northern Light search engine [Nor]. These facilities allow the user to enter part or all of an address in the USA or Canada, along with a category of interest and a search radius in miles. It appears that the tool translates the address to a map coordinate and expands the search to include other places within the specified radius, with the aid of a digital map. The Mirago search engine [Mir] provides what is referred to as a regional web search facility with a focus in the four countries of UK, Germany, France and Spain. Here the user can select individual regions of a country on which to focus their web search. Google have recently introduced a demonstration of locational web search based in the USA [Goo]. Like the Vicinity search tools it allows the user to specify the name of a place of interest using an address or zip code, which is then matched against relevant documents. A commercial enterprise devoted to geographical indexing of documents is the company MetaCarta [Met]. Some explanation of the workings of an experimental geographical search engine can be found in [Egn]. The software uses the US Bureau of the Census TIGER/Line digital mapping data to detect street addresses in a corpus of text and then converts them to geographical coordinates. These coordinates are indexed in a two-dimensional index along with a conventional keyword index of the corpus. A query processor is able to process queries that ask for documents which match certain keywords or contain addresses within a certain radius of a specified target address. The procedure for detection of geo-referencing appears to be used in conjunction with the manual registration of URLs in the GeoURL's location-to-url reverse directory database creation [Geo]. Other research which considers the specific problem of determining the geographical context of a web document is that of [DGS00]. In their work they used a gazetteer to detect to the presence of place names in a web document, before analysing their frequency. Another approach to developing location-specific referencing of web data is to associate IP addresses of domain names with telephone area codes as suggested by Buyukokkten et al [BCGM+99]. In this approach the postal address of the web site or network administrators is used to derive a zip code that can be mapped to geographical coordinates. The Stanford Research Institute (SRI) has proposed a top level domain that is based on geographical referencing. In this system, the domain name refers to a strict hierarchy of quadrilateral cells defined by latitude and longitude. Existing domain names would be able to register themselves with a *.geo* domain server which would store, for a set of cells, all registered web sites that relate to each of the given cells. An experimental system for geographical navigation of the web has been described by McCurley [McC01]. A variety of techniques is proposed for extraction of the geographical context of a web page, on the basis of the occurrence of text addresses and post codes, place names and telephone numbers. This information is then transformed to one of a limited set of point-referenced map locations. Geographic search is initiated by the user asking to find web sites that refer to places in the vicinity of a currently displayed web site.

Global Atlas search engine [BOL00] indexes maps, images and HTML documents on the Web. The indexes are maintained for the available information according to their "geoprint" in addition to the traditional keywords and categories used by most search-engines. Queries are expressed as rectangles drawn on a map together with the traditional keyword filters. The registration of document footprints into an Oracle based spatial database is done with the help of gazetteers such as the Getty Thesaurus of Geographic Names [Get].

3   Architecture of the SPIRIT Search Engine

The SPIRIT search engine consists of the following components: user interface; geographical and domain-specific ontologies; web document collection; core search engine; textual and spatial indexes of document collection; relevance ranking and metadata extraction as shown in figure 1. We now provide a summary of the functionality associated with each of these components and the interactions between the components required to support the functionality. The summary starts with the user interface to introduce the system functionality from the user's perspective.

3.1   User Interface

The user interface allows the user to specify a subject of interest and a geographical location. The initial provisions for specifying a query are a structured text interface, a free text interface and a map. It is also planned that facilities for query by sketch will be introduced. The structured text interface allows the user to specify the subject of the query, a place name and a spatial relationship to the place name. The term or terms that form the subject of the query are treated in our initial prototype as non-spatial, but they could include types of places such as "hotels", or "cities."

The spatial relationships that are to be supported initially are *inside*, *outside* and *near* (distinguishing between whether the query is to include the named place or not) as well as cardinal direction and proximity relationships, namely *within a specified distance*.

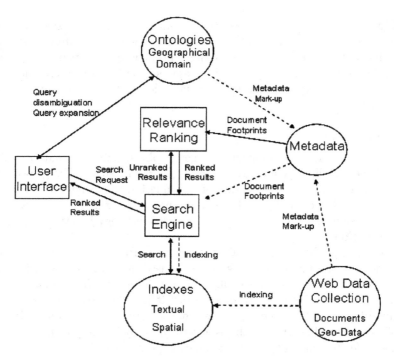

Fig. 1. SPIRIT search engine architecture.

Once the user enters a query, it is tested for ambiguity and alternative options are presented for purposes of disambiguation before the query is submitted to the search engine. The combination of the place name and spatial relationship is then used to determine a geometric query footprint, which consists of a polygon representing the interpretation of the spatial relationship applied to the place name. It might, for example, consist of the region of space inside a city or within some specified distance of a city. The shape of the footprint is echoed on the user interface map. The processes of disambiguation and query footprint generation require access to the geographical ontology. Once the user confirms the system's interpretation of the query, it is submitted to the core search engine and a relevance-ranked list of documents is returned to the user. These results will be presented both as a text list and as symbols on a map. A number of graphical visualizations are being investigated.

The present state of research in the field of spatially-aware search engines leaves many issues unresolved, concerning for example, effective user interfaces for geographical information retrieval, the most appropriate information to maintain in a gazetteer or a geo-ontology, reliable methods for determining the geographical scope of web documents, the way in which geographical relevance should be combined with thematic relevance to obtain a good ranking of retrieved documents, and the relative merits of different techniques for indexing documents with regard to both space and textual content. This paper provides one of the first published accounts of the overall architecture of a spatially-aware search engine that serves as a testbed for specific research to address these issues. It elucidates the key role of a geographical ontology in supporting geographical information retrieval functionality, and presents experimental results on a novel technique for integrating spatial and textual indexing for efficient access to web documents.

3.2 Ontologies

The primary ontology component is a geographical ontology that provides a model of the terminology and structure of geographic space. The geo-ontology plays a key role in the interpretation of user queries; the formulation of system queries, generation of spatial indexes, relevance ranking and metadata extraction. The design and implementation of the geo-ontology is described in more detail in Section 4.

In addition to the geographical ontology, the SPIRIT prototype also maintains a domain-specific ontology, which is focused on tourism. This will enable query expansion with respect to subject query terms for this domain. Thus if the user employs the term "accommodation" it will be possible to expand this to include terms such as "hotel," "guest house" etc. It will also introduce the possibility of disambiguation with respect to the subject query terms (e.g., does "surfing" refer to wind surfing or surf boarding etc).

3.3 Web Document Collection

Initial experiments with the SPIRIT search engine employ a 1 terabyte test collection of web documents (comprising 94 million web pages). Documents in this collection are structured to facilitate indexing. In order to support spatial indexing of the collection, each referenced document that contains place names is associated with one or more document footprints that are derived from the geographical ontology entries for the respective names. The SPIRIT prototype adapts an experimental text search engine GLASS [GLA] for purposes of building, maintaining and accessing the document collection and indexes. The major modification to the existing search engine functionality concerns the introduction of spatial indexing of the indexes of web documents and facilities to search for geographical context within web documents.

3.4 Indexes

The SPIRIT search engine supports both pure text indexing and spatio-textual indexing. The text index employs a conventional inverted file structure whereby each term in the index is associated with a list of the documents that include the term. Use of this index alone to process both spatial and non-spatial terms of a query is analogous to conventional search engine functionality, and will depend for its geographical effectiveness upon exact match between query place names and place names in web documents. The results of such a search can be improved in principle by expanding the query terms using the ontology, as indicated above, to refer to synonyms and neighbouring places. Note that in some cases this may result in massive proliferation of query terms, as would be the case if the name of a large region were expanded to include all contained places.

Spatio-textual indexing combines text indexing with spatial indexing of documents with respect to their document footprint. Use of a geometric query footprint to access a spatial index of documents serves the same purpose as geographical term expansion, with a term index, in that all documents relating to the region of the query should be retrieved. Use of a query footprint to access the index avoids the potentially very heavy and possibly impracticable overhead of employing high numbers of query terms resulting from term expansion. Spatial indexing will also facilitate distance-based relevance ranking, which depends upon analysis of the geometric relationships between query and document footprints. Details of spatio-textual indexing are provided in Section 5.

3.5   Core Search Engine

This component is responsible for accessing the web document collection and its text and spatio-textual indexes. It is based on a simple text retrieval system that has been enhanced to support spatial access to web documents. The component receives a query from the user interface and processes it against the collection using the textual and spatio-textual indexes as required. For experimental purposes queries may be processed either entirely by means of the text index, or using the spatio-textual index. The initial results of the query are passed to the relevance ranking component before being returned to the user interface.

3.6   Relevance Ranking

The relevance ranking component takes results retrieved from the search engine database and relevance ranks them with respect to the non-spatial and spatial elements of the query. Text relevance ranking is based on the BM25 algorithm [RWB+95], while spatial relevance is based, in the initial prototype, on measures of distance between the query footprint and the document footprint and on angular differences from cardinal directions in the case of directionally qualified queries. There are various techniques for combining textual and spatial relevance scores to produce an integrated score. In addition to combining independent text and spatial scores it is also possible to take account of the proximity in the document between the query text terms and the query spatial terms. This will be addressed in future versions of the SPIRIT prototype. It is also intended to introduce relevance ranking measures that take account of the parent geographical regions of the query footprint and the document footprint using methods such those documented in [CAT01].

3.7   Metadata Extraction

Effective spatial indexing of web documents depends upon the development of reliable techniques to identify the presence of place names in web documents and to determine their likely importance with regard to the subject matter of the document. This is the subject of ongoing research which includes the use of machine learning methods. For an example of previously published techniques see [DGS00,Li03]. Once significant place names have been detected in a document the geographical ontology can be used to provide footprints that contribute to the geographical metadata associated with the document. In the SPIRIT project, in addition to developing techniques for extraction of geographical context from web documents, work is also being pursued on the detection of features within geo-datasets, which may then be used to enrich the geographical ontology [HKS03].

4   A Geographical Ontology for Information Retrieval

When interacting with the user, the geo-ontology is used to recognise the presence of place names in a query and then to perform disambiguation. Once the user's query is formulated as a $< term, spatial relationship, place >$ expression, the ontology can be used to generate a polygonal geometric query footprint covering the spatial extent of the query region, based on the interpretation of the spatial relationship with the place. This query footprint is then used to access the spatial index of web documents. The geo-ontology could also be used to "expand" the user's query terms to include alternative names for the same place as well as the names of geographically associated places that may be inside, nearby or contain the specified place. The relevance ranking component accesses the geographical ontology to retrieve geometric footprints of places that are being compared with the query footprint, as well as with associated data providing the geographical context of a place, such as its containing and overlapping places. In the process of metadata extraction from web documents, the ontology is essential in identifying the presence of place names within text. There is also scope for enriching the ontology by including imprecise places found as a result of analysis of geo-datasets.

To support the functions above, actual and alternative place names, including multi-lingual versions need to be supported by the ontology. Geographical containment hierarchies and place types are required for query expansion and disambiguation. A geographic place is associated with possibly multiple geometric footprints. For example, detailed geometric footprints need to be used for accurate spatial indexing of documents. As indexing is a pre-processing operation, no impact on run time performance is expected. However, when generating a query footprint, access to detailed geometries can be expected to introduce processing overheads and hence there is a strong case for supporting generalised polygonal e.g. an MBR, or point-based geometries. The same reasoning applies to the use of a footprint for spatial relevance ranking of documents at query time. Consequently, the design of the geo-ontology supports multiple spatial representations of geographic places, including centre points, minimum bounding rectangles besides more faithful representations of geometries.

Several types of spatial relationships are stored and supported by the geo-ontology. Part-of relationships are used for maintaining containment hierarchies (e.g. based on different types of administrative hierarchies). Overlap and adjacency relationships are utilised by the relevance ranking component. In addition, overlap and containment relationships are used to derive similarity metrics between pairs of places. The later function is used when building or updating the ontology using different data sources. Hence, the geo-ontology supports part-of, contains, overlap and adjacency relationships between geographic places. The main components of the SPIRIT geo-ontology are shown in figure 2.

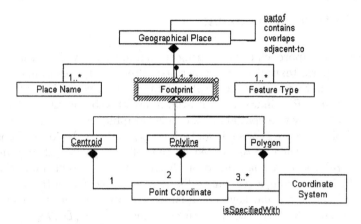

Fig. 2. Conceptual design of the geo-ontology.

The geographical ontology has been implemented using the Oracle Spatial database management system. The set of access functions developed are summarised as follows.

matchPlace: is an operation used by the user interface component to test whether a specific term in a query is a valid geographical place name.

getFeatureID: is an operation used by the geo-markup process to get the identifiers of places of specified terms.

getFootprint: is an operation to retrieve the geographical footprints for the specified place.

queryDisambiguation: is an operation to retrieve the broader geographical contexts for the name appearing in a query, using the partof relationship to derive the geographical hierarchies which are returned.

queryExpansion: is an operation that takes as arguments the disambiguated place name and the spatial relationship and derives the desired geographical search extent for the query (i.e. the query footprint).

Multiple data sources for different countries in Europe have been used to populate the geo-ontology. In particular, the EuroGeographics SABE data set [Sea] was used to extract the locations of towns and administrative boundaries and the Getty Thesaurus of Geographical Names [Get] is used to complement the ontology by providing other information such as alternative place names.

Employing multiple data sources in building the ontology is a complex task involving a pre-processing stage for checking the similarity of the data sets. Data sets may be different in many respects including the accuracy of representation, the type of spatial representation of the geographic objects as well as in the semantic classification used. Research in this area is still ongoing and will be reported elsewhere.

5   Spatio-textual Indexing

In this section an approach to integrating spatial indexing with textual indexing by means of spatio-textual keys is described. Preliminary experimental results with synthetic data compare the results of the spatio-textual indexing method with pure textual (PT) indexing. A spatial index of web documents can be created in a similar way to a spatial index of a geographic data set. Each document is allocated one or more geometric footprints, typically in the form of polygons. The footprints can then be referenced by the cells of spatial indexing methods such as a regular grid, a quadtree, or an R-tree. In the current implementation of the SPIRIT prototype, a regular grid scheme is employed. In this section, the derivation of spatio-textual keys is explained for an example document space as shown in figure 3. A collection of 16 documents, $D = \{D_1, D_2, \cdots, D_{16}\}$, is distributed over a document space R divided into 4 cells. The footprints of the documents (approximated by rectangular bounding boxes) are shown. Let SR be the document space associated with the entire set D, and the respective division for cells R_1, R_2, R_3 and R_4 be $SR1$, $SR2$, $SR3$, $SR4$.

$$SR = \{D_1, D_2, \cdots, D_{16}\}$$
$$SR1 = \{D_1, D_7, D_{12}, D_{15}\}$$
$$SR2 = \{D_{15}, D_{10}, D_{11}, D_3, D_{13}\}$$
$$SR3 = \{D_2, D_5, D_{14}, D_{12}, D_{15}\}$$
$$SR4 = \{D_{15}, D_{14}, D_9, D_6, D_{11}, D_{16}, D_4, D_8\}$$

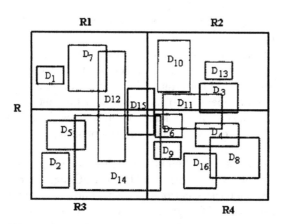

Fig. 3. An example of a document's search space.

A complete text index for SR corresponds to the index type PT. In such an index each term is associated with a list of the documents (inverted document list) in which it occurs. An example of such a list for a text term "spirit" is as follows:

$$\boxed{\text{spirit} \,|\, \{D_1, D_2, D_3, D_7, D_8, D_9, D_{11}, D_{13}\}}$$

Using the cell divisions shown in figure 3, this inverted list can be re-arranged in the form of a spatially-referenced list of documents which records for each cell those documents that contain the term:

$$\boxed{\text{spirit} \,|\, \{R1(D_1, D_7); R2(D_3, D_{11}, D_{13}); R3(D_2); R4(D_8, D_9, D_{11})\}}$$

An index with this structure can be exploited by first searching for a textual term and then using the associated spatial index of documents to filter out those meeting the spatial constraints.

A greater degree of integration of text and space, which effectively reduces the dimensionality of the problem, can be obtained by creating spatio-textual keys which join a text term to its corresponding spatial cells. Here the numeric cell identifiers ($R1$, $R2$ etc in the example) are concatenated with the text. In the example, the resulting keys would then be associated with their document lists as follows:

spiritR1	D_1, D_7
spiritR2	D_3, D_{11}, D_{13}
spiritR3	D_2
spiritR4	D_8, D_9, D_{11}

Searching this index requires text query terms to be concatenated at run time with the identifiers of spatial cells intersecting the query footprint, prior to matching the transformed query terms with the spatio-textual index terms. This indexing strategy shall be denoted spatio-textual (SP). Textual terms are used as a prefix for making the spatio-textual key. Identical results should be expected if cell identifiers were used as prefix, as the resulting spatio-textual list is of the same size and is constructed in exactly the same manner. It is possible for the spatio-textual index size to be bigger than the pure textual index, due to the introduction of multiple lists of documents for each term. The individual document lists may however be much smaller than for the PT index, provided that the terms referred to within documents in individual cells are a relatively small subset of the total number of terms in the document database. The following experiments were designed to predict the performance of the indexing strategies.

5.1   Experiments

In the absence of a very large collection of spatially indexed documents, a synthetic collection is used in which documents are assigned random footprints.

Queries are also generated synthetically. A total of six synthetic data sets were used comprising 1000 to 10,000 documents by employing techniques described in the public domain "mg" source code [MG.]. This text data generator allows the generation of data and queries with realistic properties as per their storage requirements. The document search space is divided into 1 to 60 cells to study the effect of cell size on search performance. As query time for textual queries up to 100 terms is practically undetectable, a query set comprising 1000 queries is used. The source code is written in C++ using MFC and STL in the Windows environment. A Pentium 4 PC is used (2 GHz processor and 256 MB of RAM). Standard MFC/STL arrays are used to store all indexes and table lookup is implemented with a standard binary search algorithm. A discussion of the results of the experiments are presented below.

5.2 Discussion of Results

The batch of 1000 queries was subjected to a total of 13 indexes for each of the six data sets. Table 1 lists the different index types.

Table 1. Different index types used in the experiments.

Index Type	Cells	Rows	Columns	Cell Size (%)
PT				
$PT + S$				
SP_R1_C1	1	1	1	100.0
SP_R2_C1	2	2	1	50.0
SP_R2_C2	4	2	2	25.0
SP_R5_C2	10	5	2	10.0
SP_R4_C5	20	4	5	5.0
SP_R5_C5	25	5	5	4.0
SP_R6_C5	30	6	5	3.0
SP_R7_C5	35	7	5	2.8
SP_R5_C8	40	5	8	2.5
SP_R5_C10	50	5	10	2.0
SP_R10_C6	60	10	6	1.6

In table 1, PT is a pure textual index comprising text terms only. All documents matching query terms are returned. The $PT + S$ index is the same as the PT index, but a spatial post-processing phase is employed to ensure that the only query results returned are those that fall within the query's footprint. This index is used to give an indication of query times and document hits when spatial relevance of documents is decided during query time. $SP\_R * \_C*$ are spatio-textual indexes with one or more cells.

Table 1 and figures 4 and 5 depict characteristics of the different index types. SP terms (Max) is the maximum number of SP terms obtained as a product of the total number of indexed terms (T) and the number of spatial cells (S).

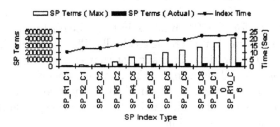

Fig. 4. SP index statistics.

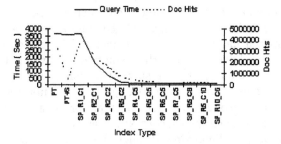

Fig. 5. Query results.

On comparing the theoretically expected maximum and actual SP terms, for each of the data sets, the actual number of text terms in the SP indexes is found to be much less than the expected maximum. Theoretically the number of SP terms should increase when the SP index has a higher number of cells. In practise this number is very low as can be seen in the graphs. This is a very positive observation in terms of memory considerations for spatio-textual indexes.

The SP indexes are constructed by post-processing the PT indexes and as such their construction times is much smaller in comparison. Figures 4 and 5 correspond to a collection size of 10,000 documents but a similar trend was also observed for other data sets.

As shown in figure 5, the query time for PT and $PT+S$ indexes is far higher than most SP indexes as the search space is the textual index of the entire collection. The only exception is the SP index SP_R1_C1, where there is only one spatial cell and the search space is identical to the PT and $PT+S$ indexes i.e. the entire collection.

The query time for the $PT+S$ index is slightly less than the PT index in all cases. Initially this may not sound logical as the $PT+S$ index performs additional calculations for filtering out the documents that do not intersect the query's footprint. However, observing the number of documents returned for both those index types it can be seen that the document hits after spatial filtering by the

$PT + S$ index are far lower than the PT index. Thus the lower query time is possibly due to fewer I/O operations used in writing the results.

The query times for the SP indexes initially begin to fall with increasing resolution of the grid but then decrease gradually. The document hits keep falling with increasing grid resolution. This suggest that excessive grid refinement can cause information loss. Comparing both time and document hits for the SP indexes with the $PT + S$ index, it appears that the best configuration of the grid is for SP_R4_C5 i.e. when the cell size is 5%. This corresponds to greater than 30 times improvement in query time.

6   Conclusions

This paper has summarised the architecture of a spatially-aware search engine and focused on the design of its component geo-ontology and on the integration of textual and spatial indexing of web documents. The prototype is currently under development and demonstrates the viability of the overall design. The geo-ontology plays a key role in providing support for query disambiguation, query expansion via the generation of geometric query footprints, relevance ranking to compare a query footprint to a document footprint, and the extraction of metadata to record the geographical context of web documents and geo-datasets. The introduction of spatial indexing, through the use of spatio-textual keys that concatenate text and spatial cell identifiers, reduced query times in excess of 30 times for large data sets for some grid resolutions. A study of the effect of the size of a spatial cell was conducted in the context of the regular grid indexing scheme and the best spatial search times were achieved when the cell size was 5% of the total area of the data sets employed.

Further work relating to the role of geo-ontologies and spatio-textual indexing will address issues that include integration of multiple data sources for construction of multi-national geo-ontologies, multi-scale representation of place footprints, representation of imprecise named places, alternative forms of spatio-textual key and experiments with large collections of geo-referenced documents. Research in parallel with that presented here focuses on issues concerned with user interface design, geo-referencing of text in web documents (to establish document footprints), geographical relevance ranking methods and metadata extraction from and enhancement of geo-datasets for purposes of web search.

Acknowledgments

This research is supported by the European Commission Framework V project SPIRIT: Spatially-Aware Information Retrieval on the Internet, funded under the Semantic Web Technologies action line.

References

[Ale]   Alexandria Digital Library Project. http://www.alexandria.ucsb.edu/.

[BCGM⁺99]   O. Buyukokkten, J. Cho, H. Garcia-Molina, L. Gravano, and N. Shiv-akumar. Exploiting Geographical Location Information of Web Pages. In *Proceedings of Workshop on Web Databases (WebDB'99) held in conjunction with ACM SIGMOD'99*, pages 91–96. ACM Press, 1999.

[BOL00]   S. Bressan, B.C. Ooi, and F. Lee. Global Atlas: Calibrating and Indexing Documents from the Internet in the Cartographic Paradigm. In *Proceedings of the 1st International Conference on Web Information Systems Engineering*, volume 1, pages 117–124, 2000.

[CAT01]   Jones C.B., H. Alani, and D. Tudhope. Geographical information retrieval with ontologies of place. In *Spatial Information Theory Foundations of Geographic Information Science, COSIT 2001*, volume LNCS 2205, pages 323–335. Springer Verlag, 2001.

[DGS00]   J. Ding, L. Gravano, and N. Shivakumar. Computing Geographical Scopes of Web Resources. In *Proceedings of the 26th Very-Large Database (VLDB) Conference*, pages 546–556. Morgan Kaufmann, 2000.

[Egn]   Egnor, D. http://www.google.com/programming-contest/winner.html.

[Geo]   GeoURL ICBM Address Server. http://geourl.org/.

[Get]   Getty Thesaurus of Geographic Names. http://www.getty.edu/research/conducting_research/vocabularies/tgn/index.html.

[GLA]   GLASS: Online Documentation. http://dis.shef.ac.uk/mark/glass/.

[Goo]   Google Local. http://local.google.com/lochp.

[HKS03]   F. Heinzle, M. Kopczynski, and M. Sester. Spatial Data Interpretation for the Intelligent Access to Spatial Information in the Internet. In *Proceedings of 21st International Cartographic Conference*, 2003.

[Li03]   H. Li. Infoxtract location normalization: a hybrid approach to geographic references in information extraction. In *Proc. of the HLT-NAACL 2003 Workshop on Analysis of Geographic References*, pages 39–44, 2003.

[Map]   Mapblast. http://www.mapblast.com.

[McC01]   K.S. McCurley. Geospatial Mapping and Navigation of the Web. In *Proceedings of Tenth International World Wide Web Conference*, pages 221–229. ACM Press, 2001.

[Met]   Metacarta. http://www.metacarta.com.

[MG.]   MG. Information Retrieval System. http://www.cs.mu.oz.au/mg/.

[Mir]   Mirago: Mirago the UK Search Engine. http://www.mirago.co.uk/.

[Nor]   Northern Light. http://www.northernlight.com/index.html.

[RWB⁺95]   S.E. Robertson, S. Walker, M.M. Beaulieu, M. Gatford, and A. Payne. Okapi at trec-4. In *Proc. of the 4th Text REtrieval Conference (TREC-4)*, pages 73–96, 1995.

[Sea]   Seamless Administrative Boundaries of Europe (SABE) dataset. http://www.eurogeographics.org/eng/04-sabe.asp.

[Vic]   Vicinity.com. http://home.vicinity.com/us/mappoint.htm.

Comparing Exact and Approximate Spatial Auto-regression Model Solutions for Spatial Data Analysis

Baris M. Kazar[1], Shashi Shekhar[2], David J. Lilja[1],
Ranga R. Vatsavai[2], and R. Kelley Pace[3]

[1] Electrical and Computer Engineering Department, University of Minnesota,
Twin-Cities MN 55455
{Kazar,Lilja}@ece.umn.edu

[2] Computer Science and Engineering Department, University of Minnesota,
Twin-Cities MN 55455
{Shekhar,Vatsavai}@cs.umn.edu

[3] LREC Endowed Chair of Real Estate, 2164B CEBA, Department of Finance
E.J. Ourso College of Business, Louisiana State University
Baton Rouge, LA 70803-6308
Kelley@pace.am

Abstract. The spatial auto-regression (SAR) model is a popular spatial data analysis technique, which has been used in many applications with geo-spatial datasets. However, exact solutions for estimating SAR parameters are computationally expensive due to the need to compute all the eigenvalues of a very large matrix. Recently we developed a dense-exact parallel formulation of the SAR parameter estimation procedure using data parallelism and a hybrid programming technique. Though this parallel implementation showed scalability up to eight processors, the exact solution still suffers from high computational complexity and memory requirements. These limitations have led us to investigate approximate solutions for SAR model parameter estimation with the main objective of scaling the SAR model for large spatial data analysis problems. In this paper we present two candidate approximate-semi-sparse solutions of the SAR model based on Taylor series expansion and Chebyshev polynomials. Our initial experiments showed that these new techniques scale well for very large data sets, such as remote sensing images having millions of pixels. The results also show that the differences between exact and approximate SAR parameter estimates are within 0.7% and 8.2% for Chebyshev polynomials and Taylor series expansion, respectively, and have no significant effect on the prediction accuracy.

1 Introduction

Explosive growth in the size of spatial databases has highlighted the need for spatial data analysis and spatial data mining techniques to mine the interesting but implicit spatial patterns within these large databases. Extracting useful and interesting patterns

M.J. Egenhofer, C. Freksa, and H.J. Miller (Eds.): GIScience 2004, LNCS 3234, pp. 140–161, 2004.
© Springer-Verlag Berlin Heidelberg 2004

from massive geo-spatial datasets is important for many application domains, such as regional economics, ecology and environmental management, public safety, transportation, public health, business, and travel and tourism [8,34,35]. Many classical data mining algorithms, such as linear regression, assume that the learning samples are *independently and identically distributed (i.i.d)*. This assumption is violated in the case of spatial data due to *spatial autocorrelation* [2,34] and in such cases classical linear regression yields a weak model with not only low prediction accuracy [35] but also residual error exhibiting *spatial dependence*. Modeling spatial dependencies improves overall classification and prediction accuracies.

The spatial auto-regression model (SAR) [10,14,34] is a generalization of the linear regression model to account for spatial autocorrelation. It has been successfully used to analyze spatial datasets in regional economics and ecology [8,35]. The model yields better classification and prediction accuracy [8,35] for many spatial datasets exhibiting strong spatial autocorrelation. However, it is computationally expensive to estimate the parameters of SAR. For example, it can take an hour of computation for a spatial dataset with 10K observation points on a single IBM Regatta processor using a 1.3GHz pSeries 690 Power4 architecture with 3.2 GB memory. This has limited the use of SAR to small problems, despite its promise to improve classification and prediction accuracy for larger spatial datasets. For example, SAR was applied to accurately estimate crop parameters [37] using airborne spectral imagery; however, the study was limited to 74 pixels. A second study, reported in [21], was limited to 3888 observation points.

Table 1. Classification of algorithms solving the serial spatial auto-regression model

	Exact	Approximate
Maximum Likelihood	Applying Direct Sparse Matrix Algorithms [25]	ML based Matrix Exponential Specification [26]
	Eigenvalue based 1-D Surface Partitioning [16]	Graph Theory Approach [32]
		Taylor Series Approximation [23]
		Chebyshev Polynomial Approximation Method [30]
		Semiparametric Estimates [27]
		Characteristic Polynomial Approach [36]
		Double Bounded Likelihood Estimator [31]
		Upper and Lower Bounds via Divide & Conquer [28]
		Spatial Autoregression Local Estimation [29]
Bayesian	None	Bayesian Matrix Exponential Specification [19]
		Markov Chain Monte Carlo (MCMC) [3,17]

A number of researchers who have been attracted to SAR because of its high computational complexities have proposed efficient methods of solving the model. These solutions, summarized in Table 1, can be classified into exact and approximate solutions, based on how they compute certain compute-intensive terms in the SAR solution procedure. Exact solutions suffer from high computational complexities and memory requirements. Approximate solutions are computationally feasible, but many of these formulations still suffer from large memory requirements. For example, a standard remote sensing image consisting of 3000 lines (rows) by 3000 pixels (columns) and six bands (dimensions) leads to a large neighborhood (**W**) matrix of size 9 million rows by 9 million columns. (The details for forming the neighborhood matrix **W** can be found in Sect. 2.) Thus, the exact implementations of SAR are simply not capable of processing such large images, and approximate solutions must be found. We choose Taylor and Chebyshev approximations for two reasons. First, the solutions are scalable for large problems and secondly these methods provide bounds on errors.

Major contributions of this study include scalable implementations of the SAR model for large geospatial data analysis, characterization of errors between exact and approximate solutions of the SAR model, and experimental comparison of the proposed solutions on real satellite remote sensing imagery having millions of pixels. Most importantly, our study shows that the SAR model can be efficiently implemented without loss of accuracy, so that large geospatial datasets which are spatially auto-correlated can be analyzed in a reasonable amount of time on general purpose computers with modest memory requirements. We are using an IBM Regatta in order to implement parallel versions of the software using open source ScaLAPACK [7] linear algebra libraries. However, the software can also be ported onto general-purpose computers after replacing ScaLAPACK routines with the serial equivalent open source LAPACK [1] routines. Please note that, even though we are using a parallel version of ScaLAPACK, the computational timings presented in the results section (Table 7) are based on serial execution of all SAR model solutions on a single processor. The remainder of the paper is organized as follows: Section 2 presents the problem statement, and Section 3 explains the exact algorithm for the SAR solution. Section 4 discusses approximate SAR model solutions using Taylor series expansion and Chebyshev polynomials respectively. The experimental design is provided in Section 5. Experimental results are discussed in Section 6. Finally, Section 7 summarizes and concludes the paper with a discussion of future work.

2 Problem Statement

We first present the problem statement and the notation in Table 2; and then explain the exact and approximate SAR solutions based on maximum-likelihood (ML) theory [12].

The problem studied in this paper is defined as follows: *Given* the exact solution procedure described in the Dense Matrix Approach [16] for one-dimensional geospatial datasets, we need to *find* a solution that scales well for large multi-dimensional geo-spatial datasets. The *constraints* are as follows: the spatial auto-regression parameter ρ varies in the range [0,1]; the error is normally distributed, that is, $\varepsilon \sim N(0,\sigma^2 I)$ *iid*; the input spatial dataset is composed of normally distributed random variables; and the size of the neighborhood matrix **W** is n. The *objective* is to implement scalable and portable software for analyzing large geo-spatial datasets.

Table 2. The notation in this study

Variable	Definition	Variable	Definition
ρ	The spatial auto-regression (autocorrelation) parameter	**I**	Identity matrix
y	n-by-1 vector of observations on the dependent variable	λ	Eigenvalue of a matrix
x	n-by-k matrix of observations on the explanatory variable	$tr(.)$	Trace of the "." matrix
W	n-by-n neighborhood matrix that accounts for the spatial relationships (dependencies) among the spatial data	π	Pi constant which is equal to 3.14
k	Number of features	$|.|$	Determinant of the "." matrix
β	k-by-1 vector of regression coefficients	$(.)^{-1}$	Inverse of the "." matrix
n	Problem size (also number of observation points or pixels)	$T_i(.)$	A Chebyshev polynomial of degree i. "." can be a matrix or a scalar number.
p	Row dimension of spatial framework (image)	k	Index variable
q	Column dimension of spatial framework (image)	$\sum(.)$	Summation operation on a matrix/vector Index variable running on i of $T_i(.)$
C	n-by-n Binary neighborhood matrix	$\prod$	Product operation on a matrix/vector
D	n-by-n Diagonal matrix with elements $1/s_i$, where s_i is the row-sum of row i of **C**	$\exp(.)$	Exponential operator i.e., $e^{(.)}$
$\widetilde{\mathbf{W}}$	n-by-n Symmetric equivalent of **W** matrix in terms of eigenvalues	$(.)^T$	Transpose of the "." Matrix/vector
ε	n-by-1 vector of unobservable error	$(.)_{ij}$	ij^{th} element of the "." matrix
σ^2	The common variance of the error ε	Σ	n-by-n Diagonal variance matrix of error defined as $\sigma^2 \mathbf{I}$
$\ln(.)$	Natural logarithm operator	$O(.)$	"O" notation for complexity analysis of algorithms
q	The highest degree of the Chebyshev polynomials	**N**	n-by-n Binary neighborhood matrix from Delaunay triangulation
Ψ	Current pixel in the spatial framework (image) with "s" neighbors	$\cos(.)$	Cosinus trigonometric operation

2.1 Basic SAR Model

The *spatial auto-regression model (SAR)* [10], also known in the literature as a *spatial lag* model or *mixed regressive* model, is an extension of the linear regression model and is given in equation 1.

$$\mathbf{y} = \rho\mathbf{W}\mathbf{y} + \mathbf{x}\boldsymbol{\beta} + \boldsymbol{\varepsilon}. \tag{1}$$

Here the parameters are defined in Table 2. The main point to note here is that a spatial autocorrelation term $\rho\mathbf{W}\mathbf{y}$ is added to the linear regression model in order to model the strength of the *spatial dependencies* among the elements of the dependent variable, **y**.

2.2 Example Neighborhood Matrix (W)

The neighborhood matrices used by the spatial auto-regression model are the neighborhood relationships on a one-dimensional regular grid space with two neighbors and a two-dimensional grid space with "s" *neighbors*, where "s" is four, eight, sixteen, twenty-four and so on neighbors, as shown in Fig. 1. This structure is also known as regular square tessellation one-dimensional and two-dimensional planar surface partitioning [12].

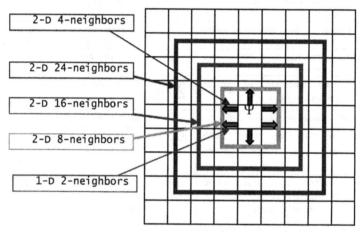

Fig. 1. The neighborhood structures of the pixel Ψ on one-dimensional and two-dimensional regular grid space

2.3 Illustration of the Neighborhood Matrix Formation on a 4-by-4 Regular Grid Space

As noted earlier, modeling spatial dependency (or context) improves the overall classification (prediction) accuracy. Spatial dependency can be defined by the relationships among spatially adjacent pixels in a small neighborhood within a spatial framework that is a regular grid space. The following paragraph explains how **W** in the SAR model is formed. For the four-neighborhood case, the neighbors of the $(i,j)^{th}$ pixel of the regular grid are shown in Fig. 2.

$$neighbors(i,j) = \begin{cases} (i-1,j) & 2 \le i \le p, 1 \le j \le q \text{ NORTH} \\ (i,j+1) & 1 \le i \le p, 1 \le j \le q\text{-}1 \text{ EAST} \\ (i+1,j) & 1 \le i \le p\text{-}1, 1 \le j \le q \text{ SOUTH} \\ (i,j-1) & 1 \le i \le p, 2 \le j \le q \text{ WEST} \end{cases}$$

Fig. 2. The four neighbors of the $(i,j)^{th}$ pixel on the regular grid

The $(i,j)^{th}$ pixel of the surface will fill in the $(p(i-1)+j)^{th}$ row of the non-row-standardized neighborhood matrix, **C**. The following entries of **C**, i.e. $\{(p(i-1)+j), (p(i-2)+j)\}, \{(p(i-1)+j),(p(i-1)+j+1)\}, \{(p(i-1)+j),(p(i)+j)\}$ and $\{(p(i-1)+j),(p(i-1)+j-1)\}$ will be "1"s and the others all zeros. The row-standardized neighborhood matrix **W** is formed by first finding each row sum (i.e., there will be pq or n number of row-sums since **W** is pq-by-pq) and dividing each element in a row by its corresponding row-sum. In other words, $\mathbf{W} = \mathbf{D}^{-1}\mathbf{C}$, where the elements of the diagonal matrix **C** are defined as $d_{ii} = \sum_{i=1}^{n} c_{ij}$ and $d_{ij} = 0$. Fig. 3 illustrates the spatial framework and the matrices. Thus, the rows of matrix **W** sum to 1, which means that **W** is row-standardized i.e., row-normalized or row-stochastic. A non-zero entry in the jth column of the ith row of matrix **W** indicates that the jth observation will be used to adjust the prediction of the ith row where i is not equal to j. We described forming a **W** matrix for a regular grid that is appropriate for satellite images; however, **W** can also be formed for irregular (or vector) datasets as discussed further in the Appendix [12].

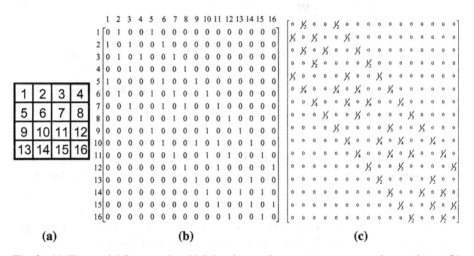

(a)   (b)   (c)

Fig. 3. (a) The spatial framework, which is p-by-q, where p may or may not be equal to q, (b) the pq-by-pq non-normalized neighborhood matrix **C** with 4 nearest neighbors, and (c) the normalized version (i.e., **W**), which is also pq-by-pq. The product pq is equal to n, the problem size

3 Exact SAR Model Solution

The estimates for the parameters ρ and β in the SAR model (equation 1) can be found using either maximum likelihood theory or Bayesian statistics. In this paper we consider the maximum likelihood approach for estimating the parameters of the SAR model, whose mechanics are presented in Fig. 4.

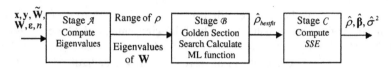

Fig. 4. System diagram of the serial exact algorithm for the SAR model solution composed of three stages ($\mathcal{A}$, $\mathcal{B}$, and C)

Fig. 4 highlights the three stages of the exact algorithm for the SAR model solution. It is based on maximum-likelihood (ML) theory, which requires computing the logarithm of the determinant (i.e., log-determinant) of the large $(\mathbf{I} - \rho\mathbf{W})$ matrix. The first term of the end-result of the derivation of the logarithm of the likelihood function (equation 2) clearly shows why we need to compute the (natural) logarithm of the determinant of a large matrix. In equation 2 "$\mathbf{I}$" denotes an n-by-n identity matrix, "T" denotes the transpose operator, "ln" denotes the logarithm operator, and σ^2 is the common variance of the error.

$$\ln(L) = \ln|\mathbf{I} - \rho\mathbf{W}| - \frac{n\ln(2\pi)}{2} - \frac{n\ln(\sigma^2)}{2} - \frac{SSE}{2\sigma^2}. \tag{2}$$

where $SSE = \left\{ (\mathbf{y}^T (\mathbf{I} - \rho\mathbf{W})^T [\mathbf{I} - \mathbf{x}(\mathbf{x}^T\mathbf{x})^{-1}\mathbf{x}^T]^T [\mathbf{I} - \mathbf{x}(\mathbf{x}^T\mathbf{x})^{-1}\mathbf{x}^T](\mathbf{I} - \rho\mathbf{W})\mathbf{y}) \right\}$.

Therefore, Fig. 4 can be viewed as an implementation of the ML theory. We now describe each stage. Stage $\mathcal{A}$ is composed of three sub-stages: pre-processing, Householder transformation [33], and QL transformation [9]. The pre-processing sub-stage not only forms the row-standardized neighborhood matrix $\mathbf{W}$, but also converts it to its symmetric eigenvalue equivalent matrix $\widetilde{\mathbf{W}}$. The Householder transformation and QL transformation sub-stages are used to find all of the eigenvalues of the neighborhood matrix. The Householder transformation sub-stage takes $\widetilde{\mathbf{W}}$ as input and forms the tri-diagonal matrix whose eigenvalues are computed by the QL transformation sub-stage. Computing all of the eigenvalues of the neighborhood matrix takes approximately 99% of the total serial response time, as shown in Table 3.

Stage $\mathcal{B}$ computes the best estimates for the spatial auto-regression parameter ρ and the vector of regression coefficients $\boldsymbol{\beta}$ for the SAR model. While these estimates are being found, the logarithm of the determinant of $(\mathbf{I} - \rho\mathbf{W})$ needs to be computed at each step of the non-linear one-dimensional parameter optimization. This step uses the golden section search [9] and updates the auto-regression parameter at each step. There are three ways to compute the value of the logarithm of the likelihood function: (1) compute the eigenvalues of the large dense matrix $\mathbf{W}$ *once*; (2) compute the determinant of the large dense matrix $(\mathbf{I} - \rho\mathbf{W})$ at each step of the non-linear optimization; (3) approximate the log-determinant term. For small problem sizes, the first two methods work well; however, for large problem sizes approximate solutions are needed.

Equation 3 expresses the relationship between the eigenvalues of the $\mathbf{W}$ matrix and the logarithm of the determinant (i.e., log-determinant) of the $(\mathbf{I} - \rho\mathbf{W})$ matrix. The optimization is of $O(n)$ complexity.

$$|\mathbf{I} - \rho\mathbf{W}| = \prod_{i=1}^{n} (1 - \rho\lambda_i) \xrightarrow[\text{logarithm}]{\text{taking the}} \ln|\mathbf{I} - \rho\mathbf{W}| = \sum_{i=1}^{n} \ln(1 - \rho\lambda_i). \qquad (3)$$

The eigenvalue algorithm applied in this study cannot find the eigenvalues of any dense matrix. The matrix $\mathbf{W}$ has to be converted to its symmetric version $\widetilde{\mathbf{W}}$, whose eigenvalues are the same as the original matrix $\mathbf{W}$. The conversion is derived as shown in Fig. 5.

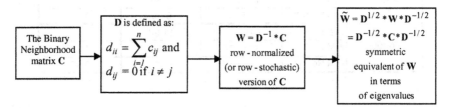

Fig. 5. Derivation of the $\widetilde{\mathbf{W}}$ matrix, the symmetric eigenvalue equivalent of the $\mathbf{W}$ matrix

The matrix $\widetilde{\mathbf{W}}$ (i.e. $\mathbf{D}^{-1/2}\mathbf{C}\mathbf{D}^{-1/2}$) is symmetric and has the same eigenvalues as $\mathbf{W}$. The row standardization can be expressed as $\mathbf{W} = \mathbf{D}^{-1}\mathbf{C}$, where $\mathbf{D}$ is a diagonal matrix with elements $1/s_i$, where s_i is the row-sum of row i of $\mathbf{C}$. The *symmetrization* subroutine is the part of the code that does this job.

Finally, stage C computes the sum of the squared error, i.e., the *SSE* term, which is $O(n^2)$ complex. Table 3 shows our measurements of the serial response times of the stages of the exact SAR model solution based on ML theory. Each response time given in this study is the average of five runs. As can be seen, computing the eigenvalues (stage $\mathcal{A}$) takes a large fraction of the total time.

Now, we outline the derivation of the maximum likelihood function. This derivation not only shows the link between the need for eigenvalue computation and the spatial auto-regression model parameter fitting but also explains how the spatial auto-regression model works and can be interpreted as an execution trace of the algorithm.

We begin the derivation by choosing a SAR model that is described by equation 1. Ordinary least squares are not appropriate to solve for the models described by equation 1. One way to solve is to use the maximum likelihood procedure. In probability, there are essentially two classes of problems: the first is to generate a data sample given a probability distribution and the second is to estimate the parameters of a probability distribution given data. Obviously in our case, we are dealing with the latter problem.

It is assumed that ε is generated from a normal distribution, which has to be formally defined to go further in the derivation. The normal density function is given in equation 4.

Table 3. Measured serial response times of stages of the exact SAR model solution for problem sizes of 2500, 6400 and 10K. Problem size denotes the number of observation points

Problem size (n)	Machine	Serial Execution Time (sec) Spent on		
		Stage A	Stage B	Stage C
		Computing Eigenvalues	ML Function	Least Squares
2500	SGI Origin	78.10	0.41	0.06
	IBM SP	69.20	1.30	0.07
	IBM Regatta	46.90	0.58	0.06
6400	SGI Origin	1735.41	5.06	0.51
	IBM SP	1194.80	17.65	0.44
	IBM Regatta	798.70	6.19	0.42
10000	SGI Origin	6450.90	11.20	1.22
	IBM SP	6546.00	66.88	1.63
	IBM Regatta	3439.30	24.15	0.93

$$N(\varepsilon) \equiv (2\pi)^{-\frac{n}{2}} |\Sigma|^{-\frac{1}{2}} \exp\left\{-\frac{1}{2}\varepsilon^T \Sigma \varepsilon\right\}, \tag{4}$$

where $\Sigma = \sigma^2 I$ and "T" means transpose of a vector or matrix. It is worth noting again that we are assuming in our derivation that the error vector ε is governed by the standard normal distribution with zero mean and variance Σ. The prediction of the spatial auto-regression model solution heavily depends on the quality of the normally distributed random numbers generated. The term $|d\varepsilon/dy|$ needs to be calculated out in order to find the probability density function of the variable y, which is given by equation 5. The notation $|x|$ denotes the determinant of matrix x. Hence, the probability density function of the observed dependent variable (y vector) is given by equation 6. When ε is replaced by $((I - \rho W)y - x\beta)$ in equation 6, the explicit form of the probability density function of y given by equation 7 is obtained. It should be noted that $|\Sigma| = |\sigma^2 I| = \sigma^{2n}$, where n is the rank (i.e., row-size and column-size) of identity matrix, I.

$$|d\varepsilon/dy| = |I - \rho W|. \tag{5}$$

$$L(y) = N((I - \rho W)y - x\beta)|d\varepsilon/dy|. \tag{6}$$

$$L(y) = (2\pi\sigma^2)^{-\frac{n}{2}} |I - \rho W| \exp\left\{-\frac{1}{2\sigma^2}\{[(I - \rho W)y - x\beta]^T [(I - \rho W)y - x\beta]\}\right\}. \tag{7}$$

$L(y)$ will henceforth be referred to as the "likelihood function." It is a probability distribution but now interpreted as a distribution of parameters which have to be calculated. Since the log-likelihood function is monotonic, we can then equivalently minimize the log-likelihood function, which has a simpler form and can handle large numbers. This is because the logarithm is advantageous, since $\log(ABC) = \log(A) + \log(B) + \log(C)$. After taking the natural logarithm of equation 7, we get equation 8, i.e., the log-likelihood function with the estimators for the variables β and σ^2, which are represented by $\hat{\beta}$ and $\hat{\sigma}^2$ respectively in equations 9a and b.

$$\ln(L) = \ln|\mathbf{I} - \rho\mathbf{W}| - \frac{n\ln(2\pi)}{2} - \frac{n\ln(\sigma^2)}{2} - \frac{1}{2\sigma^2}\left\{[(\mathbf{I}-\rho\mathbf{W})\mathbf{y} - \mathbf{x}\beta]^T[(\mathbf{I}-\rho\mathbf{W})\mathbf{y} - \mathbf{x}\beta]\right\}. \tag{8}$$

$$\hat{\beta} = (\mathbf{x}^T\mathbf{x})^{-1}\mathbf{x}^T(\mathbf{I}-\rho\mathbf{W})\mathbf{y}. \tag{9a}$$

$$\hat{\sigma}^2 = \mathbf{y}^T(\mathbf{I}-\rho\mathbf{W})^T[\mathbf{I} - \mathbf{x}(\mathbf{x}^T\mathbf{x})^{-1}\mathbf{x}^T](\mathbf{I}-\rho\mathbf{W})\mathbf{y}/n. \tag{9b}$$

The term $[(\mathbf{I}-\rho\mathbf{W})\mathbf{y} - \mathbf{x}\beta]$ is equivalent to $[\mathbf{I} - \mathbf{x}(\mathbf{x}^T\mathbf{x})^{-1}\mathbf{x}^T](\mathbf{I}-\rho\mathbf{W})\mathbf{y}$ after replacing $\hat{\beta}$ given by equation 9a with β in equation 8. That leads us to equation 10 for the log-likelihood function (i.e., the logarithm of the maximum likelihood function) to be optimized for ρ.

$$\ln(L) = \ln|\mathbf{I} - \rho\mathbf{W}| - \frac{n\ln(2\pi)}{2} - \frac{n\ln(\sigma^2)}{2} - \tag{10}$$
$$\frac{1}{2\sigma^2}\left\{\mathbf{y}^T(\mathbf{I}-\rho\mathbf{W})^T[\mathbf{I} - \mathbf{x}(\mathbf{x}^T\mathbf{x})^{-1}\mathbf{x}^T]^T[\mathbf{I} - \mathbf{x}(\mathbf{x}^T\mathbf{x})^{-1}\mathbf{x}^T](\mathbf{I}-\rho\mathbf{W})\mathbf{y})\right\}$$

The first term of equation 10 (i.e., the log-determinant) is nothing but the logarithm of the sum of a collection of scalar values including all of the eigenvalues of the neighborhood matrix $\mathbf{W}$. The first term transforms from a multiplication to a sum as shown by equation 3. That is why all of the eigenvalues of $\mathbf{W}$ matrix are needed.

$$\underset{|\rho|<1}{\text{MIN}} \sum_{i=1}^{n} \ln(1 - \rho\lambda_i) - \frac{1}{2\sigma^2}\left\{\mathbf{y}^T(\mathbf{I}-\rho\mathbf{W})^T[\mathbf{I} - \mathbf{x}(\mathbf{x}^T\mathbf{x})^{-1}\mathbf{x}^T]^T[\mathbf{I} - \mathbf{x}(\mathbf{x}^T\mathbf{x})^{-1}\mathbf{x}^T](\mathbf{I}-\rho\mathbf{W})\mathbf{y}\right\} \tag{11}$$

Therefore, the function is optimized using the golden section search to find the best estimate for ρ, as shown in equation 11, which must be solved using nonlinear optimization techniques. Equation 9a could be estimated and then equation 8 could be optimized iteratively. Rather than computing that way, there is a faster and easier way such that equation 9a is substituted directly into equation 8 to get equation 10, which is a single expression in one unknown to be optimized. Once the estimate for ρ is found, both $\hat{\beta}$ and $\hat{\sigma}^2$ can be computed. Finally, the simulated variable ($\mathbf{y}$ vectors or thematic classes) can be computed using equation 12.

$$\mathbf{y} = (\mathbf{I} - \rho\mathbf{W})^{-1}(\mathbf{x}\beta + \varepsilon). \tag{12}$$

Equation 12 needs a matrix inversion algorithm in order to get the predicted observed dependent variable ($\mathbf{y}$ vector). For small problem sizes, one can use exact matrix inversion algorithms; however, for large problem sizes (e.g., > 10K) one can use geometric series expansion to compute the inverse matrix in equation 13. (For more details see lemma 2.3.3 [11].) In the next section we show how the complexity of these calculations can be reduced using approximate solutions.

4 Two Approximate SAR Model Solutions

Since an exact SAR model solution is both memory and compute intensive, we need approximate solutions that do not sacrifice accuracy and can handle very large data-sets. We propose to use two different approximations for solving the SAR model solution, Taylor's series expansion and Chebyshev polynomials. The purpose of these approximations is to calculate the logarithm of the determinant of $(I - \rho W)$.

4.1 Approximation by Taylor's Series Expansion

Martin [23] suggests an approximation of the log-determinant by means of the traces of the powers of the neighborhood matrix, W. He basically finds the trace of the matrix logarithm, which is equal to the log-determinant. In this approach, the Taylor series is used to approximate the function $\sum_{i=1}^{n} \ln(1 - \rho \lambda_i)$ where λ_i represents the ith eigenvalue that lies in the interval [-1,+1] and ρ is the scalar parameter from the interval (-1,+1). The term $-\sum_{i=1}^{n} \ln(1 - \rho \lambda_i)$ can be expanded as $\sum_{i=1}^{n} \left[(\rho \lambda_i)^k / k \right]$ provided that $| \rho \lambda_i | < 1$, which will hold for all i if $| \rho | < \lambda^{-1}$. Equation 13, which states the approximation used for the logarithm of the determinant of the large matrix term of maximum likelihood, is obtained using the relationship between eigenvalues and the trace of a matrix, i.e., $\sum_{i=1}^{n} \lambda_i^{k} = tr(W^k)$.

$$\ln |I - \rho W|^{-1/n} = \frac{1}{n} \sum_{k=1}^{\infty} \{tr(W^k)\rho^k / k\}. \tag{13}$$

The approximation comes into the picture when we sum up to a finite value, r, instead of infinity. Therefore, equation 13 is relatively much faster because it eliminates the need to calculate the compute-intensive eigenvalue estimation when computing the log-determinant. The overall solution is shown in Fig. 6.

4.2 Approximation by Chebyshev Polynomials

This approach uses the symmetric equivalent of the neighborhood matrix W (i.e., $\tilde{W}$) as discussed in Sect. 3. The eigenvalues of the symmetric $\tilde{W}$ are the same as those of the neighborhood matrix W. The following lemma leads to a very efficient and accurate approximation to the first term on the right-hand side of the logarithm of the likelihood function shown in equation 2.

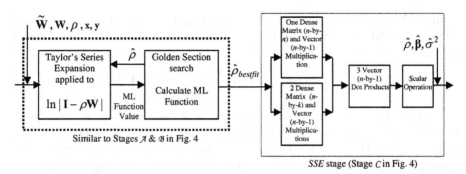

Fig. 6. The system diagram for the Taylor's Series expansion approximation for the SAR model solution. The inner structure of Taylor series expansion is similar to that of Chebyshev Polynomial except that there is one more vector sum operation, which is very cheap to compute

Lemma 1. The Chebyshev solution tries to approximate the logarithm of the determinant of $(I - \rho W)$ involving a symmetric neighborhood matrix $\widetilde{W}$ as in equation 14. The first three terms are sufficient for approximating the log determinant term with an accuracy of 0.03%.

$$\ln | I - \rho W | \equiv \ln | I - \rho \widetilde{W} | \cong \sum_{j=1}^{q+1} c_j(\rho) tr(T_{j-1}(\widetilde{W})) - \frac{1}{2} c_1(\rho). \tag{14}$$

Proof. It is available in [33].

The value of "q" is 2, which is the highest degree of the Chebyshev polynomials. Therefore, only $T_0(\widetilde{W}), T_1(\widetilde{W})$ and $T_2(\widetilde{W})$ have to be computed where:

$$T_0(\widetilde{W}) = I; T_1(\widetilde{W}) = \widetilde{W}; \ T_2(\widetilde{W}) = 2\widetilde{W}^2 - I; ...; T_{k+1}(\widetilde{W}) = 2\widetilde{W}T_k(\widetilde{W}) - T_{k-1}(\widetilde{W})$$

The Chebyshev polynomial coefficients $c_j(\rho)$ are given in equation 15.

$$c_j(\rho) = (\frac{2}{q+1}) \sum_{k=1}^{q+1} \ln[1 - \rho \cos(\frac{\pi(k-1/2)}{q+1})] \cos(\frac{\pi(j-1)(k-1/2)}{q+1}). \tag{15}$$

In Fig. 7, the maximum likelihood function is computed by computing the maximum of the sum of the logarithm of the likelihood function values and the *SSE* term. The spatial auto-regression parameter ρ that achieves this maximum value is the desired value that makes the classification most accurate. The parameter "q" is the highest degree of the Chebyshev polynomial which is used to approximate the term $\ln(I - \rho W)$. The system diagram of the Chebyshev polynomial approximation is presented in Fig. 7. The following lemma reduces the computational complexity of the Chebyshev polynomial from $O(n^3)$ to approximately $O(n^2)$.

Lemma 2. For regular grid-based nearest-neighbor symmetric neighborhood matrices, the relationship shown in equation 16 holds. This relationship saves a tremendous amount of execution time.

$$trace(\widetilde{\mathbf{W}}^2) = \sum_{i=1}^{n} \sum_{j=1}^{n} \widetilde{w}_{ij}^{2} \quad \text{where } (i,j)^{th} \text{ element of } \widetilde{\mathbf{W}} \text{ is } \widetilde{w}_{ij}. \tag{16}$$

Proof. The equality property given in equation 16 follows from the symmetry property of the symmetrized neighborhood matrix. In other words, it is valid for all symmetric matrices. The trace operator sums the diagonal elements of the square of the symmetric matrix $\widetilde{\mathbf{W}}$. This is the equivalent of saying that the trace operator first multiplies and adds the ith column with the ith row of the symmetric matrix, where the ith column and the ith row of the matrix are the same entries in a symmetric matrix. This results in squaring and summing the elements of the symmetric neighborhood matrix $\widetilde{\mathbf{W}}$. Equation 16 shows this shortcut for computing the trace of the square of the symmetric neighborhood matrix.

In Fig. 8, the powers of the $\mathbf{W}$ matrices, whose traces are to be computed, go up to 2. The parameter "q" is the highest degree of the Chebyshev polynomial which is used to approximate the term $\ln(\mathbf{I} - \rho\mathbf{W})$. The ML function is computed by calculating the maximum of the likelihood functions (i.e. the logarithm determinant term plus the SSE term). The pseudo-code of the Chebyshev polynomial approximation approach is presented in Fig. 8.

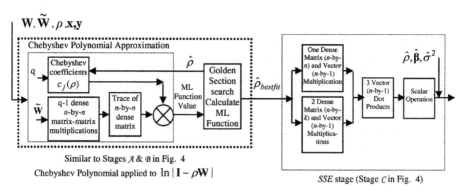

Fig. 7. System diagram of the approximate SAR model solution, where $\ln(\mathbf{I} - \rho\mathbf{W})$ is expressed as a Chebyshev polynomial. The term "q" is the degree of the Chebyshev Polynomial

5   Experimental Design

We evaluate our solution models using satellite remote sensing data. We first present the system setup and then introduce our real dataset along with the comparison metrics.

System Setup. The control parameters for our experiments are summarized in Table 4. One of our objectives is to make spatial analysis tools available to the GIS user

$Input : \widetilde{\mathbf{W}}, n, \hat{\rho}, q(= nposs)$
$Output :$ The estimate to $\ln |\mathbf{I} - \rho\mathbf{W}|$
CHEBYSHEV - APPROXIMATION - TO - LOGDET
1   $tdl \leftarrow 0$
2   $td2 \leftarrow \mathrm{TRACE}(\widetilde{\mathbf{W}}^2)$
3   $cheby\_poly\_coeffs \leftarrow [1\,0\,0 ; 0\,1\,0 ; -1\,0\,2]$
4   $nposs \leftarrow 3$
5   $seq1nposs \leftarrow \begin{bmatrix} 1 & 2 & 3 \end{bmatrix}^T$
6   $x_k \leftarrow \cos(\pi * (seq1nposs - 0.5)/nposs)$
7   for $j \leftarrow 1$ to $nposs$ do
8       $cposs[i,j] \leftarrow (2/nposs)\mathrm{sum}(\log(1 - \rho * x_k).*$
$\cos(\pi(j-1).*(seq1nposs - 0.5)/nposs))$
9   $tdvec \leftarrow \begin{bmatrix} n & tdl & td2 - 0.5*n \end{bmatrix}^T$
10 $comboterm \leftarrow cposs_{length(\rho\_vec) \times nposs} * cheby\_poly\_coeffs_{3 \times 3}$
11 $cheby\_logdet\_approx \leftarrow comboterm_{length(\rho\_vec) \times 3} * tdvec_{3 \times 1}$

Fig. 8. The pseudo code of the Chebyshev polynomial approximated $\ln(\mathbf{I} - \rho\mathbf{W})$

community. Notable solutions for the SAR model have been implemented in Matlab [18]. These approaches have two limitations. First, the user cannot operate without these packages and secondly these methods are not scalable to the application size. Our approach is to implement a general purpose package that works independently and scales well to the application size. All solutions described in this paper have been implemented using a general purpose programming language, f77, and use open source matrix algebra packages (ScaLAPACK [7]). All the experiments were carried out using the same common experimental setup summarized in Table 4.

Table 4. The experimental design

Factor Name	Parameter Domain
Language	f77
Problem Size (n)	2500,10K and 2.1M observation points
Neighborhood Structure	2-D w/ 4-neighbors
Method	Maximum Likelihood for Exact & Approximate SAR Model
Auto-regression Parameter	[0,1)
Hardware Platform	IBM Regatta w/1.3 GHz Power4 architecture processor
Data set	Remote Sensing Imagery Data

Dataset. We used real data-sets from satellite remote-sensing image data in order to evaluate the approximations to SAR. The study site encompasses Carlton County, Minnesota, which is approximately 20 miles southwest of Duluth, Minnesota. The region is predominantly forested, composed mostly of upland hardwoods and lowland conifers. There is a scattering of agriculture throughout. The topography is relatively flat, with the exception of the eastern portion of the county containing the St. Louis River. Wetlands, both forested and non-forested, are common throughout the area. The largest city in the area is Cloquet, a town of about 10,000. For this study we used a spring Landsat 7 scene, taken May 31, 2000. This scene was clipped to the Carlton county boundaries, which resulted in an image of size 1343 lines by 2043 pixels and 6-bands. Out of this we took a subset image of 1200 by 1800 to eliminate boundary zero-valued pixels. This translates to a **W** matrix of size 2.1 million x 2.1 million (2.1M x 2.1M) points. The observed variable **x** is a matrix of size 2.1M by 6. We chose nine thematic classes for the classification.

Comparison Metrics. We measured the performance of our implementation for accuracy, scalability (computational time), and memory usage. We first calculated the *percentage error* of the spatial auto-regression parameter ρ and the vector of regression coefficients β estimates from the approximate and exact SAR model solutions. Next, we calculated another *accuracy* metric using the standard root-mean-square (RMS) error. We computed the RMS error of the estimates of the observed dependent variable (**y** vectors or $\hat{\mathbf{y}}$) i.e. the thematic classes from the approximate and exact SAR model solutions. *Scalability* is reported in terms of computation (wall) time on an IBM Regetta 1.3GHz Power4 processor. *Memory* usage is determined by the total memory required by the program (which includes data and instruction space).

6 Results and Discussion

Since the main focus of this study is to find a scalable approximate method for the SAR model solution for very large problem sizes, the first evaluation is to compare the estimates from the approximate methods for the spatial auto-regression parameter ρ and the vector of regression coefficients β with the estimates obtained from the exact SAR model. Using the percentage error formula, Table 5 presents the comparison of accuracies of ρ and β obtained from the exact and the approximate (Chebyshev Polynomial and Taylor Series expansion based) SAR model solutions for the 2500 problem size. The estimates from the approximate methods are very close to the estimates obtained from the exact SAR model solution; there is an error of only 0.57% for the ρ estimate obtained from the Chebyshev polynomial approximation case and an error of 7.27% for the ρ estimate from the Taylor series expansion approximation. A similar situation exists for the β estimates. The maximum error among the β estimates is 0.7% for the Chebyshev polynomial approximation case and 8.2% for the Taylor series expansion approximation. The magnitudes of the errors for the ρ and β estimates are on the same order across methods.

Lemma 3. Taylor series approximation performs worse than Chebyshev polynomial approximation because Chebyshev polynomial approximation has a potential error canceling feature of the logarithm of the determinant (log-determinant) of a matrix. Taylor series expansion produces different error magnitudes for positive versus negative eigenvalue λ_i whereas the Chebyshev polynomials tend to produce error of more equal maximum magnitude [30].

Proof. The main reason behind this phenomenon is that Taylor series approximation does better than the Chebyshev polynomial approximation for values of ρ nearer to zero, bur far worse for extreme ρ (see Sect. 2.3 of [30]). Since the value of ρ is far greater than zero in our case, our experiments also verify this phenomenon, as shown in Table 5.

Table 5. The comparison of accuracies of ρ, the spatial auto-regression parameter, and β, the vector of regression coefficients, obtained from the exact and the approximate (Chebyshev Polynomial and Taylor Series expansion) SAR model solutions for the 2,500 problem size

Problem		ρ	β					
Size			1	2	3	4	5	6
50x50	Exact	0.4729	-2.473	-0.516	3.167	0.0368	-0.4541	3.428
(2500)	Chebyshev	0.4702	-2.478	-0.520	3.176	0.0368	-0.456	3.440
	Taylor	0.4385	-2.527	-0.562	3.291	0.0374	-0.476	3.589

The second evaluation is to compute the RMS (root-mean-square) error of the estimates of the observed dependent variable (**y** vectors or $\hat{\mathbf{y}}$) i.e., the thematic classes. The RMS error is given in equation 17 to show how we use it in our formulation. Table 6 presents the RMS values for all thematic classes. A representative RMS error value for the Taylor method is 2.0726 and for the Chebyshev method, 0.1686.

$$RMSerror_{cp} = \sqrt{\sum\left(\frac{\left(\hat{\mathbf{y}}_{cp} - \hat{\mathbf{y}}_{ee}\right)^2}{n-2}\right)} \tag{17}$$

$$RMSerror_{ts} = \sqrt{\sum\left(\frac{\left(\hat{\mathbf{y}}_{ts} - \hat{\mathbf{y}}_{ee}\right)^2}{n-2}\right)}.$$

The values of the RMS error suggest that estimates for the observed dependent variable (**y** vector or thematic classes) from the Chebyshev polynomial approximated SAR model solution are better than those of the Taylor series expansion approximated SAR model solution. This result agrees with the estimates for the spatial auto-regression parameter ρ and the vector of regression coefficients β shown in Table 5.

Table 6. RMS values for each thematic class of a dataset of problem size 2500

Training Thematic Class	RMS error value for Chebyshev	RMS error value for Taylor	Testing Thematic Class	RMS error value for Chebyshev	RMS error value for Taylor
y1	0.1686	2.0726	y1	0.1542	1.9077
y2	0.2945	2.0803	y2	0.2762	2.0282
y3	0.5138	3.3870	y3	0.5972	4.0806
y4	1.0476	6.9898	y4	1.4837	9.6921
y5	0.3934	2.4642	y5	0.6322	3.9616
y6	0.3677	2.3251	y6	0.4308	2.8299
y7	0.2282	1.5291	y7	0.2515	1.7863
y8	0.6311	4.3484	y8	0.5927	4.0524
y9	0.3866	3.8509	y9	0.4527	4.4866

Table 7. The execution time in seconds and the memory usage in mega-bytes (MB)

Problem Size (n)	Time (Seconds)			Memory (MB)		
	Exact	*Taylor*	*Chebyshev*	*Exact*	*Taylor*	*Chebyshev*
50x50 (2500)	38	0.014	0.013	50	1.0	1.0
100x100 (10K)	5100	0.117	0.116	2400	4.5	4.5
1200x1800 (2.1M)	Intractable	17.432	17.431	$\sim32*10^6$	415	415

The predicted images (50 rows by 50 columns) using exact and approximate solutions are shown in Fig. 9. Although the differences in the images predicted by the exact and approximate solutions is hard to notice, there is a huge difference between these methods in terms of computation and memory usage. As can be seen in Table 7, even for large problem sizes, the run-times are pretty small due to the fast log-determinant calculation offered by Chebyshev and Taylor's series approximation. By contrast, with the exact approach, it is impossible to solve any problem having more than 10K observation points. Even if we used sparse matrix determinant computation, it is clear that approximate solutions will still be faster.

The approximate solutions also manage to provide close estimates and fast execution times using very little memory. Such fast execution times make it possible to sale solutions for large problems consisting of billions of observation points. The memory usage is very low due to the sparse storage techniques applied to the neighborhood matrix W. Sparse techniques cause speedup since the computational complexity of linear algebra operations decrease because of the small number of non-zero elements within the W matrix. As seen from Figures 6 and 7, the most complex operation for Taylor series expansion and Chebyshev Polynomial approximated SAR model solutions is the trace of powers of the symmetric neighborhood matrix $\widetilde{W}$, which requires matrix-matrix multiplications. These operations are reduced to around $O(n^2)$ complexity by Lemma 2 given in Sect. 4.2. All linear algebra matrix operations are efficiently implemented using the ScaLAPACK [7] libraries.

We fitted the SAR model for each observed dependent variable (y vector). For each pixel a thematic class label was assigned by taking the maximum of the predicted values. Fig. 9 shows a set of labeled images for a problem size of 2500 pixels (50 rows x 50 columns). For a learning (i.e., training) dataset of problem size 2500, the prediction accuracies of the three methods were similar (59.4% for the exact SAR model, 59.6% for the Chebyshev polynomial approximated SAR model, and 60.0% for the Taylor series expansion approximated SAR model.) We also observed a similar trend on another (testing) dataset of problem size 2500. The prediction accuracies were 48.32%, 48.4% and 50.4% for the exact solution, Chebyshev polynomial and Taylor series expansion approximation based SAR models respectively. This is an interesting result. Even though the estimates for the observed dependent variables (y vectors) or thematic classes are more accurate for the Chebyshev polynomial based approximate SAR model than for the Taylor series expansion approximated SAR model solution, the classification accuracy for the Taylor series expansion approximated SAR model solution becomes better than the ones for not only the Chebyshev polynomial based approximate SAR model but also even the exact SAR model solution. We think that the opposite trend will be observed for larger size images because SAR might need more samples to be trained better. Even though we do not suggest a new exact SAR model solution, further research and experimentation is needed to fully understand SAR model's training needs and its impact on prediction accuracy with the solution methods discussed in this paper.

7   Conclusions and Future Work

Linear regression is one of the best-known classical data mining techniques. However, it makes the assumption of *independent identical distribution (i.i.d.)* in learning data samples, which does not work well for geo-spatial data, which is often characterized by spatial autocorrelation. In the spatial auto-regression (SAR) model, spatial dependencies within data are taken care of by the autocorrelation term, and the linear regression model thus becomes a spatial auto-regression model.

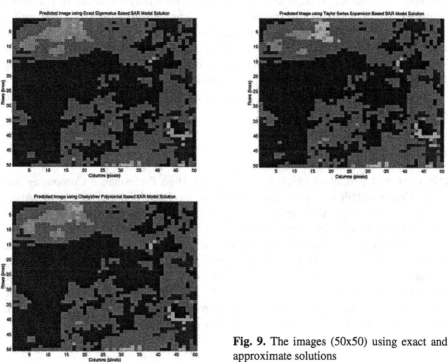

Fig. 9. The images (50x50) using exact and approximate solutions

Incorporating the autocorrelation term enables better prediction accuracy. However, computational complexity increases due to the need for computing the logarithm of the determinant of a large matrix $(\mathbf{I} - \rho\mathbf{W})$, which is computed by finding all of the eigenvalues of the $\widetilde{\mathbf{W}}$ matrix. This paper applies one exact and two approximate methods to the SAR model solution using various sizes of remote sensing imagery data i.e., 2500, 10K and 2.1M observations. The approximate methods applied are Chebyshev Polynomial and Taylor series expansion. It is observed that the approximate methods not only consume very little memory but they also execute very fast while providing very accurate results. Although the software is written using a parallel version of ScaLAPACK [7], SAR model solutions presented in this paper can be run either sequentially on a single processor of a node or in parallel on single or multiple nodes. All the results presented in Sect. 6 (Table 7) are based on sequential runs on the same (single) node of an IBM Regetta machine. It should be noted that the software can be easily ported onto general purpose computers and workstations by replacing open source ScaLAPACK routines with the serial equivalent routines in the open source LAPACK [1,13] library. Currently, LAPACK libraries can be compiled on Windows 98/NT, VAX, and several variants of UNIX. In our future release of SAR software, we plan to provide both ScaLAPACK and LAPACK versions.

In this study we focused on the scalability of the SAR model for large geospatial data analysis using approximate solutions and compared the quality of exact and approximate solutions. Though in this study we focused only on quality of parameter estimates, we do recognize that training and prediction errors were also important for these methods to be widely applied in various geospatial application domains. Towards this goal we are conducting several experiments on several geospatial data sets from diverse geographic settings. Our future studies will also focus on comparing SAR model predictions against competing models like Markov Random Fields. We are also developing algebraic cost models to further characterize performance and scalability issues.

Acknowledgments

This work was partially supported by the Army High Performance Computing Research Center (AHPCRC) under the auspices of the Department of the Army, Army Research Laboratory (ARL) under contract number DAAD19-01-2-0014. The content of this work does not necessarily reflect the position or policy of the government and no official endorsement should be inferred. The authors would like to thank the University of Minnesota Digital Technology Center and Minnesota Supercomputing Institute for the use of their computing resources. The authors would also like to thank the members of the Spatial Database Group, ARCTiC Labs Group for valuable discussions. The authors thank Kim Koffolt for helping improve the readability of this paper and anonymous reviewers for their useful comments.

References

1. Anderson, E., Bai, Z., Bischof, C., Blackford, S., Demmel, J., Dongarra, J., Du Croz, J., Greenbaum, A., Hammarling, S., McKenney, A., Sorensen, D.: LAPACK User's Guide, 3rd Edition, Society for Industrial and Applied Mathematics, Philadelphia, PA (1999)
2. Anselin, L.: Spatial Econometrics: Methods and Models, Kluwer Academic Publishers, Dorddrecht (1988)

3. Barry, R., Pace, R.: Monte Carlo Estimates of the log-determinant of large sparse matrices. Linear Algebra and its Applications, Vol. 289 (1999) 41-54
4. Bavaud, F.: Models for Spatial Weights: A Systematic Look, Geographical Analysis, Vol. 30 (1998) 153-171
5. Besag, J. E.: Spatial Interaction and the Statistical Analysis of Lattice Systems, Journal of the Royal Statistical Society, B, Vol. 36 (1974) 192-225
6. Besag, J. E.: Statistical Analysis of Nonlattice Data, The Statistician, Vol. 24 (1975) 179-195
7. Blackford, L. S., Choi, J., Cleary, A., D'Azevedo, E., Demmel, J., Dhillon, I., Dongarra, J., Hammarling, S., Henry, G., Petitet, A., Stanley, K., Walker, D., Whaley, R.C.: ScaLAPACK User's Guide, Society for Industrial and Applied Mathematics, Philadelphia, PA (1997)
8. Chawla, S., Shekhar, S., Wu, W., Ozesmi, U.: Modeling Spatial Dependencies for Mining Geospatial Data, Proc. of the 1st SIAM International Conference on Data Mining, Chicago, IL (2001)
9. Cheney, W., Kincaid, D.: Numerical Mathematics and Computing, 3rd ed. (1999)
10. Cressie, N. A.: Statistics for Spatial Data (Revised Edition). Wiley, New York (1993)
11. Golub, G. H., Van Loan, C. F.: Matrix Computations. Johns Hopkins University Press, 3rd edn. (1996)
12. Griffith, D. A.: Advanced Spatial Statistics, Kluwer Academic Publishers (1988)
13. Information about Freely Available Eigenvalue-Solver Software: http://www.netlib.org/utk/people/JackDongarra/la-sw.html
14. Kazar, B., Shekhar, S., Lilja, D.: Parallel Formulation of Spatial Auto-Regression, AHPCRC Technical Report No: 2003-125 (August 2003)
15. Kazar, B. M., Shekhar, S., Lilja, D. J., Boley, D.: A Parallel Formulation of the Spatial Auto-Regression Model for Mining Large Geo-Spatial Datasets, Proc. of 2004 SIAM International Conf. on Data Mining Workshop on High Performance and Distributed Mining (HPDM2004), Orlando, Fl. USA (2004)
16. Li, B.: Implementing Spatial Statistics on Parallel Computers, In: Arlinghaus S. (Ed.), ed. Practical Handbook of Spatial Statistics, CRC Press, Boca Raton, FL (1996) 107-148
17. LeSage, J.: Solving Large-Scale Spatial autoregressive models, presented at the Second Workshop on Mining Scientific Datasets, AHPCRC, University of Minnesota (July 2000)
18. LeSage, J. P.: Econometrics Toolbox for MATLAB. http://www.spatial-econometrics.com/
19. LeSage, J., Pace, R. K.: Using Matrix Exponentials to Explore Spatial Structure in Regression Relationships (Bayesian MESS), Technical Report (October 2000) http://www.spatial-statistics.com
20. LeSage, J., Pace, R. K.: Spatial Dependence in Data Mining, in Data Mining for Scientific and Engineering Applications, R. L. Grossman, C. Kamath, P. Kegelmeyer, V. Kumar, and R. R. Namburu (eds.), Kluwer Academic Publishing (2001) 439-460
21. Long, D. S.: Spatial autoregression modeling of site-secepific wheat yield. Geoderma, Vol. 85 (1998) 181-197
22. Marcus, M., Minc, H.: A Survey of Matrix Theory and Matrix Inequalities, New York: Dover (1992)
23. Martin, R. J.: Approximations to the determinant term in Gaussian maximum likelihood estimation of some spatial models, Communications in Statistical Theory Models, Vol. 22 Number 1 (1993) 189-205
24. Ord, J. K.: Estimation Methods for Models of Spatial Interaction, Journal of the American Statistical Association, Vol. 70 (1975) 120-126

25. Pace, R. K., Barry, R.: Quick Computation of Spatial Auto-regressive Estimators. Geographical Analysis, Vol. 29, (1997) 232-246
26. Pace, R. K., LeSage, J.: Closed-form maximum likelihood estimates for spatial problems (MESS), Technical Report (September 2000) http://www.spatial-statistics.com
27. Pace, R. K., LeSage, J.: Semiparametric Maximum Likelihood Estimates of Spatial Dependence, Geographical   Analysis, Vol. 34, No.1 The Ohio State University Press (Jan 2002) 76-90
28. Pace, R. K., LeSage, J.: Simple bounds for difficult spatial likelihood problems, Technical Report (2003) http://www.spatial-statistics.com
29. Pace, R. K., LeSage, J.: Spatial Auto-regressive Local Estimation (SALE), Spatial Statistics and Spatial Econometrics, Edited by Art Getis, Palgrave (2003)
30. Pace, R. K., LeSage, J.: Chebyshev Approximation of Log-Determinant of Spatial Weight Matrices, Computational Statistics and Data Analysis, Technical Report, Forthcoming.
31. Pace, R. K., LeSage, J.: Closed-form maximum likelihood estimates of spatial auto-regressive models: the double bounded likelihood estimator (DBLE), Geographical Analysis-Forthcoming
32. Pace, R. K., Zou, D.: Closed-Form Maximum Likelihood Estimates of Nearest Neighbor Spatial Dependence, Geographical Analysis, Vol. 32, Number 2, The Ohio State University Press (April 2000)
33. Press, W., Teukulsky, S. A., Vetterling, W. T., Flannery, B. P.: Numerical Recipes in Fortran 77, 2nd edn. Cambridge University Press (1992)
34. Shekhar, S., Chawla, S.: Spatial Databases: A Tour, Prentice Hall (2003)
35. Shekhar, S, Schrater, P., Raju, R., Wu, W.: Spatial Contextual Classification and Prediction Models for Mining Geospatial Data, IEEE Transactions on Multimedia, Vol. 4, Number 2 (June 2002) 174-188
36. Smirnov, O., Anselin, L.: Fast Maximum Likelihood Estimation of Very Large Spatial Auto-regressive Models: A Characteristic Polynomial Approach, Computational Statistics & Data Analysis, Volume 35, Issue 3 (2001) 301-319
37. Timlin, J., Walthall, C. L., Pachepsky, Y., Dulaney, W. P., Daughtry, C. S. T.: Spatial Regression of Crop Parameters with Airborne Spectral Imagery. Proceedings of the 3rd Int. Conference on Geospatial Information in Agriculture and Forestry, Denver, CO; (November 2001)

Appendix: Constructing the Neighborhood Matrix W for Irregular Grid

Spatial statistics requires some means of specifying the spatial dependence among observations [12]. The neighborhood matrix i.e., spatial weight matrix fulfills this role for lattice models [5,6] and can be formed on both regular and irregular grid. This appendix shows a way to form the neighborhood matrix on the irregular grid which is based on Delaunay triangulation algorithm [28,29]. [30] describes another method of forming the neighborhood matrix on the irregular grid which is based on nearest neighbors.

One specification of the spatial weight matrix begins by forming the binary adjacency matrix N where $N_{ij} = 1$ when observation j is a neighbor to observation i ($i \neq j$). The neighborhood can be defined using computationally very expensive

Delaunay triangulation algorithm [18]. These elements may be further weighted to give closer neighbors higher weights and incorporate whatever spatial information the user desires. By itself, $\mathbf{N}$ is usually asymmetric. To insure symmetry, we can rely on the transformation $\mathbf{C} = (\mathbf{N} + \mathbf{N}^T)/2$. The rest of forming neighborhood matrix on irregular grid follows the same procedure discussed in Sect. 3 (see Fig. 5). Users often re-weight the adjacency matrix to create a row-normalized i.e., row-stochastic matrix or a matrix similar to a row-stochastic matrix. This can be accomplished in the following way. Let $\mathbf{D}$ represent a diagonal matrix whose ith diagonal entry is the row-sum of the ith row of matrix $\mathbf{C}$. The matrix $\mathbf{W} = \mathbf{D}^{-1/2}\mathbf{D}^{-1/2}\mathbf{C} = \mathbf{D}^{-1}\mathbf{C}$ is row-stochastic (see Fig. 5) where $\mathbf{D}^{-1/2}$ is a diagonal matrix such that its ith entry is the inverse of the square root of the ith row of matrix $\mathbf{C}$. Note that the eigenvalues of the matrix $\mathbf{W}$ do not exceed 1 in absolute value as noted in Sect. 4.1, and the maximum eigenvalue equals 1 via the properties of row-stochastic matrices (see Sect. 5.13.3 in [22]). Despite the symmetry of $\mathbf{C}$, the matrix $\mathbf{W}$ will be asymmetric in the irregular grid case as well. One can however invoke a similarity transformation as shown in equation 18 (see Sect. 3, Fig. 5 of this study).

$$\widetilde{\mathbf{W}} = \left(\mathbf{D}^{-1/2}\right)^{-1} \mathbf{W} \left(\left(\mathbf{D}^{-1/2}\right)^{-1}\right)^{-1} = \mathbf{D}^{1/2}\mathbf{W}\mathbf{D}^{-1/2} = \mathbf{D}^{-1/2}\mathbf{C}\mathbf{D}^{-1/2}. \tag{18}$$

This results in $\widetilde{\mathbf{W}}$ having eigenvalues i.e., λ equal to those of $\mathbf{W}$ [24]. That is why we call $\widetilde{\mathbf{W}}$ the symmetric eigenvalue-equivalent matrix of $\mathbf{W}$ matrix. Note, the eigenvalues of $\mathbf{W}$ do not exceed 1 in absolute value via the properties of row-stochastic matrices (5.13.3 of [22]) because $\widetilde{\mathbf{W}}$ is similar to $\mathbf{W}$ due to the equivalent eigenvalues i.e., $-1 \leq \lambda_i^{\widetilde{\mathbf{W}}} \leq 1$.

From a statistical perspective, one can view $\mathbf{W}$ as a spatial averaging operator. Given the vector $\mathbf{y}$, the row-stochastic normalization i.e., $\mathbf{W}\mathbf{y}$ results in a form of local average or smoothing of $\mathbf{y}$. In this context, one can view elements in the rows of $\mathbf{W}$ as the coefficients of a linear filter. From a numerical standpoint symmetry of $\widetilde{\mathbf{W}}$ simplifies computing the logarithm of determinant and has theoretical advantages as well. (See [4,28,29,30] for more information on spatial weight matrices.)

3D GIS for Geo-coding Human Activity
in Micro-scale Urban Environments

Jiyeong Lee

Department of Geography, 7 Armstrong Hall
Minnesota State University, Mankato, MN 56001, USA
jlee68@uncc.edu

Abstract. The study of human activity and movement in space and time has been an important research area in the social sciences. However, the difficulty in collecting and analyzing the space-time activity (STA) data using current 3D location positioning techniques has limits in applying the time-geographic and activity theory to transportation and urban research and analyses. This paper develops a "3D Indoor Geo-Coding" technique to identify 3D indoor locational data for analyzing human activities. For implementing 3D positioning methods, this paper presents (1) to model micro-scale urban environments to be a frame of reference (reference data) to identify distance and direction, in order to obtain locational data, and (2) to develop a 3D indoor geo-coding method to identify locational data on individual activities based on the reference data. Finally, this paper describes the output from implementing the 3D indoor geo-coding technique to demonstrate the potential benefits of the 3D indoor geo-coding technique for improving the speed of emergency response using GIS data of the study area at Minnesota State University-Mankato, MN (USA).

1 Introduction

From the terrorist attacks at the World Trade Center in New York and the Pentagon in Washington, DC, September 11, 2001, there are many lessons to be learned for the development and implementation of geospatial technologies on responding to similar disasters on multi-level structures in urban areas in the future. These events occurred within 3D micro-spatial environments with a relatively complex internal structure that renders speedy escape and emergency response difficult. For such a disaster situation, an emergency response system needs to be built upon a highly flexible and distributed system architecture, where the 3D GIS database and decision support functionalities remain accessible to emergency personnel through multiple channels via wireless and mobile communications technologies.

Kwan and Lee [16] proposed a GIS-based Intelligent Response System (GIERS) (Figure 1). It illustrates that a GIERS is part of an Emergency Management Information System, integrating the ground transport component based on an Intelligent Transportation Systems with the route systems of the multi-level structures in an urban area through a series of Intelligent Building Systems (IBS).

In the system, emergency and rescue crews operating within multi-level structures under disaster conditions are equipped with location-aware devices and other mobile or handheld communications equipment that interact in real-time with the rescue

M.J. Egenhofer, C. Freksa, and H.J. Miller (Eds.): GIScience 2004, LNCS 3234, pp. 162–178, 2004.

command center of a GIERS. These location-aware devices provide location information that is transmitted via cell phones or mobile GIS devices. In addition, these locational devices are equipped with mobile GIS software (such as ArcPad) and can generate on-screen geo-referenced maps to support rescuers' operations on-site (e.g. providing navigational guidance). The location information generated by location-ware devices is in turn transmitted back to the rescue command center in real-time for the purpose of locating emergency crews and disaster events and conditions within the multi-level structure. Since most current 3D location positioning techniques are not appropriate for Location-Based Services (LBSs) in micro-scale urban environments (3D indoor), this paper conducted an experimental implementation of the 3D Indoor Geo-Coding technique to demonstrate the potential benefits of the 3D indoor geo-coding technique for improving the speed of emergency response.

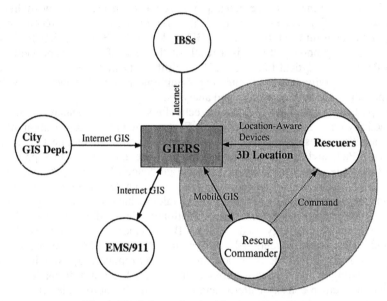

Fig. 1. The Information Architecture of GIERS

The purpose of this paper, therefore, is to develop a "3D indoor geo-coding" technique, in order to identify 3D indoor locational data for analyzing human activities in space and time. It proposes a new approach to dealing with 3D positioning methods. The requirements of a 3D indoor geo-coding technique that are presented include (1) a frame of reference (reference data) to identify distance and direction in order to obtain locational data, and (2) a 3D indoor geo-coding method to identify locational data on individual activities based on the reference data.

Section 2 of this paper reviews current possibilities and limitations for wireless 3D locational positioning systems. Section 3 proposes a 3D indoor positioning method utilizing a conventional address matching technology. The following section describes the output from implementing the 3D indoor geo-coding technique and demonstrates the potential benefits of the 3D indoor geo-coding technique for improving the speed of emergency response. The final section discusses several significant substantive insights derived from this study.

2 3D Location Positioning to Spatial Analysis

The study of human activity and movement in space and time has been an important research area in the social sciences, referred to as spatial analysis. It covers a wide range of topics, including travel behavior, migration and residential mobility, pedestrian choice behavior, as well as shopping and commuting behavior [15]. One of the earlier perspectives for the analysis of human activity patterns and movements in space and time is time-geography, introduced by Torsten Hägerstrand [10]. The time-geographic perspective shows the importance of space for understanding geography in time, and it also allows the researcher to examine the complex interaction between space and time and its effects on the structure of human activity patterns in particular areas [5].

Despite the usefulness of time geography in social sciences, very few studies have been conducted using space-time individual activity data, except for recent attempts using geographical information systems (GISs) [13-15, 25, 36]. The limited development of time-geography methods is a result of the lack of detailed individual-level activity data and analytical tools [15, 26]. The difficulty in collecting and analyzing the space-time activity data has limits in applying time-geographic and activity theory to transportation and urban research and analyses.

In the past few years, new information technology has greatly enhanced the collection of activity data. The Global Positioning System (GPS) provides location information. In addition, these locational devices are equipped with mobile GIS software (such as ArcPad) and can generate on-screen geo-referenced maps to support users' operations on-site (e.g., providing navigational guidance). However, there are limitations in using GPS within a multi-level structure due to a degradation or loss of signal in certain areas of a building. Positioning techniques have been developed in LBS, which seeks to provide better location information inside buildings. Network-based or hybrid positioning technology used by most LBS providers to achieve a positional fix faster and easier than conventional GPS-based technology. Nevertheless, the problems of loss-of-fix in LBS-derived locational data still remain [16]. New indoor positioning technologies are needed to provide more exact location information for LBS and analyzing human activity patterns and movements in micro-scale urban environments. Dürr and Rothermel [8] proposed a fine-grained Geocast location model to determine client positions within a building based on geometric and symbolic addressing.

In recent years, the positioning techniques to collect 3D locational positioning data have improved the quality and quantity of these data and reduced their cost using location-aware devices such as cellular phones, wireless PDAs, GPS receivers and radiolocation methods [30]. GPS devices are able to offer the easiest methods and a relatively accurate way of 3D positioning of the user. The receivers are becoming more portable, cheaper and convenient for general use with handheld devices.

Even though GPS-based systems provide accurate location information outdoors (3-5 meters horizontal error with Differential GPS or Wide Area Augmentation System), there are many difficulties in utilizing GPS for LBS. One main problem is that the GPS receiver has to have a line of sight to the satellites. Many applications requiring the use inside of buildings or vehicles make it difficult for GPS to provide the required service because of the relatively weak satellite signals. As a result, an Indoor GPS or Assisted GPS has been developed [23, 34] to detect highly attenuated GPS signals indoors. A-GPS, however, requires the use of modified handsets that receive the GPS signals and then send those readings to a network server.

The positioning within mobile networks using only the information related to the base network transmitter is a very cost-effective method, but it is very inaccurate and practically not applicable for LBS [38]. The only advantage, compared to GPS positioning, is the possibility of working inside buildings. None of techniques increases the accuracy by more than 50 meters.

Due to weak GPS signals and the limited accuracy of radio network solutions, neither technique is appropriate for indoor positioning. Wireless LANs (WLANs) are used to track mobile users in closed spaces as buildings and tunnels [29]. The system for location positioning using WLANs runs on a standalone server and gives the x,y-position and the floor of the mobile unit. The positioning accuracy achieved by the system is up to 1 meter. Despite providing accurate positioning for indoors, however, the WLANs have problems with implementation because they require a reference database for an average signal measurements at fixed points throughout a building [38], [28].

As mentioned above, there are problems and limitations of current 3D location positioning for LBSs in micro-scale urban environments, this paper proposes a new approach to 3D Indoor Geo-Coding methods, based on a well-known Address Matching technology [7].

3 Requirements for 3D Indoor Geo-coding

To identify and develop methods of displaying real-time 3D indoor locational data by using address matching technology, there are two important requirements. One is to model an urban entity (internal structure of a building) to serve as a reference data for geo-coding individual activities in 3D. The other is to implement a "3D Indoor Geo-Coding" method to provide the location of users for location-based services within a building.

3.1 Modeling 3D Urban Environments

3D location positioning data (location x,y,z-coordinates) is only one aspect of providing LBS. Although having the location coordinates obviously is essential in LBSs, the location would be related to other pertinent information to understand the environment and to provide users with the location of an item or place they are interested in finding. A 3D urban entity (internal structure of a building) should be modeled to serve as reference data for 3D geo-coding individual activities in micro-scale urban environments.

3D urban models have become important due to the increasing demand for a realistic presentation of the real world in GIS. The 3D models data are potentially of great importance to the understanding of urban structure and the mechanism of urban growth through visualizing urban and built environments [19, 20, 31, 32]. This paper uses ERDAS Imagine 3D Virtual Extension Module and ESRI ArcScene to display the urban features. The 3D model of the MSU-Mankato campus (Figure 2a) is based on the data model of the ESRI ArcView 3D Shapefile. The 3D model is obtained with links to GIS database including the Campus Map and floor-by-floor information from Trafton East Hall (Figure 2b).

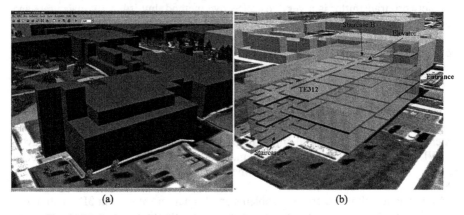

Fig. 2. 3D Geometric Representation: (a) 3D View and (b) Layout of a 3D View

In such multi-level structures, compartmentalized zones or areas (rooms) are connected by complex transport routes such as corridors shown horizontally (layer-based floor by floor). In addition, different levels of these structures are connected by vertical conduits, such as elevators and stairways. When their internal structure are represented using navigable 3D GIS data models, many GIS-based analytical techniques can be applied for implementing network-based analyses, such as pathfinding, allocation and tracing analyses in micro-spatial environments [18, 19]. Further, as the horizontal and vertical conduits within multi-level structures are ultimately connected to the ground transportation system, this will support network analyses in urban environments. In order to model the internal structures of a building, this paper abstracts the transport routes (between corridors and elevators or stairways) from the building (Figure 3) showing the transportation system within a building.

Fig. 3. Transport Routes within a Building

Instead of representing all internal structures of a building (including all rooms), the paper represents the transport routes using 3D Geometric Network Model (GNM) [19] to be a reference data for 3D geo-coding individual activities. For developing the network model, a 3D Poincaré Duality Transformation [18, 19] is utilized in order to represent the connectivity relationships between horizontal routes (corridors) and vertical conduits (elevators and stairways), and a Straight-Medial Axis Transformation (Straight-MAT) [3, 19] is used to represent corridors on the floor of the building. Both techniques are reviewed in the following section.

3.1.1 Modeling internal structures of a building.

The connectivity relationships among 3D spatial units (transport routes within a building) are the combinations of the connectivity relations in the horizontal directions (on a floor) and the connectivity relations in the vertical direction (among floors). The connectivity relationships of 3D units in horizontal directions on a floor are derived from the connectivity relationships among polygons in 2D (e.g., between corridors and stairways or elevators) [18]. The vertical connectivity relationships are defined by the locations of stairways or elevators.

In order to model the connectivity relationships among the transport routes on floor j, 3D Poincaré Duality Transformation [19] is utilized, which will be explained in the following section. The connectivity relationships are modeled as a 3D network structure, the dual graph of 3D spatial units. A graph $G = (V(G), E(G))$ consists of two sets: a finite set V of elements called vertices and a finite set E of elements called edges, $V(G) = \{v_1, ..., v_n\}$ and $E(G) = \{(v_i, v_j) / v_i, v_j \in V\}$. Each edge is identified with a pair of vertices (v_1, v_2), where $v_1, v_2 \in V$.

To represent the transport routes using the 3D GNM, a graph $Nh_j = (V(Nh_j), E(Nh_j))$, the first step is that the hallway is transformed into linear features based upon the Straight-MAT (a graph $MAT_j = (V(MAT_j), E(MAT_j))$) proposed in the previous section. After that, each node representing 3D spatial units (stairways or elevators) is projected and connected into the medial axis. The projection $p(q, e)$ of a point q onto an edge e is the intersection of the edge e and the edge perpendicular to the edge e through the point q. Suppose the edge e is $Ax + By + C = 0$ and the point $q = (x_0, y_0)$; the linear equation of the projection $p(q, e)$ is $Bx - Ay + (Ay_0 - Bx_0) = 0$. The intersection point $i = (x_i, y_i)$ can be calculated based upon Cramer's rule: $x_i = (B(Bx_0 - Ay_0) - CA) / (A^2 + B^2)$ and $y_i = (-A(Bx_0 - Ay_0) - BC) / (A^2 + B^2)$.

The graph $Nh_j = (V(Nh_j), E(Nh_j))$, representing the geometric network within floor j (Figure 4a), is combined with graph $Nv_n = (V(Nv_n), E(Nv_n))$ using a union operation to produce the graph $N_i = (V(N_i), E(N_i))$ for a building i (Figure 4(b)), which is the geometric network model of the NRS. The graph $Nv_n = (V(Nv_n), E(Nv_n))$ is a subgraph representing the connectivity relations in vertical directions between floor i-1 and floor i.

The network structure, a graph $N_i = (V(N_i), E(N_i))$, consists of a set of Nodes V and a set of Edges E. The primary classes of the model are Node, Edge, and Network, whose schema is shown in Figure 5. The class Node consists of an identifier and position data in 3D (x,y,z-coordinates), and the class Edge consists of an identifier, start node, and end node. The class Network consists of an identifier and lists of all nodes and of all edges in a network. The database schema for attribute data of a class Node and Edge is as follows:

NODE (<u>Node_ID</u>, RoomType)
EDGE (<u>Edge_ID</u>, Length, F_RAdd_l, T_RAdd_l, F_RAdd_r, T_RAdd_r,
  Traffic_Capacity, Speed, Occupant_NO, Impedance)

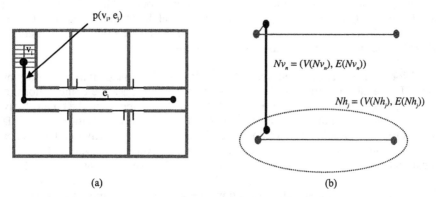

(a)   (b)

Fig. 4. 3D Geometric Network Model (GNM): (a) 2D GNM and (b) 3D GNM

```
class Node {                          class Edge {
    Int Node_ID;                          Int Edge_ID;
    Double x, y, z;                        Node initial_node;
    };                                     Node end_node;
                                           };
class Network {
    Int Network_ID;
    Node ArrayNode = new Node[];
    Edge ArrayEdge = new Edge[];
    };
```

Fig. 5. A 3D Network Data Model

Each node in the database has an identifier, room type, and room-based information that is a description of node characteristics in the 3D network. An edge has an identifier, length, occupant movement, elevator/stairway capacity, corridor capacity and traffic flow impedance. In addition, four attributes store the low and high room numbers on the edge's left side and the low and high room number on its right side, which information will be used for 3D indoor geo-coding explained in the following section.

The 3D geometric network, $N_i = (V(N_i), E(N_i))$, (an indoor network of building i) needs to be integrated with the 2D network of the ground transportation system, $S = (V(S), E(S))$, for implementing network analyses in urban environments, which is represented as a 3D network, $R = (V(R), E(R))$. For the integration, the indoor network, $N_i = (V(N_i), E(N_i))$, is represented with a hierarchical network structure by consolidating it to a high-level node, a Master_Node as seen in Figure 5(b), using graph methods [18, 22]. The connectivity relationships between 3D objects within a building are represented as a sub-network of the 3D network, $R = (V(R), E(R))$. In the multi- level network, in order to connect this sub-network to the upper level of network, the Master_Node for a building i, MN_i, should include Connect_edges, Nc_i = $(V(Nc_i), E(Nc_i))$, and data, which is defined in Figure 6.

The Connect_edges, Nc_i = $(V(Nc_i), E(Nc_i)$, represent the connectivity relationships between an indoor network for building i, N_i = $(V(N_i), E(N_i))$, and the 2D street network, S = $(V(S), E(S))$. Therefore, the initial node of Connect_edges is not an element of the node set N, and the end node of Connect_edges is an element of the node set N, or vice versa. Figure 6 shows the data structure of the Master_Node. The database schema of the attribute data of Master_Node is as follows:

MASTER_NODE (M_Node_id, Build_Name, ST_Address)

The Master_Node in the database has an identifier, building name, street address and building-based information that describe building characteristics in the 3D network of urban environment.

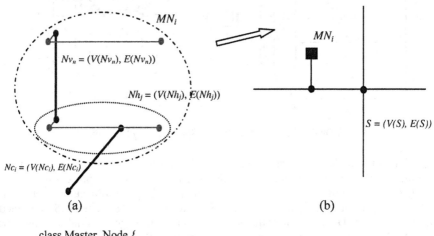

(a)   (b)

```
class Master_Node {
        Int MNode_ID;
        Network Sub_Network = new Network[];
        Connect_edges = {(ni, nj) / ( ni ∉ N ∩ nj ∈ N) ∪ ( ni ∈ N ∩ nj ∉ N )}
        };
```

Fig. 6. Hierarchical Network Data Model

3.1.2 3D Poincaré Duality Transformation. In order to represent the connectivity relationships between corridors and elevators and stairways, 3D Poincaré Duality Transformation [19] is utilized. The dual graph, an adjacency graph, has been used to represent adjacency relationships between area objects in 2D GIS [27]. The adjacency graph has a corresponding spatial partition, where the nodes in dual space are the units of the partition (polygons in 2D GIS), and two nodes are linked by an edge (in dual space) when the corresponding spatial objects are adjacent in primal space. Duality depends upon the dimensionality of the space; duality in 2D differs from duality in 3D, but the general idea is the same [35]. Since 3-cells or volumes in a 3D is manipulated and represented by their dual state, the 0-cells or point, the spatial relations between 3-cells are represented by the duality transforms of 3-cells (s_i) ↔ 0-cells (s_i') and 2-cells (f_i) ↔ 1-cells (f_i') (Figure 7).

To plan and design an evacuation network within a building in a 3D, the building is modeled as a dual graph, called a node-edge graph [4, 11, 21, 33], which is a logical network model. In this model, each room, corridor, staircase, and exit of the building is represented as a node, and each corridor from one node to another node is an edge. Each node has two attributes: maximum capacity and initial occupancy, and each edge also has two attributes: maximum capacity and travel time. This network defines a directed graph with sources and sinks, and is then used to solve network flow problems (building evacuation problems) by determining an optimal plan for minimizing the time required to evacuate the building.

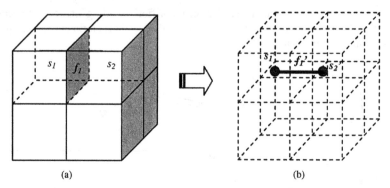

(a)   (b)

Fig. 7. Duality in 3-cells: (a) Primal Space and (b) Dual Space

Similar to the node-edge graphs, which use duality to represent space-activity interactions, the Node-Relation Structure (NRS) was developed to represent more than just adjacency and connectivity relationships among 3D spatial objects in built environments, based upon 3D Poincaré Duality Transformation [18]. The NRS is defined as a set of nodes (3D entities in primal space) with a set of edges (spatial relationships between 3D entities in primal space) that represent the topological relationships among entities in built environments.

One advantage in the duality is that all topological properties are preserved under the duality transformation. The topological relationships between 3D entities can be represented in dual space with sets of nodes and edges. By not dealing with 3D entities, therefore, the Poincaré Duality helps resolve the problems associated with data storage and computationally efficiency to keep topological consistencies of current 3D entity-based data models. This paper, therefore, represents the connectivity relationships between corridors and elevators and stairways using a dual graph based on the NRS.

3.1.3 Straight-Medial Axis Transformation (Straight-MAT). The medial axis transformation was first proposed by Blum [3] to describe an object or a shape. The medial axis of the polygon is also called the skeleton or the symmetric axis of the object because of its shape. Since the time that the notion of medial axis was introduced, a number of researchers became interested in image recognition [12] and in vectorization [37] have investigated methods for computing the medial axis by using a one-dimensional structure that represents a two-dimensional shape. The developed algorithms deal with raster data to describe an object in a digital image without representing the boundary pixel by pixel.

The computation for vector data requires the boundary of the object to be represented geometrically, such as in lines and curves instead of pictorially in raster data. The medial axis of the vector features is determined by algorithms during the computation of the Voronoi diagram because there is a close relationship between the medial axis of a simple polygon and its Voronoi diagram, which is restricted to the interior of the polygon (Figure 8) [17].

The Voronoi diagram of a simple polygon G contains parabola segments in the neighborhood of the vertices of G. As a result, in implementing the medial axis algorithms to convert raster data to vector data, or to simplify double line features to single line features for map generalizations, there are intersection problems caused by a distortion of the features near intersections. These problems are classified into three major types: T displacement, X destroying [1], and L displacement. In such applications there have been several attempts to linearize and simplify Voronoi diagrams of planar straight-line graphs [9], called the Straight-MAT of a simple polygon.

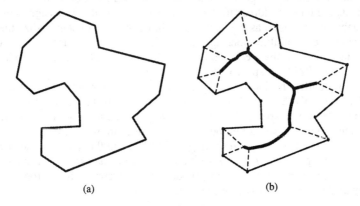

(a)   (b)

Fig. 8. Medial Axis Transformation (MAT): (a) Simple Polygon and (b) its Medial Axis

In order to remove the intersection distortions in the thinning processes, this paper implements a Straight-MAT algorithm for a simple polygon, based on Lee's Straight-MAT algorithm [19]. The straight medial-axis, constructed with the angular bisectors of each pair of neighboring edges, is used to represent the horizontal routes in modeling internal structures of a building.

3.2 3D Indoor Geo-coding Techniques

In general, most small business offices in a skyscraper in downtown or in a shopping mall have a little different address formats, called "a composite address." The address format consists of two components, which are an indoor address and a street address. For example, the address of a lawyer's office located in Minneapolis, MN is "Suite 300, 701 Fourth Avenue South, Minneapolis, Minnesota 55415-1810." "Suite 300" represents the indoor address of the office, while the rest, "701 Fourth Avenue South, Minneapolis, Minnesota 55415-1810," represents the street address. In order to geo-code the composite address, two steps are processed: one is to geo-code the address based on the street address to define the location information of the building, and the other is to define the 3D indoor positioning of the office within the building based on the suite or room number [2]. Because the outdoor address matching can be implemented by querying attribute information (building name or street address) of Master_Node, this paper presents the "3D Indoor Geo-Coding" method to define the office location within the building.

The fundamental principle of a 3D Indoor Geo-Coding method employs a traditional address matching technology developed for location positioning of outdoor phenomena. A TIGER-type reference database model can be comprised of the 3D Geometric Network Model (GNM). An edge in the 3D GNM, a hallway line segment, contains an indoor address (room number) as a range attribute, which refers to the low (F_RA_l) and high (T_RA_l) room numbers on the edge's left side and the low (F_RA_r) and high (T_RA_r) room number on its right side [2]. In general, each room in a building is uniquely named or labeled. The room names or labels provide a useful value in a database table as well as an intuitive indicator of the floor. For example, Room 200 implies a specific place on the second floor of a building. Unlike a street address, the indoor address for a room or suite is not always assigned as a number in sequence or is named as the "Liberty Room". In order to standardize the indoor address (using a sequential numbering), translation tables can be used. For example, a translation table would change "212A," "212B," and "212C" as the standard form into "212," "213," "214," respectively.

In order to identify the location of Room k in the indoor network (3D GNM) for building i, $N_i = (V(N_i), E(N_i))$, 3D Geo-Coding method will be executed in four steps. The first step is to find a specific edge (e_i) defining the target room number ($Room\_k$) by comparing it with the room number ranges of edges in 3D GNM. The second step is to calculate a distance (d_{1k}) from From_Node (FN_i), and the next is to determine the location positioning (n_k) of $Room\_k$ (with x, y, z coordinates). The final step is to apply a user-specified offset to move the node location (representing a room) a certain distance away from the hallway line segment.

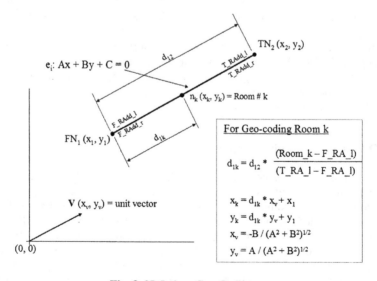

Fig. 9. 3D Indoor Geo-Coding

As seen in Figure 9, the distance d_{lk} is equal to the distance (d_{12}) of the edge (e_l) defined in the first step multiplied by the proportion of room number ranges (($Room\_k - F\_RA\_l)/(T\_RA\_l - F\_RA\_l)$). The distance between FN_1 and TN_2 can be considered as either a Euclidean distance when the edge is a straight line, or a network distance when it is a polyline. The edge e_l is characterized by a linear function $Ax + By + C = 0$. The x,y coordinates of a node n_k (representing Room k) are calculated by the functions described in Figure 9. V is a unit vector of an edge e_l. The z-coordinate of a node is determined by the room number $Room\_k$ or by the edge e_l having the same z-coordinate as the node n_k.

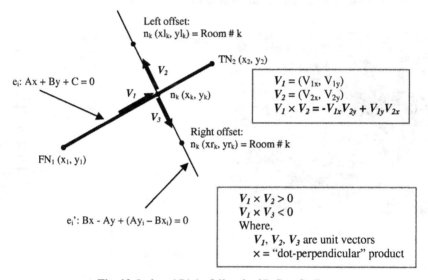

Fig. 10. Left and Right Offset for 3D Geo-Coding

As seen in Figure 10, a vector method for identifying left or right offset of the node n_k is to calculate the edge-vector cross product. This method is to determine the sign of the z component of the cross products. If the z-component turns out to be positive, the left offset of the node n_k is determined. Otherwise, the right offset of the node n_k is made. V_l is a unit vector of an edge e_l. V_2 and V_3 are unit vectors being perpendicular to an edge e_l with different directions. The cross product $V_1 \times V_2$ for two successive edge vectors is a vector perpendicular to the xy-plane, with a z-component equal to - $V_{1x}V_{2y} + V_{1y}V_{2x}$. In order to avoid working in 3D, the "dot-perpendicular" product is used, which is a signed measure of the area of the parallelogram determined by two vectors: $(V_{1x}, V_{1y}) \cdot \perp(V_{2x}, V_{2y}) = (V_{1x}, V_{1y}) \cdot (-V_{2y}, V_{2x}) = -V_{1x}V_{2y} + V_{1y}V_{2x} = z$ coordinate of $(V_{1x}, V_{1y}, 0) \times (V_{2x}, V_{2y}, 0)$ [18]. The x,y-coordinates of a node n_k (representing Room k) with right or left offset are calculated by the functions described in Figure 9.

4 Experimental Implementation of 3D Indoor Geo-coding

The system for implementing the 3D Indoor Geo-Coding technique is developed in a Visual Basic development environment, called "3D NRS Implementation Module."

The study area for this system is the surrounding area of Trafton East Hall located in Minnesota State University at Mankato. The 3D model of the MSU-Mankato campus is based on the data model of ESRI ArcView 3D Shapefile, obtained with the links to a GIS database including the Campus Map and floor layout information. Figure 11a displays the transport routes within Trafton East Hall in 3D View, and Figure 11b depicts the extracted 3D Geometric Network Data Model of the study area. The figure presents the connectivity relationships among transport routes (such as corridors, stairways and elevators) inside the building and a street network in the study area. The 3D GNM is generated by integrating the 3D internal network of the building with the 2D network of the ground transportation system.

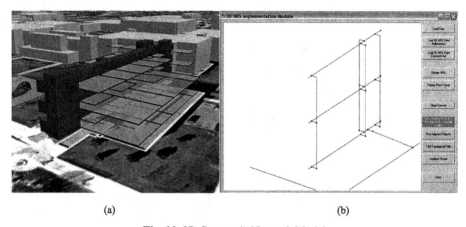

(a)   (b)

Fig. 11. 3D Geometric Network Model

Suppose that an emergency situation, for instance, a fire or chemical explosion, has occurred on the 3rd floor of Trafton East Hall at a lab (Labeled "TE312" in Figure 12a). In order to approach the disaster site quickly, emergency responders need (1) to know where the room is located within the building, and (2) to identify the optimal route from a source node (labeled "Entrance") to the destination node (room TE312 on the 3rd floor of the building). In order to geo-code the disaster site (TE312), the user selects the "Geo-Coding by Room #" button in the 3D NRS Implementation Module (Figure 12b) to send a request to the system. The input message box is displayed in the Viewer area, and the user enters the room number (or room name) of the disaster site in the input box. The disaster site is geo-coded and displayed in the Viewer area (Figure 12c). The figure also illustrates the optimal route from the Entrance to the disaster site (TE312). In order to define the optimal route in the 3D GNM, Dijkstra's priority-first search [6, 19] for finding the shortest path in graphs has been applied in this experimental implementation. The optimal route was defined on the basis of the network's geometric distance as a travel cost, because the rescuers would walk from the origin (Entrance) to the destination (TE312). The rescuers passing the entrance may turn left and walk along the hallway on the 1st floor of Trafton East Hall to use a stairway located at the south side of the building, and go up to the 3rd floor. They may turn right and walk along the hallway to the disaster site, TE312.

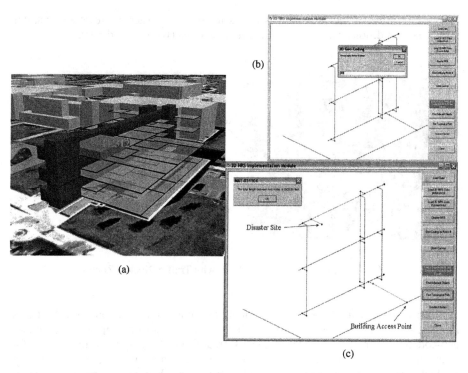

Fig. 12. 3D Indoor Geo-Coding and Optimal Route Searching

In the emergency situation, the impact will cause traffic blocks in the middle of the transport routes (labeled "Traffic Blocked Zone") and other places inside the building. There is no connectivity relationship at the zone of the transport routes. When attempting to reach the disaster site without knowing which stairways or hallways are feasible and safe, rescuers may be blocked in the middle and have to find an alternative way to proceed. The location of Blocked Zones will be transmitted via mobile or handheld communications equipment that interacts in real-time with the rescue command center. The location is identified in the 3D GNM using the 3D Geo-Coding technique (Figure 13b), and the alternative optimal route is defined from the source node to the destination node. Rescuers should use the stairway located at the east side of the building to access to the disaster site in this emergency situation, instead of using the stairway located at the south side as seen in the previous example.

5   Conclusions

From the terrorist attacks at the WTC, geographers have learned for the development and implementation of geospatial technologies on responding to similar disasters on multi-level structures in urban areas. Kwan and Lee [16] outlined the important elements of a GIS-based Intelligent Response System (GIERS), including 3D GIS network data models, real-time and distributed geographic databases, mobile GIS technologies, and analytical and modeling methods. In order to implement GIERS successfully, one of the important issues is to locate emergency crews and disaster events and conditions within the multi-level structures (within buildings).

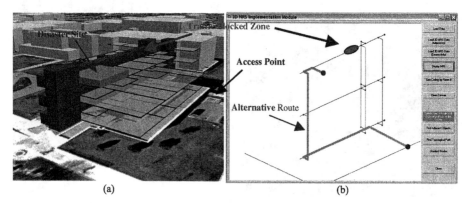

(a)   (b)

Fig. 13. Alternative Optimal Route with Traffic Blocked Zone

Since most current 3D location positioning techniques are not appropriate for LBSs in micro-scale urban environments (3D indoor), this paper developed a "3D indoor geo-coding" technique, in order to identify real-time 3D indoor locational data for analyzing human activities in space and time. For implementing 3D positioning methods, this paper presents (1) to model micro-scale urban environments to serve as a frame of reference (reference data) to identify distance and direction, in order to obtain locational data, and (2) to develop a 3D indoor geo-coding method to identify location data on individual activities based on the reference data.

While focusing on a 3D indoor geo-coding method and the usefulness of 3D network data models, this paper ignored several issues related to the implementation and deployment of the 3D indoor geo-coding techniques. First, in order to integrate with a 3D visualization system to allow the manipulation and exploration of geo-referenced virtual environments, an additional data model is needed for geometric representation to visualize geographic entities in 3D viewers. This means that dual data models are required for implementing the 3D indoor geo-coding technique in a 3D GIS. In order to maintain the data consistency between two data models, 3D GISs require the development of database designs. Second, the comprehensive GIS data to implement the 3D geo-coding technique raise serious concerns about issues of data security, as the data can be misused by unauthorized people. Means for preventing access to the data by such people should be considered before employing the system.

Lastly, the 3D indoor geo-coding technique, however, is not the most appropriate solution for the new generation of location-based services. The positioning technique requires the users to manually or verbally enter location information (e.g., room number) into the system (first generation technique of an LBS). Further research is therefore required to develop new positioning techniques to determine the user's location information without the assistance of the mobile users to provide user-friendly LBSs.

References

1. Aichholzer, O. and Aurenhammer, F.: Straight Skeletons for General Polygonal Figures in the Plane, In Cai, J.-Y. and Wong C.K. (eds.) *Proceedings of the Second Annual International Conference, COCOON '96, Hong Kong, Computing and Combinatorics, Lecture Notes in Computer Science 1090*, Berlin, Springer-Verlag: 117-126 (1998)
2. Beal, J.: Contextual Geolocation: A Specialized Geographic Information System for Improving Location Awareness in Urban and Indoor Network Environments, Master Thesis, Department of Computer Sciences, Minnesota State University-Mankato (2004)
3. Blum, H.: A Transformation for Extracting New Description of Shape, *Symposium of Models for Perception of Speech and Visual Forms*, MIT Press, Cambridge, MA, 362-380, (1967)
4. Chalmet, L.G., Francis, R.L. and Saunders, P.B.: Network Models for Building Evacuation, *Management Science*, 28(1), 86-105 (1982)
5. Cullen, I., Godson, V., Major, S.: The structure of activity patterns. In: Wilson, A.G., (Ed.), *Patterns and Processes in Urban and Regional Systems*. Pion, London, 281-296 (1972)
6. Dijkstra, E.W.: A Note on Two Problems in Connection with Graphs, *Numerical Mathematics*, Vol. 1: 269-271 (1959)
7. Drummond, W. J.: Address Matching: GIS Technology for Mapping Human Activity Patterns. *APA Journal* (Spring), 240-251 (1995)
8. Dürr, F. and Rothermel, K.: On a Location Model for Fine-Grained Geocast. In Dey, A.K., Schmidt, A. and McCarthy, J.F. (eds.) *Proceedings of the Fifth International Conference on Ubiquitous Computing* (UbiComp 2003), 18-35 (2003)
9. Flanagan, C., Jennings, C. and Flanagan, N.: Automatic GIS data capture and conversion, In Worboys M (ed.) *Innovations in GIS 1*. Taylor & Francis: Bristol, PA, 25-38 (1994)
10. Hägerstrand, T.: What about people in regional science? *Papers of Regional Science Association* 24, 7-21 (1970)
11. Hoppe, B and Tardos, E.: The Quickest Transshipment Problem, In *Proceeding of SODA: ACM-SIAM Symposium on Discrete Algorithms*: 433-441 (1995)
12. Kim, M.: Medial Axis Transform, Unpublished paper from Department of Computer Science, Johns Hopkins University. http://www.cs.jhu.edu/~bishop/ vision/medial.htm, (1998)
13. Kwan, M-P.: Space-time and integral measures of individual accessibility: a comparative analysis using a point-based framework. *Geographical Analysis* 30(3), 191-216 (1998)
14. Kwan, M-P.: Interactive geovisualization of activity-travel patterns using three-dimensional geographical information systems: A methodological exploration with a large data set. *Transportation Research C* 8, 185-203 (2000)
15. Kwan, M-P. and Lee, J.: Geovisualization of human activity patterns using 3D GIS: A time-geographic approach. In M. Goodchild and D. Janelle, (eds.): *Spatially Integrated Social Science: Examples in Best Practice*. Oxford University Press, Oxford 48-66 (2004)
16. Kwan, M-P. and Lee, J.: Emergency response after 9/11: the potential of real-time 3D GIS for quick emergency response in micro-spatial environments, *Computers, Environment and Urban Systems* (in press)
17. Lee, D.T.: Medial Axis Transformation of a Planar Shape, *IEEE Transactions on Pattern Analysis and Machine Intelligence*, PAMI-4 (4): 363-369 (1982)
18. Lee, J.: A 3D Data Model for Representing Topological Relationships Between Spatial Entities in Built-Environments, Ph.D. Dissertation, Department of Geography, The Ohio State University (2001)
19. Lee, J.: A Spatial Access Oriented Implementation of a Topological Data Model for 3D Urban Entities, *GeoInformatica* (in press)
20. Longley, P.A. and Batty, M. (eds.): *Advanced Spatial Analysis: The CASA Book of GIS*, Redlands: ESRI Press, Redlands, CA (2003)

21. Lu, Q., Hung, Y. and Shekhar, S.: Evacuation Planning: A Capacity Contrained Routing Approach, In Chen, H., Zeng, D.D., Demchak, C. and Madhusudan, T. (eds.) *Proceedings of the First NSF/NIJ Symposium on Intelligence and Security Information (ISI)*, Tuson, AZ, 111-125 (2003)
22. Mainguenaud, M.: Modeling the network component of geographical information systems. *International Journal of Geographical Information Systems*, **9**, 575-593 (1995)
23. Mallick, M.: *Mobile and Wireless Design Essentials*, Wiley Publishing (2003)
24. Mäntylä, M.: *An Introduction to Solid Modelling*, Computer Science Press, Rockville, MD (1988)
25. Miller, H.J.: Measuring space-time accessibility benefits within transportation networks: basic theory and computational procedures. *Geographical Analysis* 31(2), 187-212 (1999)
26. Miller, H.J.: What about people in geographic information science? In: P. Fisher and D. Unwin (eds.) *Re-Presenting Geographic Information Systems*, John Wiley, (in press)
27. Molenaar, M.: *An Introduction to the Theory of Spatial Object Modelling for GIS*, Taylor & Francis, New York (1998)
28. Pahlavan, K. and Li, X.: Indoor geolocation science and technology: Nextgeneration broadband wireless networks and navigation services. *IEEE Communications Magazine* (February), 112-118 (2002)
29. Prasithsangaree, P., Krishnamurthy, P. and Chrysanthis, P.K.: *On indoor position location with wireless LANs*. Telecommunications Program and Department of Computer Science, University of Pittsburgh. Pittsburgh, PA. (2001)
30. Samet, H.: Position Paper for Location-Based Services Meeting: *Santa Barbara Conference on Location-Based Services*, Center for Spatially Integrated Social Science, Santa Barbara, CA, http://www.csiss.org/events/meeting/location-based/ (2001)
31. Shiode, N.: 3D urban models: Recent developments in the digital modeling of urban environments in three-dimensions, *GeoJournal*, 52, pp. 263-269 (2001)
32. Smith, A.: *Adding 3D Visualization Capabilities to GIS*, http://www.casa.ucl.ac.uk/venue/3d_visualisation.html (1998)
33. Smith, J.M.: State Dependent Queueing Models in Emergency Evacuation Networks, *Transportation Science: Part B*, 25B (6): 373-389 (1991)
34. Vittorini, L.D. and Robinson, B.: Receiver Frequency Standards-Optimizing Indoor GPS Performance, *GPS World* 14(11), 40-48 (2003)
35. White, M.: Tribulations of Automated Cartography and How Mathematics Helps, *Cartographica*, 21, 148-159 (1984)
36. Weber, J., and Kwan, M-P.: Bringing time back in: A study on the influence of travel time variations and facility opening hours on individual accessibility. *Professional Geographer*, 54 (2), 226-240 (2002)
37. Zhao, Z., Saalfeld, A. and Ramirez, R.: A General Line-Following Algorithm for Raster Maps, In *Proceedings of GIS/LIS '96*, Denver, CO: 267-265 (1996)
38. Zlatanova, S. and Verbree, E.: Technological Developments within 3D Location-based Services, In *Proceedings of International Symposium and Exhibition on Geoinformation*, Shah Alam, Malaysia, 153-160 (2003)

Arc_Mat, a Toolbox for Using ArcView Shape Files for Spatial Econometrics and Statistics

James P. LeSage[1] and R. Kelley Pace[2]

[1] University of Toledo, Department of Economics,
Toledo, OH 43606, USA
`jlesage@spatial-econometrics.com`
[2] Lousiana State University, LREC Endowed Chair of Real Estate,
Department of Finance, Baton Rouge, LA 70803, USA
`kelley@pace.am`

Abstract. The ability to use statistical functionality for spatial modeling and analysis in conjunction with a mapping interface in the same environment has received a great deal of attention in the spatial analysis literature. We demonstrate the feasibility of extracting map polygon and database information from ESRI's ArcView shape files for use in statistical software environments. Specifically, we show that information containing map polygons can be used in these environments to produce high quality mapping functionality. Improvements in recent computer graphics hardware and software allow basic plotting functionality that is part of statistical software environments to produce mapping functionality based on the high quality ArcView map polygons.

1 Introduction

Users of spatial data naturally wish to both analyze and visualize their data. Unfortunately, the software packages such as ArcView that focus on visualization do not lend themselves to advanced spatial econometric analysis. Moreover, packages that have advantages in performing spatial econometric analysis such as Matlab and R/Splus do not immediately provide GIS functionality.

There are several means of marrying GIS functionality and spatial statistical analysis. One could add functionality to ArcView through dynamically linked libraries. However, this requires writing spatial econometrics code for compiled languages and is most natural for programs running in a menu as opposed to a batch format. In addition, ArcView is a demanding program and this may diminish resources available for statistical analysis. Alternatively, one could develop a program using compiled languages that has both GIS and statistical functions. For example, GeoDA (Anselin et. al, 2002) is a freestanding program that does not require a specific GIS. GeoDA runs under the Microsoft Windows operating system and provides an interactive environment that combines maps with statistical graphics, using dynamically linked windows and statistical modeling functionality through the use of C++ classes with associated methods.

M.J. Egenhofer, C. Freksa, and H.J. Miller (Eds.): GIScience 2004, LNCS 3234, pp. 179–190, 2004.

Finally, the approach we discuss here involves adding GIS functionality to existing statistical software. This approach has the advantage that thousands of user-contributed functions for statistical and spatial statistical modeling already exist for software environments such as Matlab or R/Splus. What these environments need is a means of visualizing the results. This paper discusses such an interface for Matlab that we have labelled *Arc_Mat*.

Arc_Mat consists of an underlying public domain library of C/C++ language Application Programming Interface (API) functions known as *shapelib* accessed via the Matlab C/C++ language interface labelled *C-MEX* combined with ordinary Matlab commands for conducting such operations as Moran scatterplots, mapping of residuals, and choropleth maps of regression coefficients. The *shapelib* functions include one API for processing, manipulating, and extracting map polygon information from ESRI's shape files, and another API for processing, manipulating and extracting database information from associated files. The associated metadata contains sample data observations for each region or polygon area on the map, which can be used in spatial econometric modeling.

The majority of the *Arc_Mat* toolbox consists of Matlab programs that can be developed and debugged in an interactive programming environment, avoiding the edit-test-debug development loop common to compiled language program development. Specifically, a set of Matlab development tools and support functions for graphical user interface (GUI) program development, was used to produce an interface to the underlying Matlab mapping functions. While this is specific to the Matlab environment, similar GUI functionality exists in the other statistical software. Our implementation calls a single function using the command-line, which creates a GUI interface. This interface presents the map as well as pull-down menus and other GUI tools for changing map properties, such as the color scheme, legend, variables mapped, as well as zooming in and out on sub-regions of the map. Readers familiar with the ArcView program will recognize that this same type of functionality is available in ESRI's ArcView software environment.

Another aspect of the tools described here is that they can be used to extract GIS information for use in other mapping software that presently exists for the Matlab environment. An extensive set of functions labelled *GeoXP* described in Heba, et al. (2002) has been developed for both Matlab and R/Splus software environments. The polygon coordinates produced by the C/C++ language interface of Matlab described here can be used to provide support for this library of functions.

Section 2 of this paper describes software issues related to the underlying file-transfer mechanism used to extract both mapping and database information from the ArcView files. The Matlab mapping software functions and issues pertaining to the design of the GUI interface are set forth in Section 3, as well as issues related to usage of the mapping functions and spatial econometric analysis. Focus is on situations where information from the ArcView shape files is useful in spatial econometric and spatial statistical modeling. Section 4 draws conclusions.

2   Underlying Algorithms and Design for Extracting and Importing Shape File Information into Matlab

The Matlab programming environment consists primarily of a command-line interface, where commands input using the keyboard or files containing command instructions are processed by a combined interpreter and just-in-time compiler. As the name Matlab suggests, the programming environment is based on matrix and vector constructs. User-written functions represent a key feature of the software environment, and these can be written in the Matlab programming language as well as external languages such as C/C++, or Fortran. Other statistical software environments such as R/Splus, Octave, and SciLab operate in a similar fashion and provide an external language interface. External language programs must be compiled and linked to the Matlab program execution environment using an interface gateway known as C-MEX (for C/C++) or F-MEX (for Fortran). C-MEX files are written in the C/C++ language, and in the case of our file interface to the ArcView *shape files*, they represent an intermediate file that translates input arising from a function call executed within Matlab into appropriate calls to a public domain library of functions known as *shapelib*. This library provides a series of API functions for reading, writing and extracting information from ArcView shape files. The term shape file is typically used to refer to a number of files that are generated by ArcView mapping projects, leading us to use the plural term shape files. These files can contain information on sample data stored in a database (*dbf*) file format, map projections, map polygons, and a host of other things related to ArcView projects. The *shapelib* functions provide an API for manipulating, reading and writing both mapping and database information from the ArcView formatted files.

We rely on a single Matlab function written in the native programming language that is named **shape_read** and takes a single argument containing a string with the basename of the shape files to be imported into Matlab. As already noted, the term "shape file" is a misnomer since this is a reference to at least three different files with the same basename but different filename extensions (.shp, .shx and .dbf). There are other possible files such as those containing map projection information that have a .prj extension, that are not currently used in our implementation.

The **shape_read** function calls underlying C-MEX functions that call the *shapelib* API functions to: (1) read polygon information and (2) extract database information. In addition to functionality for extracting information, the *shapelib* API also contains functions for performing calculations using information from the ArcView formatted files to calculate polygon centroid coordinates, which represent a key variable used to construct spatial weight matrices in spatial modeling.

A specific example of the simplest possible way to produce a map is shown below:

```
filename = '..\shape_files\china';
results  = shape_read(filename);
arc_histmap(results.data,results);
```

This would produce a mapping GUI containing a map of China that presents the first variable vector in the matrix stored in results.data, a matrix returned by the **shape_read** function. The GUI provides a pull-down menu that allows other variable vectors to be presented on the map. These variables are labelled with names constructed from the 'results.vnames' vector obtained from the database (.dbf) component of the shape files.

The objective of *Arc_Mat* is not cartographic reality, but rather comparative visualization of sample data relationships that may exist between regions. Visualizing a cluster of residuals that take on similar values to those from neighboring observations does not require high precision projections. Rather, emphasis is on preserving the aspect ratio of the initial projection so as not to disorient the user when zooming in or out on the map. In theory, map projection transformations of the polygon coordinates could be incorporated as an option in *Arc_Mat*, but this is not currently implemented. An exploration of the time required to plot a large number of polygons suggests a trade-off between speed and cartographic reality. Changing from the initial map projection to a new one would slow map zooming operations by a factor of five times.

3   The GUI Interface for Mapping

The Matlab programming environment provides a host of functions that can be used to produce a graphical user interface. A program that utilizes the GUI functions can be constructed to allow a single function call from the Matlab command window, or a file containing Matlab commands to create the GUI interface. This *Arc_Mat* design decision was motivated by the need to allow users to pass matrices or vectors returned by functions for spatial econometric analysis to produce a map. A number of different mapping GUI's exist as separate functions that produce different types of maps relevant for spatial statistical analysis.

Calling one of the mapping GUI functions turns control over to the mapping GUI, allowing various pull-down menus to be used to control characteristics of the map presentation. For example, pull-down menus allow the user to change: color schemes, variable selection from the input matrix, zooming of sub-regions on the map, choice of the number of legend categories, and labelling of the map polygons. On exit from the mapping GUI, control is returned to Matlab where further spatial econometric analysis or calls to another another type of GUI mapping function could be undertaken.

The basic mapping function is named **arc_histmap**, which produces a map based on an n by k matrix input, where n denotes the rows in the matrix which represent regions or observations, and k denotes the number of columns in the matrix which contain variable vectors. The mapping GUI produces a map of the first variable vector (column) of the input matrix in one figure window along with a histogram showing the distribution of the observation values in another figure window. Figure 1 provides an illustration of this type of map, with the associated legend shown in Figure 2. The map displays population growth of provinces in China over the 1980 to 1995 period, allowing easy identification of high versus low growth provinces.

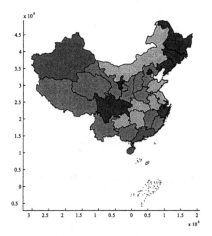 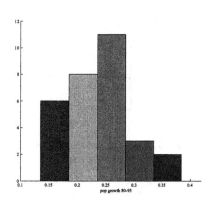

Fig. 1. China map of population growth rates 1980-1995

Fig. 2. Legend for China map of population growth rates

A second type of mapping GUI is the Moran scatterplot, which depicts a variable y on the horizontal axis with the average of neighboring region values for the variable y on the vertical axis, as a scatter of points. The averages for neighboring regions are constructed using an n by n spatial weight matrix W, that contains non-zero entries in the i, jth position if observation j represents a neighboring region, and zeros on the main diagonal. For the case of neighboring regions defined using a spatial contiguity, the matrix W could be constructed by placing values of unity in positions i, j, where j indicates regions that have borders touching region i. This matrix is then row-standardized to have row-sums of unity. The matrix product Wy then produces an average of the values from regions meeting this definition of neighbors. Various functions for producing spatial weight matrices based on contiguity, nearest neighbors, and distance are a part of the *Spatial Econometrics Toolbox* and the *Spatial Statistics Toolbox*. Since Matlab represents a programming environment, users are free to construct spatial weight matrices based on other criterion that appropriately define the structure of connectivity relationships between regions for their particular problem.

Figure 3 shows the Moran scatter plot where the population growth rates in the vector y described above have been transformed to deviation from the means form. This allows the population growth rates shown on the horizontal axis for each region to be expressed as positive if they above the overall average

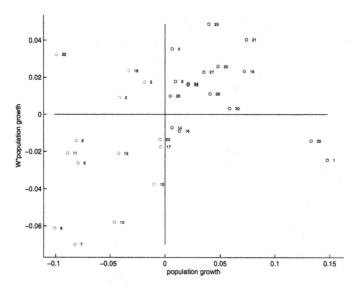

Fig. 3. Moran scatter plot for Chinese provincial population growth rates

population growth rate, or negative if they are below the average. The vertical axis was constructed using a first-order contiguity spatial weight matrix that produces average growth rates of neighboring provinces, again in deviation from the means form. The four color-coded quadrants in the scatter plot depict (1) regions where higher than average growth rates are associated with an average of neighboring region growth rates that are also higher than average (the points in the upper right-hand quadrant), (2) regions where lower than average growth rates are associated with an average of neighboring region growth rates that are also lower than average (the points in the lower left-hand quadrant), (3) regions where lower than average growth is associated with neighboring regions having higher than average growth (the points in the lower right-hand quadrant), and (4) regions where higher than average growth is associated with lower than average growth in neighboring regions (the points in the upper left-hand quadrant).

The map (Figure 4 depicts regions using the color coding from the scatter plot, and numerical labels can be provided as an option in both figures to allow identification of the growth rates associated with each region versus that of neighboring regions. This type of scatter plot is frequently used to assess the extent to which sample data observations exhibit spatial dependence. The presence of a large number of points in the upper right-hand quadrant are indicative of strong spatial association since provincial growth rates shown on the horizontal axis are positively associated with the average growth rate of neighboring provinces. Similarly, a large number of points in the lower left-hand quadrant point to positive spatial association since they indicate that the value of the variable on the horizontal axis is positively associated with the average value from neighboring regions. In contrast, points in the upper left-hand and lower right-

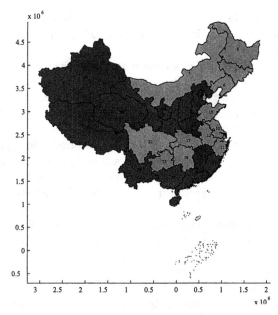

Fig. 4. China map linked to the Moran scatter plot

hand quadrants reflect negative spatial association, since these are observations
that are inversely related to the average of neighboring observations.

A third use for the mapping GUI functions is something labelled **sar_map**,
where SAR stands for a spatial autoregressive econometric model. The spatial
autoregressive model represents a variant of simple regression: $y = X\beta + \varepsilon$ which
adds an additional explanatory variable to the model, Wy, that reflects a spatial
lag: $y = \rho Wy + X\beta + \varepsilon$.

This function allows the user to select sub-regions on the map for which
spatial autoregressive model estimates of the parameters β and ρ as well as the
noise variance estimate σ_ε^2, log-likelihood function value, R^2 measure of fit, and
other information will be produced. The estimates for a sub-region are presented
alongside those for the entire sample. This allows an assessment of local variability in the nature of the underlying linear spatial autoregressive relationship
of interest.

This type of GUI mapping interface illustrates a distinct advantage of Matlab
over more conventional GIS mapping environments, in that spatial econometric methods can be used as part of the map interface. Solving for maximum
likelihood estimates of basic spatial regression models provides some computational challenges when the problem involves large samples. This makes it difficult
to provide estimation functionality in mapping software, which would need to
provide sparse matrix functionality as well as algorithms for numerical linear
algebra. In contrast, this functionality is already provided by the public domain *Spatial Econometrics Toolbox* (LeSage, 1999) and *Spatial Statistics Toolbox*

(Pace, 2002) functions for the Matlab environment. Similar spatial econometric modeling functionality exists in *spdep* (Bivand, 2002) for the R/Splus software environment.

The toolbox functions utilize sparse matrix algorithms available in Matlab as well as computationally efficient algorithms (Barry and Pace, 1999) for computing the log-determinant of an n by n matrix, and a vectorized approach to producing maximum likelihood estimates (Pace and Barry, 1997). Taken together, these allow maximum likelihood estimates to be produced in one or two seconds for problems involving over 3,100 US counties. This speed allows the initial global estimates for a model to be shown in a graphics window and rapid updating of estimates for sub-regions selected using the GUI interface. Single regions can be added or subtracted from the sub-region selection set to test for influential observations.

The GUI for the mapping functions provide a series of pull-down menus along the bottom of the map figure that allow user selection of map characteristics. For example, in the case of the **arc_histmap** function, these menus allow selection of the colormap, a zoom menu, that allows sub-regions on the map to be selected using a drag-rectangle, the number of categories for the histogram, the variable/column of the input matrix to be selected for viewing on the map, an option to place numerical labels on the map polygons that correspond to sample data observation numbers, and finally an option for exiting the mapping GUI and returning to the Matlab command line.

Optional input arguments are provided to the mapping functions that allow user input of variable names, color schemes and an indicator vector for missing values that will be coded differently on the map. There are also switches to add top menus to the map and legend figures respectively. These menus contain standard Matlab functions for operating on graphical figures, such as a file menu item that allows the figure window to be printed or exported in numerous graphic file formats. Other top menu items allow the user to add lines and text annotations to the figure windows, and a host of other graphic functionality.

The ability of the spatial econometric routines to provide maximum likelihood estimates for the parameter ρ in the first-order spatial autoregressive model provides a nice complement to the Moran scatter plot mapping GUI. Another aspect of operating in an environment where spatial econometric routines are available is the ability to generate sets of different spatial weight matrices that can be used as an input argument to the Moran scatter plot function. For example, one could examine the scatter plot and associated map using a spatial weight matrix constructed using first-order neighbors based on distance or contiguity, as well as higher-order neighbors based on contiguity or distance. Functions to construct a variety of spatial weight matrices represent a standard part of the *Spatial Econometrics* and *Spatial Statistics* toolboxes are well as the R/Splus spatial functions *spdep*.

In addition to maximum likelihood estimation procedures, algorithms exist for robust Bayesian estimates for this family of models that allow for nonconstant variance across the observations. Using Markov Chain Monte Carlo

(MCMC) estimation, variance scalar estimates can be produced for each observation using methods described by Geweke (1993) for the case of linear least-squares and LeSage (1997) for this family of spatial regression models. The ability to plot the variance estimates for each region/observation on a map would help identify regions where the model may be inadequate. For example, if urban regions exhibit higher noise variances, this may be an indication of an inadequately modeled regression relationship.

Other approaches to exploring spatial stability that have been proposed are geographically weighted regression (Brunsdon et. al., 1996) and Bayesian variants (LeSage, 2004). These models produce locally linear non-parametric estimates for each observation, based on a sub-sample of observations nearby. Mapping these parameter estimates can be useful in exploring systematic patterns between the dependent and explanatory variables over space. Pace and LeSage (2004) suggest a spatial autoregressive locally linear estimation procedure they label SALE, that relies on a recursive estimation scheme to produce estimates for the SAR model for every observation. We use parameter estimates from this model constructed using a function from the *Spatial Statistics Toolbox* to demonstrate the value of mapping locally linear parameter estimates.

A model from Pace and Barry (1997) of the relationship between voter participation in the 1980 presidential election and variables measuring population eligible to vote, home ownership, educational attainment and per capita income was estimated using 3,107 observations and the SALE methodology. This produces 3,107 parameter estimates for each explanatory variable in the model. Figure 5 shows the map and legend figures for the estimates of the 'home ownership' variable in the model. The histogram legend shows that this variable exerts a predominately positive impact on voter participation rates. From the map we see that home ownership exerts a larger impact on voter participation in the south and midwest counties, and a smaller impact for counties on the east and west coasts.

Finally, we note the need to sort spatial econometric data sets to match the order of the polygon and data information read from the shape files. One approach to using the *Arc_Mat* toolbox would be to rely on ArcView shape files for the mapping polygons and perhaps centroids of the polygons, while relying on a separate data file for spatial econometric analysis. This requires that a geography key be included in the econometric analysis data set. It is likely such a code would exist in any spatial econometric sample data.

For example, consider a sample of 1,008 Ohio zip code areas and a set of shape files that contain the polygon information for the zip code areas. An example of the Matlab code needed to sort the spatial econometric data sample into an order consistent with the polygon ordering in the shape file is provided below. In this example the shape files are named ohio.dbf and ohio.shp, while the spatial econometric data file is named odata.txt. We assume that the zip code identifier is contained in the first column of the results.data matrix obtained from the database file. We also assume that a zip code identifier exists in the first column of the data file odata.txt.

```
filename = 'ohio';
results = shape_read(filename);
nobs = results.npoly; % should equal 1,008 the # of observations
map_zips = results.data(:,1);
load odata.txt;
dat_zips = odata(:,1);
% loop over the map_zips and find an index corresponding to
% each of these zip codes in the odata matrix
out = zeros(size(odata)); % allocate a matrix for the results
missing = zeros(nobs,1);     % keep track of missing values
for i=1:nobs
zipi = map_zips(i,1);
ind = find(zipi == dat_zips);
   if length(ind) > 0
    out(i,:) = odata(ind,:);
   else
    missing(i,1) = 1;
   end;
end;
options.missing = missing;
arc_histmap(out,results,options);
```

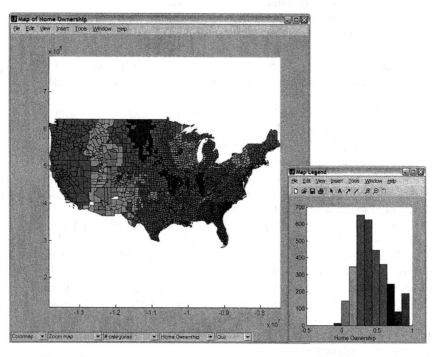

Fig. 5. SALE parameter estimates for the home ownership variable

The code loops over all zip code areas using the order from the shape files. Each zip code area is extracted into the scalar variable zipi, and this is used in the Matlab **find** function that searches the vector dat_zips for a matching zip code. If a match is found, an index into the vector dat_zips is returned and this index is used to extract a row from the matrix odata and place it in the matrix 'out'. If no match is found, we record this in a vector 'missing' that could be used as an input option to the mapping GUI function. Note that in this case the matrix 'out' would contain a row of zero values used to initialize this matrix. This would adversely affect the histogram legend in the **arc_histmap** function, which would interpret these as actual values without the optional options.missing vector used as an input to the function.

We note that any data information contained in the structure variable 'results.data' could be combined with the sorted version of the spatial econometric sample data matrix. Because these are now sorted by zip code area, Matlab would combine these two matrices with the simple command:

```
combined = [results.data out];
```

Since regression estimates are unaffected by the ordering of the observations, it would seem that sorting the spatial econometric data to match the order of the shape file information should occur at the outset. This would allow users to map residuals, predicted values and other information using 'results' structure information returned by the estimation functions in the *Spatial Econometrics Toolbox*.

4   Conclusions

We demonstrate the feasibility of extracting map polygon and database information from ArcView shape files for use in statistical software environments. The motivation for extracting database information for use in a statistical programming environment seems quite evident. We demonstrate that information containing map polygons can also be used in these environments to produce high quality mapping functionality. Improvements in recent computer graphics hardware and software have arisen from interest in computer gaming. This allows rapid rendering of map polygons on almost all recent desktop and laptop computers using basic plotting functionality that is part of statistical software environments. A byproduct of this is that mapping functionality based on the high quality ArcView map polygons can be created in a statistical software environment in support of spatial econometric and statistical analysis. Although our focus was on a particular implementation of these ideas in the Matlab software environment, a similar approach would be feasible in a number of other statistical software environments such as R/Splus, Octave, and SciLab.

Acknowledgments

The first author would like to acknowledge support from the National Science Foundation, BCS-0136229 and the second author acknowledges National Science

Foundation support from BCS-0136193. Both authors would like to thank Darren Hayunga for his helpful comments.

References

Anselin, L., I. Syabri, and O. Smirnov: Visualizing Multivariate Spatial Correlation with Dynamically Linked Windows, L. Anselin and S. Rey (eds.), Proc. CSISS Workshop on New Tools for Spatial Data Analysis, Santa Barbara, CA, Center for Spatially Integrated Social Science, (2002) CD-ROM (pdf file, 20 pp.)

Barry R. and R.K. Pace: A Monte Carlo Estimator of the Log Determinant of Large Sparse Matrices. Linear Algebra and its Applications, **289** (1999) 41–54

Bivand, R. S.: Spatial econometrics functions in R: Classes and methods. Journal of Geographical Systems **4** (2002) 405–421

Brunsdon, C., A. S. Fotheringham, and M.E. Charlton: Geographically weighted regression: A method for exploring spatial non-stationarity. Geographical Analysis **28** (1996) 281–298

Geweke J.: Bayesian Treatment of the Independent Student t Linear Model. Journal of Applied Econometrics **8** (1993) 19–40

Heba, I., E. Malin, and C. Thomas-Agnan: Exploratory spatial data analysis with GEOXP. European Regional Science Association conference papers

LeSage, J. P.: Bayesian Estimation of Spatial Autoregressive Models. International Regional Science Review **20** (1997) 113–129

LeSage, J. P.: The Theory and Practice of Spatial Econometrics. (1999) pdf file, 296 pp. available at www.spatial-econometrics.com

LeSage, J.P.: A Family of Geographically Weighted Regression Models. Advances in Spatial Econometrics, L. Anselin, J.G.M. Florax and S.J. Rey (eds.) Springer-Verlag (to appear)

Pace, R. K.: Spatial Statistics Toolbox 2.0. (2002) pdf file, 36 pp. available at www.spatial-statistics.com.

Pace, R. K. and R. Barry: Quick Computation of Regressions with a Spatially Autoregressive Dependent Variable. Geographical Analysis **29** (1997) 232–247

Pace, R.K. and J. P. LeSage: Spatial Autoregressive Local Estimation. Recent Advances in Spatial Econometrics. J. Mur, H. Zoller, and A. Getis (eds.), Palgrave Publishers (2004) 31–51.

A Predictive Uncertainty Model for Field-Based Survey Maps Using Generalized Linear Models

Stefan Leyk and Niklaus E. Zimmermann

Swiss Federal Research Institute WSL, Zuercherstrasse 111,
CH-8903 Birmensdorf, Switzerland
{stefan.leyk,zimmerma}@wsl.ch

Abstract. In this paper we present an approach for predictive uncertainty modeling in field-based survey maps using Generalized Linear Models (GLM). Frequently, *inherent* uncertainty, especially in historical maps, makes the interpretation of objects very difficult. Such maps are of great value, but usually only few reference data are available. Consequently, the process of map interpretation could be greatly improved by the knowledge of uncertainty and its variation in space. To predict *inherent* uncertainty in forest cover information of the Swiss topographic map series from the 19th century we formulate rules from several predictors. These are topography-dependent variables and distance measures from old road networks. It is hypothesized that these rules best describe the errors of historical field work and hence the mapping quality. The uncertainty measure, the dependent variable, was derived from local map comparisons within moving windows of different sizes using a local community map as a reference map. The derivation of local *Kappa coefficient* and *percent correctly classified* from these enlarged sample plots takes the local distortion of the map into account. This allows an objective and spatially oriented comparison. Models fitted with uncertainty measures from 100m windows best described the relationship to the explanatory variables. A significant prediction potential for local uncertainty could thus be observed. The explained deviance by the Kappa-based model reached more than 40 percent. The correlation between predictions by the model and independent observations was $\rho=0.76$. Consequently, an improvement of the model to 47 percent, indicated by the G-value, was calculated. The model allows the spatial-oriented prediction of *inherent* uncertainty within different regions of comparable conditions. The integration of more study areas will result in more general rules for objective evaluation of the entire topographic map. The method can be applied for the evaluation of any field-based map which is used for subsequent applications such as land cover change assessments.

1 Introduction

Historical spatial information frequently represents essential input for landscape change analysis in geographical information systems (GISs). In this context, historical maps are unique witnesses of past landscape configurations. They provide histori-

M.J. Egenhofer, C. Freksa, and H.J. Miller (Eds.): GIScience 2004, LNCS 3234, pp. 191–205, 2004.
© Springer-Verlag Berlin Heidelberg 2004

cal information on a landscape before aerial photography. The potential of historical documents has recently been recognized by various authors from different fields such as landscape management [5], landscape reconstruction, ecology or forestry [21] and GIS [3, 24].

The objects, in particular natural objects, and their delineations represented in such historical documents usually contain uncertainty to a high degree. This can be referred to as *inherent uncertainty* [24] or *production-oriented uncertainty* [19]. This type of uncertainty is complex and contains different concepts, such as *vagueness*, *ambiguity* and *error*. Different perspectives have contributed to these concepts such as philosophy [28], information theory [18], remote sensing [12], and GIS [10, 31].

Inherent uncertainty manifests itself in a spatial deviation from reality for whatever reason. Often in historical maps only this pattern of deviation can be assessed without any additional knowledge. This is due to the fact that detailed historical background information for thematic interpretation is hardly accessible. One way to assess the accuracy of a map with regard to a reference map is to carry out a quantitative map comparison. Thereby traditional techniques of accuracy assessment using error or confusion matrices produce statistical accuracy measures. There is an abundance of literature describing these techniques in detail [8, 11, 26], for a review in land cover classification). Recently these approaches were extended to assess area estimates from generalized confusion matrices [20] or to derive local summary statistics for geo-statistical analysis [4]. Furthermore combinations of confusion matrix approaches and fuzzy set operations [16, 17, 30] were shown to be flexible methods for map comparison if spatial entities have vague definitions or multi-memberships.

The above mentioned techniques allow the evaluation of map accuracies in general. But the spatial distribution of accuracies can only be considered within an area for which a reference map exists. Such combined methods provide no detailed knowledge of the uncertainty distribution over larger areas, which are not entirely covered by the reference map. In many cases the amount of *inherent* uncertainty varies greatly throughout the considered area depending on local conditions [25]. For the incorporation of historical maps into land cover change assessments, such knowledge would greatly improve the evaluation process. It would support the informed interpretation of the historically mapped objects. To enable this, the knowledge of the *inherent uncertainty* and its variation in space needs to be extrapolated to the whole historical map. This knowledge can be derived from a limited number of local reference maps. Conceptually, this can be done if rules or models can be developed, which link the local uncertainty to a set of generally available predictors. These linkages can be extrapolated into space or to areas not covered by the reference maps.

We present the development of a predictive model for mapping the *inherent* uncertainty of the forest cover of a historical national topographic map of Switzerland originating from the 19th century. These maps were surveyed in extensive field campaigns. Thus we hypothesize that the errors of such mapping efforts can be related to a set of topography-dependent predictors. These are slope, aspect, elevation, as well as distance and visibility from old road networks. The chosen explanatory variables represent factors that are assumed to have been limiting factors for field based surveys, such as visibility or accessibility. By doing so, we aim at finding more general rules, which best describe the quality and the associated errors of topographic field mapping during the 19th century.We use Generalized Linear Models (GLMs) and moving windows to explain the uncertainty found at moving window positions from a set of terrain-derived variables.

GLMs are mathematical extensions of ordinary least-square regression models that do not force data into unnatural scales. Thereby they allow for non-linearity and non-constant variances [22]. This is an important prerequisite, since our dependent variables rate the amount of error in a moving window at a scale ranging from 0 (=perfect fit) to 1 (no agreement at all). Thus the dependent variable is bounded, and cannot easily be treated with ordinary least-square regression. GLMs specifically address this problem and represent thus a suitable approach for our study. These statistical methods are successfully applied to different fields such as predictive habitat modeling in ecology [2, 14, 15], or forest inventories based on remote sensing [23].

2 Materials and Methods

2.1 Study Area and Historical Maps

The historical Swiss National topographic map evaluated here is the first edition of the so-called Siegfried map series. It was published between 1870 and 1920 at scales of 1:25,000 for the Swiss Plateau and 1:50,000 for the mountain regions. As a new feature compared to earlier maps the forest cover was delineated during the field survey. Thus this map is the only source representing forest cover throughout the entire area of Switzerland at this time.

Our study area is the community of Pontresina (Canton Grisons), which is located in an interior valley of the eastern Swiss Alps (Figure 1). As reference data were scarce, the choice of the area was limited. We found only a few areas for which reference data exist. The available reference map for the community of Pontresina belongs to a map series that preceded the official cadastral mapping in Switzerland at that time. They were now mapped as community maps, produced at a scale of 1:5,000 or 1:10,000 with a high degree of detail. These plans are the most reliable spatial representations of landscape objects for the time considered and the instructions for data survey were much more detailed (including forest). Thus we used the forest cover of these maps as a reference to test against the forest cover information of the Siegfried map. Yet there may be an unknown disagreement in the thematic meaning of *forest* leading to a certain spatial effect of uncertainty. In order to exclude additional sources of error, we chose community maps where the date of publication or production did not differ more than 5 years from those of the Siegfried maps.

Both maps were scanned and geo-referenced on the basis of the present-day Swiss national topographic map. Local distortions made the correct registration of the Siegfried map rather difficult. The resulting RMSE using the affine transformation procedure for geo-rectifying was 9.2m. The forest cover was extracted from both datasets to be treated in vector or raster format. The spatial resolution of the raster form of the maps was 1.25m.

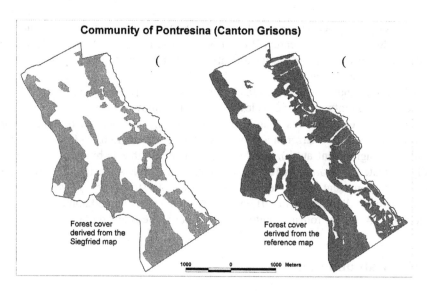

Fig. 1. The forest cover information from (b) the local community map of the study area (Pontresina) was used as a reference for map comparison with (a) the forest cover from the Siegfried map

2.2 Uncertainty Measures: The Dependent Variables

We first derived global measures, such as *percent correctly classified* (PCC) and the *Kappa coefficient* (κ) (based on [11]), from the confusion matrix by pixel-wise comparisons. To examine whether there are trends of uncertainty in relation to topographic gradients, such as slope or elevation, we derived the same measures within different strata (i.e., classes of slope). We found that uncertainty increased along these gradients, which indicated an existing relationship. The measures had to be weighted by the proportion of forest boundary per forest area unit within each stratum since uncertainty became apparent along these lines. This was necessary, because in strata where uncertainty was increasing, larger forest patches (with smaller proportions of forest boundary) occurred.

We had two options to generate a spatially distributed measure of uncertainty: (1) a pixel by pixel evaluation resulting in binary codes of 0 (both maps agree), or 1 (the two maps disagree in representing forest/non-forest), and (2) an aggregated evaluation indicating the degree of uncertainty within a moving window as illustrated in figure 2 (with continuous values between 0 to 1). We decided to use the second approach. Thus we calculated PCC and κ within rectangular moving windows of four different dimensions: 30x30m, 60x60m, 100x100m, and 180x180m. Uncertainty was then defined as 1-PCC and 1-κ. By this, the locally generated statistics provide gradual measures of local disagreement between the two maps (Figure 2).

This allowed us to test the power of variables expressing the topographic configuration in explaining the degrees of uncertainty and to test how the explanatory power varies with spatial aggregation.

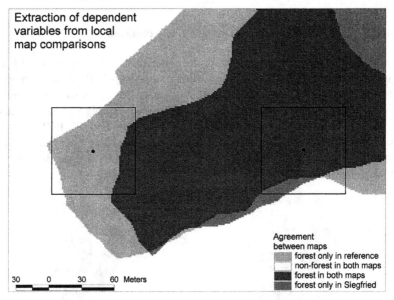

Fig. 2. The dependent variables were locally extracted from pixel-wise map comparisons within moving windows

The equations we used to derive PCC and κ from a confusion matrix of q classes occurring in both the Siegfried and the reference maps have the forms (Eqn. 1 and 2):

$$PCC = P_C = \sum_{i=1}^{q} p_{ii} \qquad (1)$$

and

$$\kappa = \frac{P_c - P_e}{1 - P_e} = \frac{P_c - \sum_{i=1}^{q} p_{i+} p_{+i}}{1 - \sum_{i=1}^{q} p_{i+} p_{+i}}, \qquad (2)$$

with
$$p_{i+} = \sum_{k=1}^{q} p_{ik} \qquad p_{+i} = \sum_{k=1}^{q} p_{ki},$$

where q is the number of classes considered for map comparison, p_{ii} is the proportion of class i that is correctly classified, P_e is the expected proportion of agreement due to chance, p_{i+} is the marginal proportion of row i and p_{+i} is the marginal proportion of column i of the confusion matrix.

In order to prevent pseudo-replication and to reduce spatial autocorrelation, we sampled the grids of the dependent variables at regular distances of at least the length of the moving window. The mesh distances of the sample locations and the resulting sample sizes, which were considerably reduced with increasing window size, were: 30m (9946 points), 60m (2481 points), 100m (1109 points) and 180m (277 points) (Table 2).

2.3   Independent Variables from Topographic Conditions

In order to find explanatory variables for the prediction of the occurrence of uncertainty in field-mapped forest cover the situation of topographic field survey more than 100 years ago has to be considered. Those conditions that had a strong influence on the quality of mapping will represent the variables with the highest explanatory potential. We included topographic conditions such as elevation (ELEV) (Figure 3b) and elevation difference from the lowest point within the study area (ELEVD) from the 25m DEM of Switzerland and its derivatives slope (SLP) (Figure 3a) and aspect (ASP).

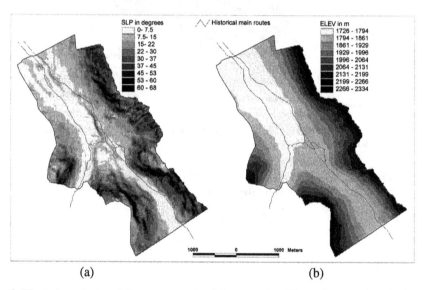

(a)   (b)

Fig. 3. The independent variables were extracted from topographic gradients such as (a) SLP or (b) ELEV and distances to routes or forest boundaries

In order to link the pixel values to the moving windows analyzed, we calculated the standard deviation of these variables within each window position using focal functions of the ArcGrid module. Thus we obtained measures of variability for elevation (ELEVCH), slope (SLPCH) and aspect (ASPCH). A next variable derived was the "pathdistance" from any location to the closest main route (STRD), which was accessible with the required equipment. We hypothesize that the distance of locations along a historical road to the considered location contains high predictive potential. We also included a measure of visibility of a location (VIS) from the positions of the main routes. Further, we derived the distance from any location in the study area to the closest forest boundary of the reference map (FORD(ref)). This variable is assumed to hold predictive power since it indicates the distance to the main source of uncertainty or, in other words, to the location where spatial patterns of uncertainty can be assessed. Since this variable is only available in the test area, the resulting model is rather an *analytical* model. We derived this measure also from the Siegfried map (FORD) in order to enable a *predictive* model for other areas. We expect that the latter will result in slightly lower model accuracies. All topographic information was derived at a resolution of 25m. A summary of all independent variables used is given in table 1. The four sets of sample points of the dependent variable were intersected with each explanatory variable in a GIS.

Table 1. Description of the independent variables used in the analytical and predictive uncertainty model

Variable	Description
ELEV	Elevation from DEM 25
ELEVCH	Standard deviation of elevation within moving window
SLP	Slope (degree) as derivative of the DEM
SLPCH	Standard deviation of slope (degree) within moving window
ASPCH	Standard deviation of aspect (degree) in moving window
STRD	Pathdistance from moving window center to nearest road
VIS	Visibility of locations from street positions
ELEVD	Elevation difference between moving window center and lowest location
FORD	Distance from moving window center to nearest forest boundary in Siegfried
FORD(ref)	Distance from moving window center to nearest forest boundary in the reference map

With the explanatory variables listed in table 1, we intended to reflect factors that may have influenced the generation of uncertainty during the historical field survey. The main routes within accessible zones such as valleys with flat terrain are assumed to be the locations the topographer passed during the survey. With increasing horizontal and vertical difference, with increasing slope and with changes of aspect from these main routes the uncertainty in mapping the landscape objects (including forest cover) is expected to grow considerably. Reasons are the limited accessibility, the reduced visibility and consequently the increased difficulty in identifying landscape objects (e.g., forest borders). These were due to the conditions of field work at the time: limited transportation capabilities and heavy equipment. Examination of the distance from the closest forest boundary (FORD) showed that locations close to the boundary carry a much higher probability of *inherent* uncertainty than those further away. This measure is expected to have significant predictive power. This is due to the fact that visibility and other terrain features mentioned are not relevant if there is no forest at all or if the location is within a very large forest patch, far away from any boundary.

We do not know from which position the cartographers worked, and many cartographers possibly used different mapping approaches. Therefore, we assume that the resulting predictive model retains a significant unexplained variation of the mapped uncertainties. The remaining variation may include several additional aspects. Apart from others these are weather conditions, seasonal differences, subjective preferences and interests of the topographer and his expertise. Furthermore, the quality of the manual map transcription and reproduction may be responsible for unexplained variation.

2.4 The Statistical Model

In the following we present the formulation of the generalized linear model. The predictor variables X_j $(j=1, ..., p)$ are combined to produce a linear predictor (LP), which is related to the expected mean value $(\mu = E(Y))$ of the response variable Y through a link function $g()$:

$$g(E(Y)) = LP = \alpha + X^T \beta , \tag{3}$$

where α is a constant called the regression intercept, X represents a vector of p predictor variables $(X_1,...,X_p)$ of any possible power T, and β denotes a vector of p regression coefficients $(\beta_1,..., \beta_p)$ which are determined for each predictor.

Prior to selecting an appropriate model, the empirical probability distribution of the response variable has to be tested and compared to the theoretical distribution. Our dependent variable follows a binomial distribution since the original measures were given in presence/absence terms of errors from a pixel-by-pixel evaluation. With increasing window size these measures are increasingly weighted due to their local environment and will be included as weighted response variables in the model.

The link function we thus used was the logistic link often termed logit regression model [22]. This link can be described as:

$$g(E(Y)) = \log(\mu/(1-\mu)) , \tag{4}$$

which is the logarithm of odds, a model widely used for binomial data [7]. In GLMs the linear combination of predictive or explanatory variables is related to the mean of the response variable through this link function to transform them to linearity and to maintain the prediction values within the range of coherent values for the response [15]. Thus the general logistic regression model we used has the form:

$$logit = \log(\mu/(1-\mu)) = \alpha + X^T \beta . \tag{5}$$

We included linear terms as well as quadratic powers and interactions of the explanatory variables. A maximum-likelihood (ML) estimator is used for fitting the model. We plotted the standardized residuals against the fitted values to identify unexpected patterns in the deviance. All calculations were performed using Splus.

We first fitted a full model, using all explanatory variables (linear and quadratic terms and interactions), then we applied stepwise regression–allowing for both backward and forward selection–in order to optimize the final model. For the analysis of the significance of eliminating or adding terms, the Akaike information criterion (AIC) was used. Thereby, given a fitted model object, individual terms of the model are removed and the respective effect is assessed in comparison to the previous model where the aim is to minimize the AIC [14].

We used χ^2 approximations [22] and AIC measures to test the deviance reduction associated with each variable for significance at a given confidence level (0.05). In addition we tested whether the model coefficients differ significantly from zero using the standard error associated with the estimated model coefficients for a Student t-test. We used the D^2 value (percent deviance explained) to evaluate the model fit, calculated as (Null.Deviance – Residual.Deviance) / Null.Deviance, where the Null.Deviance is the deviance of the model with the intercept only, and the Residual.Deviance is the deviance that remains unexplained after all final variables have been included. This is equivalent to the R^2 of a linear least-squares regression model. In addition we used the adjusted D^2 (derived from adjusted R^2 [29]) to take into account the number of observations (n) used for fitting the model and the number of explanatory variables (p):

$$D^2_{adj.} = 1 - ((n-1)/(n-p)) \times (1 - D^2).$$ (6)

To produce an uncertainty distribution map the fitted model has to be cartographically represented. For implementation of the model in GIS the inverse of the link function has to be applied to transform the values back to the scale of the original response variable. In our case the inverse logistic transformation is required, which has the form:

$$p(y) = \exp(LP)/(1 + \exp(LP)),$$ (7)

where LP is the linear predictor fitted by logistic regression (Eq. 4). In applying this retransformation we obtained values between 0 and 1, which is the same as the original response values. This inverse relationship has bias, which can be corrected by using a Taylor series approximation. However, this correction is rarely done in environmental applications. In cases where the mean of the estimated variable is within the range which can be considered fairly linear after transformation (between 0.2 and 0.8) the bias is expected to be very low. Nevertheless, the bias correction has to be done if this condition is not fulfilled.

In order to test the predictive power of the models we split the data set into two parts following the split-sample approach [27]. One part (50%) was used for calibration (training data), the other one (50%) to evaluate the model predictions and to measure the adequacy of the model (test data). Goodness-of-fit measures can then be used to evaluate the fit between the predictions and observations of the evaluation data set. We calculated the non-parametric rank correlation coefficient of Spearman (ρ Rho) and tested it for significance. Furthermore the G-value [1] was used to test the relative improvement of the model over a null model (i.e., the mean of the dependent variable within the calibration data subset).

First, we developed analytical models for PCC- and κ-based uncertainty in which the variable FORD(ref) was included as an independent variable in order to examine its explanatory power. Next, we calibrated predictive models applicable to areas other than the test site. Thus we replaced FORD(ref) by FORD for both PCC- and κ-based *predictive* models. All models were developed from split-sample data sets.

3 Results

In our study area, the *analytical* models from the 100m windows best describe the relationship between *inherent* uncertainty and the explanatory variables (Figure 4) when compared with the test data. At smaller window sizes, we did not find equally good explanations from the predictors. The model based on the 160m window data set is also inferior (see Table 2 for a comparison of the results). We therefore only present detailed results for the *predictive* models for this 100m window data set, comparing these models with the *analytical* models, respectively.

Table 2. Comparison of PCC-based predictive models calibrated from different window sizes

Window size for PCC extraction	pixel-wise	30m	60m	100m	180m
mean PCC	0.78	0.78	0.78	0.78	0.78
Sample size	9946	9946	2481	1109	277
Goodness of fit: D^2 (if model is significant)	0.16	0.21	0.25	0.31	0.29

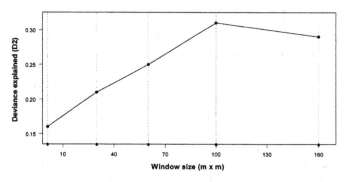

Relationship between window size and explained deviance of a model

Window size (m x m)

Fig. 4. Relationship between moving window size and deviance explained (D^2) by the fitted model (example for PCC-based models); models were optimized for 100m windows

The *analytical* models for the 100m window (Table 3) indicated significant explanatory power for the uncertainty measures. The step-wise regression resulted in a total deviance reduction of 41 percent for PCC and 46 percent for κ (Table 3, D^2 and *adjusted* D^2). The following variables contributed to the explanation of the fractions of uncertainty in the presented examples: FORD(ref), SLPCH, STRD for PCC-based uncertainty and FORD(ref), SLPCH, ELEVCH, ELEVD and STRD for κ-based uncertainty. When compared with the test data, we received a high agreement for predicted uncertainty (PCC: $\rho=0.76$, κ: $\rho=0.77$) within the test site, and the G-values indicated significant improvements over the null model (PCC: $G=37\%$, κ: $G=47\%$, Table 3). Due to the use of distance to the reference forest boundary the results are not applicable to areas outside of the study area.

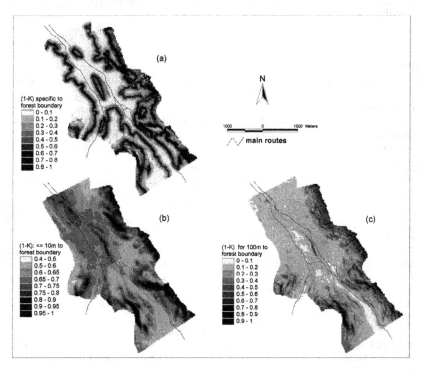

Fig. 5. Predictive uncertainty model for 1-κ as response: (a) site-specific uncertainty, with FORD as predictor, (b) uncertainty predictions with assumed constant values of FORD of 10m, and (c) the same as (b) but for FORD=100m

The calibration of a *predictive* PCC-based model resulted in a slight decrease in model qualities ($D^2=0.35$, *adjusted* $D^2=0.34$) and model accuracies (ρ, G) when compared to the left-out 554 points. Even though the FORD variable was assumed to be of significantly lower predictive power than the FORD(ref) variable, the *predictive* model still shows similar trends and accuracies ($\rho=0.71$, and $G=0.27$). The results of the PCC- *predictive* uncertainty model are presented in Table 3. The following variables contributed significantly to this model: FORD, ELEVD, ELEVCH, SLPCH and STRD.

The κ–based *predictive* uncertainty model revealed a much better model quality $D^2=0.41$, *adjusted* $D^2=0.40$), and the model accuracy is almost comparable to the κ-based *analytical* model. Testing the model using the test data resulted in comparably high ρ and G values ($\rho=0.75$, $G=0.47$, see Table 3 for an overview). The variables with significant contribution in this case were: FORD, ELEVD, SLP and SLPCH.

Table 3. Results of uncertainty modeling; the *analytical* models (M(pcc(ref)) and M(kap(ref))) and the *predictive* models (M(pcc) and M(kap)) were calibrated on 555 points using uncertainty from 100x100m moving windows, and evaluated on the second subset of 554 points

	Analytical model M(pcc(ref))	Predictive model M(pcc)	Analytical model M(kap(ref))	Predictive model M(kap)
Model quality				
D^2	0.41	0.35	0.46	0.41
Adjusted D^2	0.41	0.34	0.46	0.40
AIC	1359.1	1136.2	1595.9	1442.7
Model test				
Spearman's rank correlation ρ	0.76	0.71	0.77	0.76
G-value	0.37	0.27	0.47	0.47
Model parameters				
Constant	-4.884E-01	-6.210E-01	2.590E-01	8.749E+01
FORD(ref)	-3.368E-02	-	-3.881E-02	-
(FORD(ref))^2	-	-	1.977E-05	-
FORD	-	-1.996E-02	-	-3.157E-02
(FORD)^2	-	2.476E-05	-	2.948E-05
ELEVD	-	-3.890E-03	-7.253E-03	-9.375E-03
(ELEVD)^2	-	1.150E-05	-	1.521E-05
ELEVCH	-	9.262E-03	-	-
SLP	-	-	-	2.985E-02
SLPCH	1.694E-02	4.195E-03	1.776E-01	8.452E-02
STRD	3.446E-04	6.244E-04	3.715E-03	-
FORD(ref) : SLPCH	1.539E-03	-	-	-
FORD(ref) : ELEVD	-	-	5.974E-05	-
FORD(ref) : STRD	2.373E-05	-	-	-
FORD : SLPCH	-	8.292E-04	-	-
FORD : ELEVD	-	-	-	3.361E-05
ELEVCH : STRD	-	-	-1.295E-04	-
ELEVCH : ELEVD	-	-	2.925E-04	-
SLPCH : STRD	-	-	-1.974E-04	-

As expected, FORD was associated with the most significant deviance reduction in all cases. ELEVD and STRD were highly correlated, which occasionally resulted in a rather random mutual exclusion of one of the two. The same selection behavior was observed for ELEVCH and SLP in all models, since they naturally reflect similar topographic features. ASPFOC and VIS were excluded from all calibration processes since they did not contribute significantly to deviance reduction.

Figure 5 shows the spatial predictions of κ-based uncertainty models. In Figure 5a, we included FORD as predictor. Thus the map represents the spatial uncertainty inherent to this particular map explained by a range of topographic predictors. It is apparent that certain forest boundary zones show less uncertainty than others. In Figures 5b and 5c, predictions are displayed with assumed constant values of FORD (10m and 100m, respectively). By this, the predictions reveal the nature of the modifying effect of the terrain on the forest boundary as mapped in the 19th century and its associated *inherent* uncertainty.

4 Discussion

Overall, the four models calibrated from the 100m sample raster all resulted in models capable of explaining spatial pattern in uncertainty *inherent* in historical and other field based maps. We noticed that the κ-based uncertainty model proved superior both in quality and accuracy when compared to the PCC-based model. This is in agreement with recent research based on confusion matrix evaluations. PCC overestimates the agreement between maps in not accounting for chance agreement [6]. The Kappa coefficient measures the improvement of the *proportion correctly classified* over mere chance agreement [6]. However, κ fails to do so if one class far exceeds the others [9]. PCC showed higher overall agreement across the maps (mean of PCC=0.77) than did κ (mean of κ =0.62). This reflects the over-optimistic map accuracies, thus resulting in generally lower uncertainties and less well spread values across the scale from 0 to 1, compared to κ. This combination of drawbacks for PCC may have caused the difference between the two models.

The inclusion of VIS and ASPCH as explanatory variables did not provide any valuable contribution. The assumptions behind and the extraction processes need to be further tested. All other variables contributed significantly to the models. ELEV and STRD showed comparably high correlations, which is due to the fact that streets climb along the valley. The inclusion of FORD as predictor constrained the uncertainty to the source of error (i.e., the presence of forest boundary, irrespective of terrain features). Figure 5 demonstrates that SLP and SLPCH had high explanatory power since the predicted uncertainty distribution shows a trend along the slope gradient (compare with Figure 3).

We observed single prediction values that clearly deviated from the observed uncertainty in the evaluation subset. This is mainly caused by the many unknown historical factors which resulted in mapping uncertainty. A surveyor may occasionally have added unexpected detail or neglected easy to observe features. Thus, our model only captures the general trends of uncertainty as influenced by topographic features. Despite the rather general approach, the results provide useful insights into the historical mapping process and represent valuable evaluation support for modern use of historical maps. The choice of using window-based summary statistics proved to reflect the disagreement between both maps better than the pixel-wise global comparison. This was the result of changing the window size to optimize the predictive power of the model. One important reason for this may be that windows of 100m are less affected by (sub-pixel) geo-referencing errors. Additionally, the effects of the terrain upon accuracy seem to only be visible at a minimal area considered.

5 Conclusions

When developing the models, we aimed for simplicity and generality. We are aware that, by using additional variables, the model could have been fitted with higher accuracies within the study area. However, such models would be less easily applicable to other areas covered by the Siegfried map. From testing the model against independent data we conclude that the model is applicable to areas of similar topographic nature. Preliminary map comparisons within non-mountainous regions, such as the Swiss Plateau, indicated much higher accuracies, which is what we expect from the fitted models in mountainous terrain. Still, it remains to be tested, how well the existing models perform in non-mountainous terrain.

Historical maps contain valuable landscape information. However, there is an unknown uncertainty inherent in them. To date, only few attempts were made to analyze this uncertainty [21, 24]. In order to make informed use of such documents for land use or land cover change assessment or modeling, it is mandatory, though, to know the quality and inherent uncertainty of these historical data sources. In this paper we have developed a method to predictively map such inherent uncertainty in space.

This method can be applied to any map developed from field surveys under challenging conditions. The topography-related explanatory model variables showed satisfying predictions when the window size was optimized. The spatial predictions of *inherent* uncertainty can thus be used for the evaluation of historical data within different regions by defining fuzzy membership functions and expressing the "possibility of uncertainty" at any given location. Thus the method we presented is very well suited for incorporation into subsequent applications in a larger context to increase the objectivity of the research. We conclude that GLMs represent a very flexible tool for a range of applications. It remains to be tested, whether additional explanatory variables have the potential to improve such models, or if different uncertainty measures such as the NMI [13] are easier to be modeled spatially.

References

1. Agterberg F.P.: Trend surface analysis. In: G.L. Gaile and C.J. Willmott (eds.): Spatial Statistics and Model. Reide, Dordrecht, The Netherlands (1984) 147–171
2. Austin M.P.: Spatial prediction of species distribution: An interface between ecological theory and statistical modeling. Ecological Modelling 157 (2002) 189-207
3. Brown D.G.: Classification and boundary vagueness in mapping pre-settlement forest types. International Journal of Geographical Information Science 12(2) (1998) 105-129
4. Brunsdon C., Fotheringham S. and Charlton M.: Geographically weighted local statistics applied to binary data. In: M.J. Egenhofer and D.M. Mark (eds.): Geographic Information Science, GIScience 2002, Boulder, LNCS 2478. Springer, New York (2002) 38-51
5. Cissel J.H., Swanson F.J., Weisberg P.J.: Landscape management using historical fire regimes: Blue River, Oregon. Ecological Applications 9 (1999) 1217-1231
6. Congalton R.G.: A review of assessing the accuracy of classifications of remotely sensed data. Remote Sensing of Environment 37 (1991) 35-46
7. Dobson A.J.: An Introduction to Generalized Linear Models. Second Edition, Chapman and Hall/CRC, New York, (2002)
8. Fielding A.H.: How should accuracy be measured? In: A. Fielding (ed.): Mashine Learning Methods for Ecological Applications. Kluwer Academic Publishers (1999) 209-223
9. Fielding A.H. and Bell J.F.: A review of methods for the assessment of prediction errors in conservation presence/absence models. Environmental Conservation 24(1) (1997) 38-49

10. Fisher P.: Models of uncertainty in spatial data. In: Longley P., Goodchild M.F., Maguire D. and Rhind D. (eds.): Geographical Information Systems: Principles, Techniques, Management and Applications (1). Wiley & Sons, New York (1999) 191-205
11. Foody G.M.: Status of land cover classification accuracy assessment. Remote Sensing of Environment 80 (2002) 185-201
12. Foody G.M. and Atkinson P.M. (eds.): Uncertainty in Remote Sensing and GIS. Wiley (2002)
13. Forbes A.D.: Classification algorithm evaluation: five performance measures based on confusion matrices. Journal of Clinical Monitoring 11 (1995) 189-206
14. Guisan A., Edwards T.C. and Hastie T.: Generalized linear and generalized additive models in studies of species distribution: setting the scene. Ecological Modelling 157 (2002) 89-100
15. Guisan A. and Zimmermann N.E.: Predictive habitat distribution models in ecology. Ecological Modelling 135 (2000) 147-186
16. Hagen A.: Fuzzy set approach to assessing similarity of categorical maps. International Journal of Geographical Information Science 17(3) (2003) 235-249
17. Jäger G. and Benz U.: Measures of classification accuracy based on fuzzy similarity. IEEE Transactions on GeoScience and Remote Sensing 38 (2000) 1462-1467
18. Klir G.J. and Wierman M.J.: Uncertainty-Based Information. Springer (1998)
19. Leyk S., Boesch R. and Weibel R.: A conceptual framework for uncertainty investigation in map-based land cover change modelling. Transactions in GIS (2004) forthcoming
20. Lewis H.G. and Brown M.: A generalised confusion matrix for assessing area estimates from remotely sensed data. International Journal of Remote Sensing 22 (2001) 3223-3235
21. Manies K.L., Mladenoff D.J. and Nordheim E.V.: Assessing large-scale surveyor variability in the historic forest data of the original U.S. Public Land Surveys. Canadian Journal of Forest Research 31 (2001) 1719-1730
22. McCullagh P. and Nelder J.A.: Generalized Linear Models. Second Edition, Chapman and Hall, London (1989)
23. Moisen G.G., Edwards T.C.: Use of generalized linear models and digital data in a forest inventory of northern Utah. Journal of Agricultural, Biological and Environmental Statistics 4 (1999) 372-390
24. Plewe B.: The nature of uncertainty in historical geographic information. Transactions in GIS 6(4) (2002) 431-456
25. Steele B.M., Winne J.C. and Redmond R.L.: Estimation and mapping of misclassification probabilities for thematic land cover maps. Remote Sensing of Environment 66 (1998) 192-202
26. Stehman S.V. and Czaplewski R.L.: Design and analysis for thematic map accuracy assessment: fundamental principles. Remote Sensing of Environment 64 (1998) 331-344
27. Van Houwelingen J.C. and Le Cessie S.: Predictive value of statistical models. Statistics in Medicine 9 (1990) 1303–1325
28. Varzi A.C.: Vagueness in geography. Philosophy and Geography 4(1) (2001) 49-65
29. Weisberg S.: Applied Linear Regression. Wiley, New York (1980)
30. Woodcock C.E. and Gopal S.: Fuzzy set theory and thematic maps: accuracy assessment and area estimation. International Journal of Geographical Information Science 14(2) (2000) 153-172
31. Zhang J. and Goodchild M.F.: Uncertainty in Geographical Information. Taylor & Francis, London (2002)

Information Dissemination
in Mobile Ad-Hoc Geosensor Networks

Silvia Nittel[1], Matt Duckham[2], and Lars Kulik[3]

[1] National Center for Geographic Information and Analysis
University of Maine, Orono, ME 04469, USA
nittel@spatial.maine.edu
[2] Department of Geomatics
University of Melbourne, Victoria, 3010, Australia
mduckham@unimelb.edu.au
[3] Department of Computer Science and Software Engineering
University of Melbourne, Victoria, 3010, Australia
lkulik@cs.mu.oz.au

Abstract. This paper addresses the issue of how to disseminate relevant information to mobile agents within a geosensor network. Conventional mobile and location-aware systems are founded on a centralized model of information systems, typified by the client-server model used for most location-based services. However, in this paper we argue that a decentralized approach offers several key advantages over a centralized model, including robustness and scalability. We present an environment for simulating information dissemination strategies in mobile ad-hoc geosensor networks. We propose several strategies for scalable, peer-to-peer information exchange, and evaluate their performance with regard to their ability to distribute relevant information to agents and minimize redundancy.

1 Introduction

Increasing decentralization is a widespread feature of information system architectures, made possible by the advances in computer networks over the past two decades. Decentralized information systems are acknowledged as offering several advantages over centralized architectures, including improved reliability, scalability, and performance [1].

In the context of mobile and location-aware systems, centralization remains the dominant system architecture today. For example, location-based services, which aim to provide more relevant information to mobile users based on information about their geographic location, typically adopt a centralized architecture [2]. In a conventional location-based service, each mobile user accesses information services from remote service providers, which perform the task of capturing, managing, and updating any information relevant to their application domain.

In such a system, the centralized remote service provider can act as a weak point in the system. The bottleneck of a single access point decreases system

M.J. Egenhofer, C. Freksa, and H.J. Miller (Eds.): GIScience 2004, LNCS 3234, pp. 206–222, 2004.
© Springer-Verlag Berlin Heidelberg 2004

reliability and performance. The system is not scalable, because additional load from new users and services must be borne by the service provider. With respect to the domain of geographic information services, the service provider will also face the problem of regular maintenance and update of complex spatiotemporal data, such as environmental conditions or traffic events.

One topic where decentralized architectures are already a vital feature is *geosensor networks*. Advances in sensor technology and the development of inexpensive small-form, general-purpose computing platforms have lead to the study of *sensor networks*. Sensor networks comprise multiple miniature "PCs", each of which contains a CPU, volatile and stable memory, short-range wireless communication, battery power, and attached sensors. The on-board sensors are used to collect information about the physical world, like temperature, humidity, or the current location of objects. Sensor nodes can be deployed in high density within the physical world and enable the continuous measurement of phenomena in unprecedented detail. A geosensor network is defined as a sensor network that monitors phenomena in geographic space [3].

In this paper we apply the paradigm of geosensor networks to the problem of decentralized location-based services. Geosensor networks rely on decentralized architectures, but their sensor nodes are currently rarely mobile and primarily concerned with information capture rather than information service provision. By contrast, location-based services do provide information services to mobile users, but rely on a centralized architecture. The key question facing a decentralized location-based service is how to disseminate relevant geospatial information to spatially dispersed mobile users. The core focus of this paper is to examine this question in more detail.

In Section 2 we cover the related work and the background to the problem, with a particular focus on geosensor networks and mobile ad-hoc networks. In Section 3, we discuss the problem setting and motivation in more detail. Section 4 presents three related strategies for geospatial information dissemination in a geosensor network, and describes a simulation environment developed to enable the performance of different information dissemination strategies to be tested. Section 5 presents the initial results of using the simulation environment. We conclude with suggestions for further research in Section 6.

2   Background

The continuing development of ubiquitous wireless communication technology, including miniaturization of computing platforms and the development of nano-scale sensors, is enabling new computer applications that would not have been possible in the past. Location-based services are one of the most recent, influential application domains; users with hand-held devices employing location-sensing and wireless communication are able to retrieve up-to-date information related to their immediate geographic environment (see [4] for a review of location-sensing techniques). In this section we review the relevant literature within two fields: geosensor networks and mobile ad-hoc networks.

2.1   Geosensor Networks

Recent and projected advances in small, low-cost microelectronic and mechanical systems (MEMS) with limited on-board processing and short-range wireless communication capabilities are also changing the way that we collect and process information about the physical world [5, 6]. Small, inexpensive sensors can be attached to physical objects or embedded in the environment, and sense as well as process the information that is collected. Today, networked sensors can be constructed by using open source and commercial components at the size of an inch or smaller such as the Berkeley/Intel Mica Motes [7, 8]. Other examples of powerful sensor network platforms are the MIT Cricket [9] and the UCLA WINS systems [10].

Large collections of untethered, battery powered sensors with various sensing functions can be distributed over a geographic area, and measure traffic and road conditions, environmental indicators, or seismic activity at a fine-grained temporal and spatial scale that was not possible in the past [11–13]. Since such sensor nodes are tiny and the limited battery capabilities allow only for short range wireless communication, they must communicate with peer sensor nodes in their spatial proximity. Projecting the continued miniaturization, it is not expected that sensor nodes connect to a centralized computing server to upload or stream data directly; they might, however, communicate with a local "base station," i.e. a more powerful sensor node with larger processing, storage, communication and energy capabilities. In general, information is routed via multiple network "hops" to a centralized server and applications [14–16], or the information is processed in the local geographic environment of sensor node.

Integrating both location-aware systems and sensor network technology, we can envision sensor nodes that are aware of their geographic location, equipped with diverse sensors, mobile, and can communicate with nodes in close spatial proximity concerning information that they have collected. Sensor nodes can be mobile by either being self-propelled or by being attached to moving objects, like automobiles, USPS packages, or even humans. As defined above, a geosensor network is a network of sensor nodes that are aware of their geographic location, and capture information about environmental phenomena via on-board sensors [3].

2.2   Mobile Ad-Hoc Geosensor Networks

Efficient information routing is a significant research challenge in geosensor networks, and sensor networks at large [14, 17]. Preserving battery life is the key design criteria for designing communication protocols in order to maximize system lifetime once a sensor network is deployed. Communication is a significantly higher drain on the battery than processing information locally on a sensor node. Furthermore, once nano-sensors are deployed in the environment the battery packs cannot be replaced.

In contrast to information routing in many of today's communication networks, which is address-based (IP addresses), routing in sensor networks is data-centric. The goal is to distribute information only to sensor nodes that need the information or that can be a source of information [16,18]. Another aspect of data-centric routing is scalability: if the number of nodes in a sensor network increases to thousands or millions of nodes, a decentralized, peer-to-peer information dissemination and data collection strategy can provide efficient information distribution and eliminate the bottlenecks of a centralized database or service architecture.

The geosensor networks envisioned in this section can be seen as a type of *mobile ad-hoc network* (MANET, a self-configuring wireless network of autonomous mobile nodes). To emphasize this connection, in the remainder of this paper we refer to these geosensor networks as MAGNETs: *mobile ad-hoc geosensor networks*. Efficient location-based information dissemination strategies are highly relevant for MAGNETs [19]. Geosensor nodes in a MAGNET capture information that is relevant in a geographically constrained context, i.e., in close proximity to the event (e.g., a hazard warning). The location-dependency of information in a MAGNET contrasts strongly with generalized computer networks, such as the Internet, in which the storage location of information may be entirely independent of any locations to which that information refers.

Since sensor nodes often leave the neighborhood of an event, an important research question is how to achieve efficient sharing of relevant information between sensor nodes. This question has already begun to be answered by [20,21]. The information dissemination strategy explored in this paper, where agents communicate with one another whenever they happen to be in close spatiotemporal proximity, is similar to the *opportunistic exchange* described in [21,22].

Traditional MANET information routing strategies with the objective to route information from a specific source to a destination can be categorized into two classes: *table-driven* and *on-demand*. In table-driven routing, each node proactively maintains up-to-date information about the routing paths between every pair of nodes in the MANET. In on-demand routing, a source node reactively generates a route to a destination node when required, based on responses to a query that floods the network. Table-driving routing strategies are generally thought to be more suitable for larger MANETs with high levels of mobility [23, 24] (e.g., MAGNETs) although this result has been called into question by some other studies [25]. Opportunistic exchange is orthogonal to the MANET routing strategy, and may be used in addition to, or independently of, table-driven or on-demand routing. In this paper we assume a "routing-free" model, where information dissemination is purely opportunistic, and geosensor nodes do not need to explicitly query other nodes for information. Instead, the spatiotemporal location of a geosensor node can be thought of as an *implicit query* for any information that concerns locations in close spatiotemporal proximity.

3   Problem Statement

The previous sections have provided a background to geosensor networks and argued that collaboration strategies of geosensor networks represent a new paradigm for location-based services. For example, consider a conventional location-aware navigation system, where a centralized location-based service provider stores and manages real-time information about traffic congestion. Drivers accessing the service might expect to receive continuously updated information from a centralized service provider concerning the locations of traffic jams. In order to provide this service, the service provider would need to address all the problems of scalability, reliability, and performance discussed previously. Computing and communicating relevant customized information for each LBS client will be a significant performance problem for a centralized service with an increasing number of clients.

However, by adopting the geosensor networks paradigm, each vehicle may be thought of as a mobile geosensor node, able to sense information about its own location and local traffic conditions, process this information, and communicate this information to other vehicles in its neighborhood. Potentially, this decentralized location-based service could offer improved reliability and performance, because there exists no centralized service provider acting as a bottleneck to information dissemination and processing. Scalability issues are also positively affected, since rather than increasing the pressure on a single service provider, adding new vehicles to the system increases the number of available nodes for information dissemination and processing. As we shall see in later section, increasing the nodes within a decentralized location-based service can actively *improve* rather than degrade system performance.

In addition to the general advantages of decentralized information system architectures (scalability, reliability, and performance), we can identify at least two further advantages that are peculiar to the domain of location-based services:

1. Many types of geographic information, such as information about traffic conditions and weather, are highly *volatile* in the sense that they can change rapidly, both spatially and temporally. Using a centralized architecture, where all updates must be processed through the service-provider bottleneck, makes it harder for a service to respond rapidly to changes in volatile information.
2. Communicating rapidly changing spatial and temporal information to a centralized service provider may introduce considerable redundancy into the system. Information that refers to a very specific location and time will not be relevant to the vast majority of service users. Therefore, processing such information using the same channels as information with wide spatiotemporal relevance is not an efficient use of limited communication and computational resources.

Continuing the example of the location-aware navigation system, changes in traffic congestion or road conditions will be continually detected by vehicles moving through the environment. If all such updates must be submitted to a centralized service provider, this will further increase the performance bottleneck of the system. For rapidly evolving phenomena, such as traffic congestion or taxi cab passengers looking for a ride, information may become outdated fast, possibly before the centralized service provider is able to make updated information available to other vehicles. Further the information may only be relevant to a small percentage of the overall service users, yet it may require as much centralized communication, processing, and storage resources as information that is vital to many or even all service users.

The example of a decentralized traffic information system used above is not so futuristic. Location-aware sensors are common in modern vehicles, as are environmental sensors able to track environmental conditions (like temperature, humidity, or light conditions). A vehicle may even be able to derive road conditions from its on-board sensors such as windshield wipers, brakes, speed, and so forth [26]. Furthermore, the number of such sensor-enabled devices in the environment is steadily increasing as the technology costs decrease. With hundreds, thousands, or even millions of such sensor-enabled devices in the environment, the boundary between location-based services and geosensor networks is set to blur.

Conversely, from the perspective of geosensor networks, there is increasing interest in sensor node mobility. For example, a key application area for geosensor networks is microclimate monitoring. Hundreds or thousands of sensors distributed throughout a region can continuously sense environmental conditions, for example, temperature and humidity throughout a wine-growing region or for precision agriculture. In the future it may also be possible for such geosensor networks to reconfigure, with sensor nodes moving to the optimal location for responding to particular user queries.

A major issue facing these decentralized, ad-hoc, geosensor networks is the formulation of an efficient *information dissemination strategy*. Different information dissemination strategies vary in the way they are able to address questions such as: (1) If new information becomes available, how long does it take before the network stabilizes? (2) What is the size of the optimal distribution radius for the information origination? (3) How long does the sensed information persist in the network and what parameters can ensure persistence of information?

In the following section we distinguish between three different strategies. *Flooding* is where each geosensor node that encounters an event or receives a message about an event passes on the information to every other node within its communication range. The second approach is referred to as an *epidemic*, in which each node only informs *n* other nodes about events. The third approach is *location-constrained*, in which information is only passed on in proximity to the event, and then discarded.

4   Approach and Simulation Model

In the discussion above, we argued that it will be an advantage for the next generation of location-based services to adopt a decentralized architecture, akin to geosensor networks. In this section we begin to explore the precise nature of efficient information dissemination strategies based on localized communication between agents in a geosensor network. In this context, an agent is defined as autonomous system that is situated within an environment, senses its environment, and acts on its sensed knowledge of its environment [27]. Specifically, we are concerned with:

> mobile location-aware agents, able to sense information about their immediate geospatial environment and communicate with other agents in their neighborhood.

The goal of a MAGNET is to ensure the *efficient* communication of *relevant* information between such agents. In this context, relevance refers to the pertinence of information to the task or tasks in which an agent is engaged [28]. Efficiency (or more precisely inefficiency) can be decomposed into two key features: *ignorance* and *redundancy*. Ignorance concerns the situation where an agent fails to receive relevant information in time. Redundancy concerns the situation where an agent successfully receives irrelevant information. We will return to issues of relevance and efficiency in the following section.

To begin to investigate the potential of MAGNETs, we have developed a prototype simulation testbed, using Java. The reason for favoring a simulation approach, at least initially, is that the wide variety of possible application domains and geosensor network configurations currently precludes a more analytical approach. The prototype is intended to allow researchers to gain insight into the possible effects of manipulating different parameters and their effect on information dissemination strategies in geosensor networks.

The simulation testbed enables the manipulation of six broad classes of parameters, outlined below:

– *Environment*: A variety of different environments can be used in the testbed. The environments can be simulated or derived from information about real geographic environments. Currently, only one- and two-dimensional spatial environments (e.g, graphs and the Cartesian plane, respectively) are supported by the software, although support for three-dimensional dynamic environments would be possible with limited modifications.
– *Communication*: Limited communication capability is a fundamental feature of agents in geosensor networks, which constrains an agent's access to information. Limitations on communication range, frequency, and latency can all be modelled and manipulated within the simulation testbed. Further, different communication protocols can be implemented, discussed in more detail in the example simulation that follows.
– *Sensors*: The variety of sensors available to an agent determines the types of information that can be sensed. Other characteristics that may be modelled that include the sensor reliability, accuracy, and sensor granularity.
– *Mobility*: An agent's mobility characteristics limit the environments and neighborhoods an agent can access, and so the opportunities for sensing and communication. Mobility parameters such as speed of movement, patterns of movement (e.g., goal-directed movement or random walks), and constraints to movement (e.g., by the environment or agent congestion) are available simulation parameters.

- *Tasks*: The task an agent is performing determines whether received or sensed information is relevant to the agent. Further, an agent's task will affect other aspects of an agents behavior, such as patterns of mobility. It can also influence other agents in the system.
- *Agents*: In addition to agent mobility and tasks, discussed above, other characteristics of the agents that may be manipulated using the testbed include agent memory, agent information processing capabilities (e.g., spatial analysis or interpolation), and agent life cycle (e.g., when and how agents enter and exit the simulation). The constraints of limited agent memory have already been explored in [21].

The parameters described above may be varied both spatially and temporally. For example, an agent's speed or patterns of mobility can vary spatiotemporally, e.g., modelling a car caught in city center traffic congestion during rush hour. In the following section we examine some initial results of using the simulation testbed.

5   Example Simulation

The wealth of parameters that can affect a MAGNET leads to a large potential solution space for optimal information dissemination, with parameters that may vary according to the specific application domain. Given limited communication resources (primarily bandwidth and communication range) a vital question facing the agents in a geosensor network is "Under what circumstances should an agent transmit information to another agent in its neighborhood?" We classify three distinct classes of communication strategy in the following way:

1. *Flooding*: an agent always informs all other agents within its communication range of all information about "events" it has collected (either sensed or received from other peers).
2. *Epidemic*: an agent informs only the first n agents it encounters within its communication neighborhood after discovering an event.
3. *Proximity*: an agent informs other peers within its communication neighborhood only as long as the agent is within a certain threshold distance d from the event location.

5.1   Traffic Hazard Warning Simulation

To make our proposed approach more concrete, we consider a traffic hazard warning system based on a MAGNET. We assume road users (mobile agents) are driving around in a road network (environment) in location-aware vehicles equipped with sensors that are able to detect hazards (e.g., the presence of icy road conditions, road work, accidents, or traffic congestion). We also assume vehicles possess short-range peer-to-peer communication capabilities. Although battery power is clearly not an issue for geosensors attached to a significant power source like a vehicle, communication capabilities will in general still be limited, for example, by bandwidth. Consequently, the general form of our problem statement, that agents can only communicate with other agents in their immediate neighborhood, remains valid.

The goal of the geosensor network is then to enable relevant information about road hazards to be efficiently communicated to agents as they travel on their journey. In order to simulate this application we make several simplifying assumptions (which we will begin to relax later). First, we assume a simulation environment comprising a fixed regular grid of connected roads and a fixed single point-location hazard located fairly centrally within the grid. Second, we assume fixed and constant speed of movement as well as communication range for all vehicles. Third, we assume road users are engaged in goal directed movement: users begin and end their journeys at randomly selected locations within the environment, but take the shortest path between those locations. Fourth, the number of agents within the simulation is fixed and constant: as one agent ends its journey and leaves the simulation, another begins its journey and enters the simulation.

Figure 1 shows four of the initial iterations for this simplified simulation. The simulation is discrete, so at each iteration agents move a fixed distance along the shortest path toward their destination. Figure 1 depicts a part of the environment (the road network shown as a gray grid), the location of a simulated hazard (indicated with an exclamation mark), and the locations of agents that know about the hazard (black dots). Agents discover information about the hazard either by sensing it directly or by being informed about the hazard by other agents. The spread of information about the hazard shown in Figure 1 was produced using the flooding communication strategy.

5.2 Simulation Results

In addition qualitative simulation maps (e.g., as shown in Figure 1), we used the simulation testbed to generate quantitative information about the performance of the simulated information dissemination strategies in the geosensor network.

Ignorance. First, we tested the levels of ignorance in the different chosen information dissemination approaches. Figure 2 depicts the levels of ignorance within the system over the first 200 iterations following the initial discovery of the hazard (averaged over 5 independent simulation runs). For this simulation, 200 iterations were sufficient for the system to stabilize. In Figure 2, agent ignorance i is measured as:

$$i = 100. \left(1 - \frac{|E \cap K|}{|E|}\right)$$

where E is the set of agents who encounter the hazard and $E \cap K$ is the set of agents who encounter the hazard knowing about it in advance. Thus, the ignorance measure i varies as a percentage from 0 (total ignorance, no agents

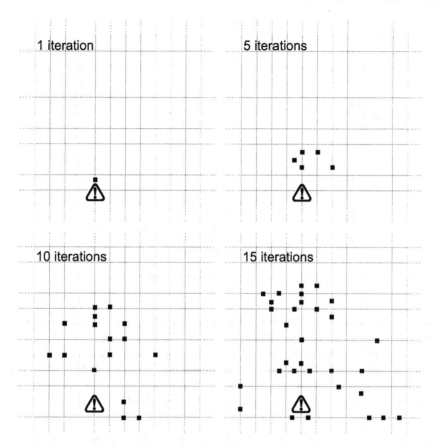

Fig. 1. The spread of information: Location of agents which know about the hazard after 1, 5, 10, and 15 iterations (using the flooding communication strategy for part of the environment)

who encountered the hazard knew about it in advance) to 100 (total knowledge, all agents who encountered the hazard knew about it in advance).

To produce Figure 2 we used a fixed number of 100 agents in the system, 736 road network nodes, and a communication range for an agent of 10 nodes, on average. Although the communication distance was fixed, the exact number of nodes within communication range varies slightly depending on the position of the agent in the network (e.g., fewer accessible nodes near the edge of the road network). Together the number of agents, environment size, and communication range can be combined to yield the probability that at any given point in the simulation an individual agent will be within communication range of at least one other agent (discussed in more detail later).

Figure 2 shows the differences in information dissemination performance across the three different communication strategies. In terms of agent ignorance, the flooding communication strategy clearly offers the best and most robust

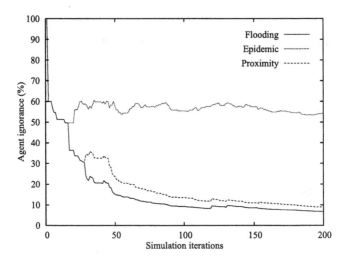

Fig. 2. Agent ignorance measure

performance: more than 90% of agents acquire the knowledge they need by the 200th iteration. The proximity strategy performs almost as well, while the epidemic strategy initially performs well, but reaches a performance ceiling at about 50% agent ignorance, i.e. 50% of the vehicles that encounter the hazard had not received the warning information from other agents ahead of time. As mentioned previously, in all our initial simulations hazards are fixed and static.

Redundancy. As discussed above, redundancy is another performance parameter in geosensor network efficiency, especially with regard to system scalability. In Figure 3, redundancy r is measured as:

$$r = 100.\left(1 - \frac{|E \cap K|}{|K|}\right)$$

where K is the set of agents who know about the hazard and $E \cap K$ is the set of agents who encounter the hazard knowing about it in advance (as above). The figure shows high levels of redundancy across all communication strategies, although the proximity strategy equilibrates at slightly lower levels of redundancy than the other two communication strategies.

Cost of Redundant Messages. Based on the evidence presented so far, the flooding strategy appears to be the primary choice for a dissemination strategy, since it achieves the lowest level of ignorance and a level of redundancy marginally worse than the proximity strategy. These parameters are initial guides, but not the only possible measures of efficiency with regard to geosensor networks. Since a large percentage of agents in geosensor networks operate in a battery- and communication bandwidth-restricted environment, the cost of

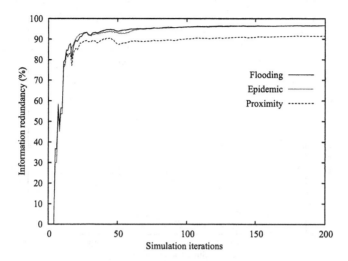

Fig. 3. Information redundancy

sending redundant messages to inform an agent about an event is a significant performance measure for a proposed information dissemination strategy. For instance, it is sufficient that an agent knows about a hazard, and is informed by one other peer (best case). However, some strategies such as flooding produce the effect that an agents "hears" of a hazard over and over again. An alternative measure of redundancy, then, is the *total number* of messages sent by agents using each communication strategy (Figure 4). These results show that the flooding communication strategy produces a significantly larger amount of messages than the proximity strategy, which in turn produces more messages than the epidemic strategy. Thus, robustness is achieved at the cost of a high degree of message passing overhead. After equilibration (following the first 50 iterations) the number of messages increases roughly linearly, suggesting that the rate of message increase could be another useful redundancy index for simulations in a steady-state.

Other Parameter Variations. Another factor that strongly affects efficiency is the degree of peer-to-peer connectivity. The number of agents, the size of the environment, and the communication range of agents can be combined to yield the probability that, at any particular iteration, an arbitrary agent is within communication range of at least one other agent. For simplicity, we refer to this probability in the following text simply as the *probability of communication*, or $P(C)$. The probability of communication is the preferred measure of connectivity, since it provides a measure of peer-to-peer connectivity that is independent of the specific context of the simulation. For a population, a, of 100 agents, an environment size, e, of 736 nodes and a connection range, c, of on average 10 nodes, the probability of communication can be computed as approximately 0.75 using the following formula:

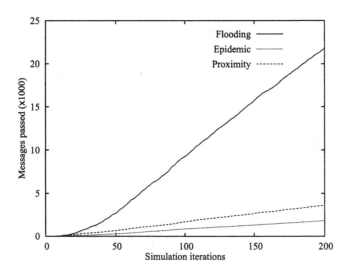

Fig. 4. Messages sent over initial 200 iterations

$$P(C) = 1 - \left(\frac{e - c}{e}\right)^a$$

Within the simulation testbed, connectivity can be investigated by varying the probability of communication and measuring the steady-state levels of agent ignorance in the system. Figure 5 shows the steady-state levels of agent ignorance achieved by the three different communication strategies at different levels of peer-to-peer connectivity. The probability of communication was calculated using the formula above, by varying the number of agents in the simulation, and maintaining a constant environment size and communication range. The agent ignorance level shown in Figure 5 is the average over 100 iterations of 5 simulation runs after each simulation has equilibrated. The error bars in Figure 5 show the variability of one standard deviation for this population of observations.

The figure shows that the proximity strategy is the most sensitive to changes in connectivity, performing almost as well as the flooding strategy at the highest levels of connectivity. The figure also indicates that higher levels of connectivity generally lead to lower variability in ignorance. This in turn may be taken to indicate that higher levels of connectivity lead to more stable geosensor networks, where performance is liable to vary less.

5.3   Summary

Overall, the proximity communication strategy seems to provide a favorable compromise in terms of MAGNET efficiency. The levels of agent ignorance achieved using the proximity strategy are comparable to those of the flooding strategy. At the same time, the proximity strategy does not lead to as high levels of information redundancy as the flooding strategy.

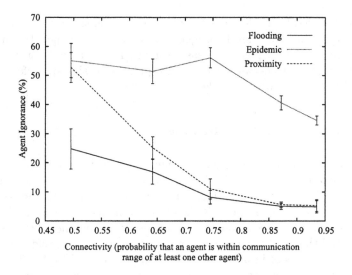

Fig. 5. Connectivity (probability of communication) and ignorance for steady-state simulations

However, it was not the aim of this section simply to suggest that the proximity protocol is necessarily a better choice for geosensor networks within the application domain of traffic networks. Instead, the preceding discussion has indicated how the simulation testbed can be used to begin to identify key strategies and factors affecting MAGNET performance. In turn, we expect that such investigations can help researchers and application domain experts begin to understand the behavior of dynamic geosensor networks. The preliminary results are a first attempt at evaluating different communication strategies, but in order to derive general recommendations further work is needed to ensure the results reflect more realistic environments and agent behavior. The following list summarizes a few related results of further simulations in this particular application domain:

- The proximity communication strategy is, as expected, sensitive to the threshold d, the distance beyond which agents using the proximity strategy will not inform other agents about a hazard. Further investigations indicated that d can be relatively small, in comparison with the environment, and still achieve effective communication.
- Simulations where agents employ a mix of different strategies often achieve high levels of efficiency. In particular, a small proportion of agents employing the flooding communication strategy mixed into a majority of agents using proximity or epidemic strategies can achieve low levels of agent ignorance and redundancy.
- Using a regular grid as a simulation environment is clearly an oversimplification with respect to real transportation networks. Simulations using environments without such uniform transportation network yield different results.

Hazards placed in hard-to-reach regions of the transportation network generally result in higher levels of ignorance and require longer to equilibrate. Conversely, hazards placed in easy-to-reach high traffic-density regions of the transportation network stabilized rapidly with almost perfect information dispersal (extremely low levels of ignorance).

6   Discussion and Conclusions

In this paper, we presented a simulation environment for testing information dissemination strategies in MAGNETs. We proposed and evaluated several strategies for scalable peer-to-peer information exchange (i.e., flooding-based, epidemic, and location-constrained). Strategy performance was measured based on the level of ignorance, redundancy, and degree of redundancy. Our simulation results showed that the proximity communication strategy provides an efficient compromise in terms of information dissemination in MAGNET efficiency. The levels of agent ignorance achieved using the proximity strategy are comparable to those achieved using the flooding strategy. At the same time, the proximity strategy does not lead to as high levels of information redundancy as the flooding strategy.

Several important issues have not been addressed in this paper, and will need to be the subject of further research. We identify below three core areas of related future research:

- Issues of *privacy* have not yet been addressed. The spatiotemporal reference associated with sensed information could be used to infer the location and movements of the agent(s) that sensed that information. Where the agents in our MAGNET are people, the location of those people becomes personal information that should not be widely disseminated. The development of strategies for safeguarding personal privacy in a decentralized geosensor network is a core issue for future research in this area.
- In this research we have assumed that the information generated by a geosensor is precise and accurate. In reality, *uncertainty* is an endemic feature of spatial and spatiotemporal information. The situation is further complicated by the possibility of malicious agents deliberately spreading misinformation throughout the geosensor network. To be practical, MAGNETs must be robust enough to continue to operate in the face of uncertainty arising from whatever source. In particular, the ability to resolve inconsistencies between multiple contradictory items of information from different sources is vital.
- While the decentralized model offers many advantages for mobile and location-aware systems, it may not be suitable for all types of location-based services. For example, safety critical applications may only be able to tolerate zero or minimal levels of ignorance. A decentralized ambulance routing location-based service, for example, could not operate if ambulances received the information they required only 90% of the time. In such an application, the minimal ignorance afforded by a centralized architecture would be vital,

possibly even at the cost of decreased system scalability and performance. Future research will need to address the suitability of the MAGNET architecture to specific application domains, perhaps developing hybrid approaches, where critical information is disseminated using a centralized server, while non-critical information is disseminated using a decentralized model.

Acknowledgments

This work was partially supported by the National Science Foundation under NSF grant number EPS-9983432 and the National Geospatial-Intelligence Agency under NGA grant number NMA201-01-1-2003. The authors would like to thank the anonymous reviewers for their helpful and constructive comments.

References

1. Kurose, J.F., Ross, K.W.: Computer Networking. 3rd edn. Addison Wesley, Reading, MA (2004)
2. Mennecke, B.E., Strader, T.J.: Mobile Commerce: Technology, Theory and Applications. Idea Group Publishing (2003)
3. Nittel, S., Stefanidis, A., Cruz, I., Egenhofer, M., Goldin, D., Howard, A., Labrinidis, A., Madden, S., Voisard, A., Worboys, M.: Report from the first workshop on geo sensor networks. Volume 33. (2004) 141–144
4. Hightower, J., Borriello, G.: Location systems for ubiquitous computing. Computer 34 (2001) 57–66
5. Weiser, M.: The computer of the 21st century. Volume 265. (1991) 94–104
6. Committee on Networked Systems of Embedded Computers, National Research Council: Embedded, Everywhere: A Research Agenda for Networked Systems of Embedded Computers. National Academies Press, Washington, DC (2001)
7. Crossbow Technology Inc.: MICA2DOT specification. http://www.xbow.com (2004) Last accessed: 03 July 2004.
8. Kahn, J., Katz, R., Pister, K.: Next century challenges: Mobile networking for smart dust. In: Proceedings of the 5th International Conference on Mobile Computing and Networking (MOBICOM), Seattle, Washington, USA. (1999) 271–278
9. Priyantha, N.B., Chakraborty, A., Balakrishnan, H.: The cricket location-support system. In: Proceedings of the 6th Annual ACM International Conference on Mobile Computing and Networking (MOBICOM00), Boston, MA. (2000) 32–43
10. University of California Los Angeles: WINS: Wireless integrated network sensors. http://www.janet.ucla.edu/WINS (2004)
11. Mainwaring, A., Polastre, J., Szewczyk, R., Culler, D., Anderson, J.: Wireless sensor networks for habitat monitoring. In: Proceedings of the 1st International Workshop on Wireless Sensor Networks and Applications (WSNA02), Atlanta, GA. (2002) 88–97
12. Ailamaki, A., Faloutos, C., Fischbeck, P.S., Small, M.J., VanBriesen, J.: An environmental sensor network to determine drinking water quality and security. ACM SIGMOD Record 32 (2003) 47–52
13. Imielinski, T., Goel, S.: DataSpace: Querying and monitoring deeply networked collections in physical space. In: Proceedings of the 1st ACM International Workshop on Data Engineering for Wireless and Mobile Access (MobiDE'99), Seattle, Washington, USA. (1999) 44–51

14. Intanagonwiwat, C., Govindan, R., Estrin, D., Heidemann, J., Silva, F.: Directed diffusion for wireless sensor networks. ACM/IEEE Transactions on Networking **11** (2002) 2–16

15. Braginsky, D., Estrin, D.: Rumor routing algorithm for sensor networks. In: Proceedings of the 1st International Workshop on Wireless Sensor Networks and Applications (WSNA02), Atlanta, GA. (2002) 22–31

16. Madden, S., Franklin, M., Hellerstein, J., Hong, W.: TAG: Tiny AGgregate queries in ad-hoc sensor networks. In: Proceedings of the USENIX Symposium on Operating Systems Design and Implementation, Boston, MA. (2002) 131–146

17. Estrin, D., Govindan, R., Heidemann, J.: Next century challenges: Scalable coordination in sensor networks. In: Proceedings of the 5th International Conference on Mobile Computing and Networking (MOBICOM99), Seattle, Washington, USA. (1999) 263–270

18. Krishnamachari, B., Estrin, D., Wicker, S.: Modelling data-centric routing in wireless sensor networks. Computer Engineering Technical Report CENG 02-14, University of Southern California (2002)

19. Schiller, J., Voisard, A.: Information handling in mobile applications: A look beyond classical approaches. In Stefanidis, A., Nittel, S., eds.: GeoSensor Networks. Taylor and Francis (2004) Forthcoming.

20. Wolfson, O., Ouksel, A., Xu, B.: Resource discovery in disconnected mobile ad-hoc networks. In: Proceedings of the International Workshop on Next Generation Geospatial Information, Boston, MA. (2003)

21. Xu, B., Ouksel, A., Wolfson, O.: Opportunistic resource exchange in inter-vehicle ad-hoc networks. In: Proceedings of the IEEE International Conference on Mobile Data Management (MDM'04), Berkeley, CA, USA. (2004) 4–12

22. Wolfson, O., Xu, B., Sistla, A.P.: An economic model for resource exchange in mobile peer to peer networks. In: Proceedings of the 16th International Conference on Scientific and Statistical Database Management (SSDBM'04), Santorini, Greece. (2004) 235–244

23. Broch, J .and Maltz, D.A., Johnson, D.B., Hu, Y.C., Jetcheva, J.: A performance comparison of multi-hop wireless ad hoc network routing protocols. In: Proceedings of the 4th annual ACM/IEEE International Conference on Mobile Computing and Networking (MOBICOM), Dallas, Texas, USA. (1998) 85–97

24. Johansson, P., Larsson, T., Hedman, N., Mielczarek, B., Degermark, M.: Scenario-based performance analysis of routing protocols for mobile ad-hoc networks. In: Proceedings of the 5th annual ACM/IEEE International Conference on Mobile Computing and Networking (MOBICOM). (1999) 195–206

25. Raju, J., Garcia-Luna-Aceves, J.: A comparison of on-demand and table driven routing for ad-hoc wireless networks. In: Proceedings of the IEEE International Conference on Communications (ICC 2000). Volume 3. (2000) 1702–1706

26. Der Spiegel: Sprechende Autos: Zu schlau fuer den Stau. http://www.spiegel.de/auto/werkstatt/0,1518,267313,00.html (2003) Last accessed: 03 July 2004.

27. Franklin, S., Graesser, A.: Is it an agent or just a program? Proceedings of the 3rd International Workshop in Agent Theories, Architectures and Languages (1996) 21–35

28. Sperber, D., Wilson, D.: Relevance: Communication and Cognition. Blackwell Publishers, 2nd Edition (1995)

Public Commons of Geographic Data:
Research and Development Challenges

Harlan Onsrud[1], Gilberto Camara[2], James Campbell[1],
and Narnindi Sharad Chakravarthy[3]

[1] Department of Spatial Information Science and Engineering
5711 Boardman Hall, University of Maine
Orono, ME 04469-5711, USA
{onsrud,campbell}@spatial.maine.edu
[2] Image Processing Division, National Institute for Space Research
Av. Dos Astronautas, 1758 –12227-001
São José dos Campos , SP, Brazil
gilberto@dpi.inpe.br
[3] GlaxoSmithCline
North Carolina, USA
narnindisharad@hotmail.com

Abstract. Across the globe individuals and organizations are creating geographic data work products with little ability to efficiently or effectively make known and share those digital products with others. This article outlines a conceptual model and the accompanying research challenges for providing easy legal and technological mechanisms by which any creator might affirmatively and permanently mark and make accessible a geographic dataset such that the world knows where the dataset came from and that the data is available for use without the law assuming that the user must first acquire permission.

1 Introduction

Geospatial data analysts require as much data as possible about geographic features to make informed judgments about their "meaning" in a particular frame of reference. While automated systems may queue satellite images or sensor data and identify potentially interesting selections for analysts to focus on, the analyst must take those queued images and put attributes to them in order to make sense of them and place their meanings in a larger context.

No matter how elegant an aerial or satellite image might be, it can only show, for example, the physical presence of power lines, not what the attributes of those lines are in terms of age, carrying capacity, interconnection links, where they run underground, or other non-visual data. An aerial photo may show a house but will not show its assessed value, the age of the roof shingles or the number of inhabitants. In short, geographic imagery requires geographic attributes to become fully useful.

How does the analyst quickly find the "on the ground" geographic attribute data corresponding to an area depicted in an aerial or satellite image that enables the ana-

M.J. Egenhofer, C. Freksa, and H.J. Miller (Eds.): GIScience 2004, LNCS 3234, pp. 223–238, 2004.

lyst to complete an assessment? Obtaining access to data appropriate to the question at hand is often difficult, at best. Yet, across the globe, local governments, small companies, non-profit organizations and individuals often generate detailed local geographic information. These parties, however, seldom expend the significant effort required to make that information available to others. It often sits on a local server, unknown outside of the organization and effectively hidden from anyone else.

The goal of the Public Commons of Geographic Data, using open-source and open-access technology, is to remove technical and legal barriers facing the tens of thousands of GIS users (e.g., researchers, local government agencies, nonprofit organizations, field scientists, and individual citizens) that wish to contribute, access, and use locally generated geographic information. This approach has the potential to help free up currently unavailable information generated by non-federal and non-professional sources, and make it available to the widest possible range of potential users. Although not all local governments, private companies, non-profits or individuals will want to provide access to any or all of their geographic data files in a "commons licensing" environment, more people will participate once a user-friendly capability is available. The historical development of the web itself demonstrates that fact.

The "public commons" incorporates both *public domain* and *open access* works. The body of scientific and technical data currently within the *public domain* is significant and the ability of researchers and others to freely use this material has contributed to the economic, social, cultural, and intellectual vibrancy of the entire world [1-5]. Geographic resources in the public domain are comprised of geographic data and information provided by U.S. federal government agencies that cannot, by law, hold copyrights; information that may have once been copyrighted but on which the copyright has expired; information that is not subject to copyright (e.g., facts); and material placed affirmatively in the public domain by its creators that would otherwise have been subject to copyright. Works within the public domain are completely free of any intellectual property restrictions.

Open access works, while still copyrighted, also allow use without obtaining prior permission since a general license is granted ahead of any specific use, provided any attached conditions of use are met. Open-access works typically invoke copyright law and licensing restrictions to help ensure that they remain freely available. Thus, software, data files, and journal articles, for example, distributed under open-source or open-access licenses contribute to the "public commons," but are not by typical legal definition within the "public domain." Examples of such licenses include the General Public License (GPL) (http://www.gnu.org/copyleft/gpl.html), Creative Commons licenses (http://www.creativecommons.org), and the Public Library of Science Open-access License (http://www.publiclibraryofscience.org).

A primary goal of the Public Commons of Geographic Data is to create a broad and continually growing set of freely usable geographic data and information products (i.e., no monetary charges for data use) similar in effect to the public domain datasets and works created by federal agencies. The overarching objective is to provide easy legal and technological mechanisms by which any creator may affirmatively and permanently mark and make accessible a geographic dataset such that the world knows where the dataset came from, and that the data is available for use without the law assuming that the user must first acquire permission.

National governments throughout the world are involved in developing spatial data infrastructures that will better facilitate the availability and access to spatial data for all citizens. A key premise in most of the initiatives is that national governments, including the government of the United States, will be unable to gather and maintain more than a small percentage of the geographic data that users want and need in the digital age. Thus, it will become increasingly important to overcome obstacles and construct ways for non-federal geographic data providers who wish to do so to make their data available to the public.

For researchers, nonprofit organizations, citizen groups, local government bodies, and others who collect and use geographic information, the implementation of a Public Commons of Geographic Data could remove many obstacles they currently face in sharing the geographic data and information they have produced, and in gaining access to the information others have produced. In the remainder of this article we outline implementation objectives for a Public Commons of Geographic Data, discuss research and development challenges in achieving those objectives, and provide some conclusions.

2   Implementation Objectives

Many who generate digital geographic information would be more than willing to make their spatial datasets and information freely available if (1) creating metadata was much easier to do, (2) creators could reliably retain credit and recognition for their contributions to the public commons, (3) creators could acquire substantially increased liability protection from uses by others, (4) creators could reap benefits, such as having their data evaluated by peers and made "visible" and widely available to potential users, and (5) creators could have their data stored in a long-term archive they would not have to maintain. We propose a conceptual model for a public commons that addresses all of these impediments.

The envisioned system, implemented through open access content support and open-source software, should:

- enable simple straightforward construction of contributor-defined open-access licenses using a check-box system suitable for use even by non-professionals,
- enable simple, straightforward construction of machine readable, standards compliant metadata using a menu driven system suitable for use even by non-professionals,
- allow non-removable identity information to be embedded in contributed files,
- track data lineage and improve the ability to find data meeting specific criteria, and
- provide access to a powerful peer-based evaluation system that is simple to use.

From the perspective of a searcher for data, the commons database and search software should be designed to allow a user to (1) locate data for a spatial region and content of interest, (2) avoid or solve data formatting and semantic translation problems, and (3) obtain detailed explanatory information about found data. Ultimately the system should also support users in extraction of subsets of information from files contained in the Commons.

To illustrate how the Public Commons of Geographic Data could interact with a person desiring to contribute data to or search the Commons, a mock-up of an interaction session with the Commons is available at http://www.spatial.maine.edu/ geodatacommons (Figure 1). Steps that users would go through in contributing data to the Commons are also summarized in Table 1.

Fig. 1. Public Commons of Geographic Data Web Mock-Up

In the conceptual model we assume that the typical data contributor, although perhaps an expert in another domain such as epidemiology or ecology, is unlikely to ever become an expert in geographic information technologies or geographic metadata creation. Further, we assume that the typical contributor has gathered digital data from several existing databases or other digital sources, collected some of her own field data, completed an analysis, and now desires to make the resulting digital geographic work product available to others.

3 Research and Development Challenges

Several research and development challenges must be addressed in order to move from concept to effective implementation.

3.1 Intellectual Property

Open-access seeks to clarify the legal status of digitally available works by enabling creators who choose to do so to make an affirmative statement that they are allowing access to and use of their work under only some (or no) conditions of current copyright law without requiring further permission for use on the part of the user.

Table 1. Conceptual Model for a Public Commons of Geographic Data: Operational Characteristics

---

1. A non-expert user creates a GIS dataset or a dataset locatable in space that he or she wants preserved and accessible to the rest of the world.

2. The user accesses a web site that automatically generates an open access license and facilitates the creation of a metadata record in response to a web interview transcript.

2.1 *Open Access License Creation*: In responding to the transcript, the contributor agrees to (1) dedicate the file to the public domain or (2) choose among a limited selection of "open access" license provisions to apply to the dataset. The basic concept of an "open access" license is that any subsequent user may freely use the data file without asking for permission yet the license also can ensure that (1) the originator and all value-adders have a legally enforceable right to credit for their work, (2) liability exposure for the data contributor may be substantially reduced, and (3) the efforts of the originator and value-adders may be protected from capture as the intellectual property of others.

2.2 *Metadata Creation*: The metadata record is created in a semi-automated fashion. The user is walked through a series of questions with limited choice responses. Portions of the transcript form are automatically filled in based on previous responses provided through the user registration and license creation processes. Other portions of the transcript automatically change depending on responses to initial questions. That is, the system guides the user by asking the data contributor to select among a fixed set of definitions for some of the terms the user selects. Those definitions, along with ontologies appropriate to the primary theme designated for the data file by the contributor, are then used to predict and simplify subsequent metadata selection choices.

This is a very different approach from current metadata approaches that have been designed for flexible use by specialists. Non-expert users probably will never take a course in how to create metadata for geographic data files, nor are they likely to have familiarity with many technical geographic terms. Therefore, open-ended questions with free-form responses need to be minimized. Yet the system also needs to be responsive to a variety of disciplines using the language and classification schemes of those disciplines.

3. The transcript responses and the actual data file to which the responses apply are submitted to an automated processing facility. An encrypted identifier is automatically embedded in the geographic data file. The identifier does not interfere with the file nor is it stripped from the file through standard GIS operations. Through the availability of freely downloadable client software, any user may readily determine the status of legal rights and metadata for any standard format geographic data file that the user has in her possession. This approach varies from the current commercial approaches in which metadata is maintained separate from the data file. If properly designed, the originator and the string of value-adders to a dataset would always be known when a file is processed in this manner. Appearance of the identifier information would provide legal evidence that a user is allowed to use the file in accordance with the license provisions without impinging on intellectual property rights.

4. The system would return a copy of the "marked" geographic data file back to the originator incorporating the embedded metadata information. In an optimal system, all files so processed also would be permanently and publicly archived. Whether maintained on the open web or maintained in a long-term commons electronic archive, anyone would be able to search for, access, and legally download and use any such datasets.

Today, there are many gray areas in what a user may or may not do with copyrighted works in the digital domain. Current U.S. Copyright law does not make it possible to copyright facts or even obvious arrangements of facts such as an alphabetical listing in a telephone directory (Feist Publications, Inc. v. Rural Tel. Service Co., 499 U.S. 340 (1991)). However, the threshold for "originality" that would make arrangements of facts (such as raw geographic data) copyrightable is very low. Thus the typical user must assume that an interest of another may exist in the vast majority of geographic data compilations openly found on the web or elsewhere, and the law thus presumes that permission must be acquired even if the compilations are mostly factual. This presumption will become even stronger if, as many are predicting, the U.S. moves closer to the database protection regimes enacted in the European Union and advocated by the World Intellectual Property Organization [6]. Providing a way for local geographic information originators to affirmatively state that their work is open-access will eliminate any present or future doubts as to its status in the eyes of potential users.

The geographic data commons concept extends from the open-source licensing model. Currently, the law assumes that geographic data creators have all proprietary rights (e.g., copyright) in the datasets they produce. A common alternative to this approach is to place the data in the public domain with no rights reserved. Emerging open access licensing approaches, derived from the open-source licensing model, provide a middle ground that allows access and use of data for wide-ranging productive purposes but with "some rights reserved."

The most prevalent current open access license approach, which is the one we advocate for use with the geographic data commons, is that developed by the Creative Commons project (http://www.creativecommons.org). Through this middle ground approach, data producers are able to specify whether future users must provide attribution, are allowed to modify the data, are allowed to modify the data as long as the users apply the same license to any derivative works, or may not use the data for commercial purposes. With the exception of these possible constraints, users are granted affirmative permission to use data drawn from the commons.

An interesting and perhaps critical aspect of the open access license model is that data and product producers have the option, if they so choose, of charging for the service of transferring their work to others and charging for support services. That is, many parties are generating substantial revenues by making their works available through open access and open source licensing approaches. This licensing approach is viewed as supporting a relatively new mode of economic production where individual contributors are organized neither in response to price signals nor by explicit firm managers [7]. However, Adam Smith's more traditional notion of "enlightened self-interest" aptly describes the motivations of many businesses and individuals contributing to open source and open access efforts. Thus such licensing models may be viewed as supporting basic free-market principles whereby claims of property right are used to distribute work efforts in furtherance of competition, creativity and enterprise. For certain products and parties this new form of production works well and even better than price signal or hierarchical management arrangements. In other instances, traditional means of marketplace production are likely to remain more efficient.

One core area for research investigation is whether the current semi-automated licensing options used by the Creative Commons project might be improved to be more responsive to the needs of the scientific and technical community. Further, what are the conditions and limits under which open access economic models for supplying geographic data succeed or fail relative to competing models? Answering this second research question will require first the development of an operational Public Commons of Geographic Data.

3.2 Metadata Generation

One of the major barriers for non-specialists who wish to offer their data to a larger audience is the generation of metadata. There are currently many competing metadata standards in use [8] and even professionals have difficulty staying current with them. In addition, using any current metadata system, for example the ISO 19115 Metadata Standard (ultimately ISO 19139 in XML), requires systematic study and practice. Non-professionals in geographic information, no matter how competent in their own areas of expertise, hesitate to wrestle with any of the current geographic metadata systems, and even many GIS professionals find metadata generation burdensome and do as little as possible. Currently, in fact, metadata fields typically are minimally populated, and there is a lack of depth in the meanings of the data submitted.

The ISO 19115 standard will soon replace the Federal Geographic Data Committee Metadata Content Standards [9], one of the most widely used metadata systems. Both the FGDC and ISO standards are geared toward professionals. Historically the FGDC system has been not fully utilized by local professionals and has almost never been used by non-professionals.

The Public Commons of Geographic Data speaks directly to this problem by creating a minimal metadata set and options for extending it to the full ISO 19115 standard. The Dublin Core specifies 15 elements that must be included to conform to its standard [10]. Most of these also map to a subset of the current FGDC standard and the new ISO 19115 standard (see http://www.spatial.maine.edu/geodatacommons/metadatadublin.html). Using the Dublin Core elements as a minimal set for metadata generation, it would be possible to provide sufficient metadata for geographic information to make it accessible to today's search engines as well as to the semantic web search tools of the future which will be based on XML, and which will recognize and parse Dublin Core elements.

With this in mind, the Public Commons of Geographic Data should incorporate ways to generate ISO compliant metadata that meets Dublin Core standards. This should be accomplished using pull down menus and other user-interactive "choose-one" techniques that will make it reasonably simple for non-professionals in geographic information to generate usable metadata without taking a course in how to do so. Professionals may also choose to utilize the proposed system in generating metadata for ease of use in populating all of the ISO 19115 fields.

As users of the system select terms and affiliated definitions appropriate to their data from pull-down menus, the Public Commons of Geographic Data system will need to be able to provide subsequent branching menus based on the "meaning" of the previous user input. We hypothesize that selected responses in a pull-down menu by a respondent may be used along with formal specifications for potential domains of interest (i.e., ontologies) to predict and simplify metadata choices. That is, existing ontologies may be affiliated with each ISO 19115 data topic category (e.g., MD_DataIdentification.topicCategory). Provision of menu choices that change based on earlier choices should speed up metadata creation for infrequent contributors and make the typical completion of metadata much more comprehensive. If thousands of users make pull-down menu choices according to an initial ontology (e.g., for "transportation"), we hypothesize that those responses may be used to automatically develop an improved ontology that reflects the primary understanding and usage of the community, as opposed to reflecting the logic of classification specialists [11]. This adaptive ontology then may be used to continuously optimize the system for future metadata submissions by the community.

To make the commons of greatest use across a variety of domains, occasional users should be encouraged, but not required, to complete more comprehensive metadata fields corresponding to the data topic categories they have selected. For instance, if contributors selected "biota" under the ISO19115 data topic category, they might be led to complete the remaining ISO19115 elements using a broad vegetation metadata profile (e.g., FGDC-STD-005). However, if they further selected under "biota" a subcategory such as "wetlands" they might be led to a different selection of pull-down menu options based on the terms and classifications used in developing metadata, for example, by the U.S. National Wetlands Inventory. Similarly, if they had chosen under "biota" a further subcategory of "flora" they might be led to complete the remaining ISO19115 elements using the classifications established by the "Darwin Core" element set. The goal is to federate the system across disciplines, and thus be responsive to the widest range of potential contributors of geographic data files. The additional depth of documentation and meaning included in the metadata should then contain sufficient text to allow inferences to be made in future semantic web environments. To be effective, the system must be designed in such a way that each additional request for information should extend for no more than a page, and take only a few minutes for the typical user to complete.

In order for a pull-down menu system to work, there will need to be a powerful dictionary underpinning it. At present, there is no standard dictionary for geographic information suitable for this use, although there are efforts underway that may be adaptable. In Scotland, the Association for Geographic Information [12] in collaboration with the University of Edinburgh maintains an online GIS Dictionary. Several commercial providers, such as ESRI, also offer dictionaries. The Alexandria Digital Libraries [13] project has developed a Thesaurus with 210 preferred terms and 946 non-preferred terms (non-preferred terms refer the user back to the preferred term, for instance, "ditch" refers the user to "canal"). Use of controlled vocabularies, such as WordNet (http://www.cogsci.princeton.edu/~wn/), may have application when terms are specified by contributors that are not contained in the standard geographic dictionary. All of these initiatives offer elements that may be able to be adapted for a commons environment, and the limited scope of the ADL Thesaurus indicates that the scope of this task is manageable. Today's methods for finding and using information on the web are often insufficient. Yet, if semantic web methods are to be able to draw inferences from text, such as the text in metadata, that metadata must exist in the first place, and be at a level of detail far greater than is currently being provided. Further, for the "Spatial Semantic Web" to reach its full potential, automated searches must be able to reach and explore actual geographic datasets [14-16] as well as their metadata. Without full access to the dataset, data semantics cannot be used to find and assess the suitability of a geographic data file for an explicit need. Additionally, searches that rely on data similarity assessments require access to the data rather than just metadata.

A further concern is that metadata entry must be extremely efficient for the occasional contributor of datasets. For example, the typical user of local level geographic data does not know the bounding latitude and longitude coordinates of the dataset they are using. Many local geographic datasets throughout the world are not tied to universal coordinate systems. Therefore, an efficient tool should be supplied to provide the approximate bounding coordinates for the data file. For metadata and search purposes, the bounding coordinates do not need to be precise. The user should be able to type in the name of the location of concern, be presented with an image centered on the coordinates of the place, and then zoom in or out on a high resolution image to allow a box to encompass the area of concern as precisely as possible. The bounding coordinates of the drawn box should then be used to automatically populate the required metadata fields for coordinates. This online process must take less than a minute to avoid frustrating contributors. The National Map Viewer (http://nmviewogc.cr. usgs.gov/viewer.htm) and the Alexandria Digital Library Gazetteer (http://fat-albert.Alexandria.ucsb.edu:8827/gazetteer) partially illustrate this capability.

Research questions include: (1) Will responding to the Dublin core set of elements as the minimum set take too much time to elicit widespread responses from the broad user community? Would this element set or fewer elements provide insufficient information for effective future searches? (2) We hypothesize that provision of menu choices that change based on earlier choices should speed up metadata creation for infrequent contributors and make the typical metadata entry far more comprehensive. Will this prove true in practice? (3) We hypothesize that initial ontologies for specific domains or data themes may be automatically revised through thousands of submissions to reflect the primary understanding and usage of the community, and that this adaptive ontology might then be used to continuously optimize the system for future metadata submissions by the community. What specific approaches might be used to promote enhanced efficiency for individual users, users reporting metadata within a specific domain, and for all users on average?

In summary, once users are familiar with the commons interface, it should take them only a few minutes to create a license, complete an accurate and sufficient metadata script, and submit their geographic data file. An initial user interface mock-up is available at http://www.spatial.maine.edu/geodatacommons.

3.3 Tracking Data File Lineage

In a commons environment, tracking of license provisions makes more sense than controlling access by the methods of more traditional Intellectual Property Rights Management Systems [17-19]. A unique encrypted identifier would be embedded in each submitted file, but should not interfere with subsequent use of the file nor should the identifier be stripped from the file through standard GIS operations. Current commercial GIS software does not provide this capability. The goal is to discourage license breakers, but not ban them. Getting credit "most of the time" is probably sufficient for most contributors to the commons. There is little incentive for those downloading to strip unobtrusive IDs, even if software becomes available to do so, since users may use the files for free anyway and license infringers may still be identified if contributors additionally use the more traditional methods of embedding false objects or watermarks in their files.

A range of methods have already been developed for embedding encrypted IDs in the most commonly used file formats, including raster files [20]. At least one vector steganography approach shows great promise as well [21]. To make the tracking system operational, open-source software should be developed to embed identifiers in all of the primary formats of files likely to be delivered to the commons (see http://www.spatial.maine.edu/geodatacommons/toload.html for a sample listing). Using known techniques, it is typically possible to embed numerous copies of an identifier throughout a single geographic data file so that even if only a small portion of a large file is extracted for use, that small part will continue to carry the ID in most instances. Thus, such methods would be used to automatically generate ID's, encrypt them, and embed them in any file delivered to the commons.

Software would also need to be developed for identifying data files that have been processed through the Public Commons of Geographic Data. If a hidden commons identifier is detected in a file on a person's desktop through use of the free software, the core license provisions are exposed and a link is provided to the complete metadata file and license in the archive.

Similarly, when a file is uploaded by a contributor to the central server, the system checks to see if there is one or more hidden identifiers in the submitted file. If found, this means the submitted file is a derivative of other files previously processed by the system. Metadata fields would be populated automatically for the new file showing that it is derived from those other files and direct links will be provided to the parent files. In this manner, any file may be traced back in time through the successive generations of other files that were used to construct it. This capability also should allow the automatic enforcement of certain license provisions, such as the "share alike" or "copyleft" provision, through successive generations of use.

A frequently suggested alternative to the identifier tracking system just described is the technique of "hashing."[22] Hashing transforms a string of characters into a shorter fixed length value or key that represents the original string. While most frequently used as a technique for increasing the efficiency of recalling information from a database or in implementing encryption processes, hashing may also be used as a check on similarities among files. Hashing is insufficient for the system envisioned since, for the most part, all hashing can indicate is that a file is likely to have a derivative or ancestry relationship to another file but cannot adequately resolve which of the files came first in time. Hashing might be used, however, to make the file lineage system more robust. In the light of the large number of standard file formats for geographic data and the range of useful steganographic approaches available, the optimal technical means for file lineage tracking is still an open question.

3.4 Archiving

Archiving ensures a backup for commons licensed data files and is a major benefit for contributors since contributors will always be able to find and copy their previously submitted files from the long-term archive. Data files for the commons would otherwise be distributed among thousands of machines that inevitably are subject to broken links and lost data over time. Similar to CiteSeer (http://www.citeseer.com), we envision that the system should generate and make accessible several standard and interchange formats of each submitted file, all containing the hidden ID, so that future users will not need to accomplish such conversions. Providing several standard formats for a file also lessens the likelihood of loss or obsolescence of datasets over time as the popularity of some data formats wane.

The Creative Commons web site already has a reference repository in place in which creators may list their work that is available under a Creative Commons license. Sites such as Geospatial One-Stop (http://www.geo-one-stop.gov) and GI-gateway (http://www.askgiraffe.org.uk/metadata) provide repositories and search capabilities for metadata for geographic data files. The Public Commons of Geographic Data system should automatically submit references of all files contributed to it to the Creative Commons repository and provide a link back to the Public Commons of Geographic Data metadata archive. Similarly the commons should automatically forward its metadata files, if allowed, to other major geographic metadata file depositories. Thus, in addition to accessing directly the Public Commons of Geographic Data metadata and geographic data repositories, potential users of open-access geographic information could have several other entry points for finding datasets available in the commons.

3.5 Data Storage and Search Optimization

Primary contributors to the commons are envisioned as individuals and organizations that produce geographic datasets only occasionally or for limited areas. As such, storage capacity limitations and file transfer times across networks should be less problematic than faced by many current data centers handling very large files. It is assumed that organizations managing large databases would continue to manage and archive them in-house. Further, while organizations like NASA have a strong need for capabilities that would allow extraction and transfer of portions of large files, such a need would be less critical for a commons environment. If such capabilities eventually emerge, they should be incorporated in the commons system environment but are unlikely to be critical to its success.

Assuming that an efficient location specification tool can be supported for contributors in creating their metadata, this opens the possibility of organizing and storing both geographic metadata and geographic datasets by their spatial extents on one or numerous servers. For example, Figure 2 illustrates the current decentralized and distributed FGDC Clearinghouse Metadata Model Approach. Under this configuration metadata are not stored by location and, therefore, a typical query must search all nodes on the network, which may number in the thousands, in order to be comprehensive. That is, a server in Sao Paulo, Brazil might contain metadata and associated geographic data files pertaining to a location in Maine, and vice versa. Figure 3 suggests an alternative arrangement for storing metadata gathered through the Public Commons of Geographic Data system. Depending on the bounding coordinates provided in the metadata and the size of geographic area encompassed by those coordinates, metadata could be automatically categorized and stored by geographic location (e.g., assume organization through equal size grid cells covering the globe), extent of coverage (i.e., this would result in approximate groupings of files containing images or maps of similar scales), and by primary topic (i.e., theme) of the file. Searches germane to any specific bounded region might be made much more efficient through this arrangement.

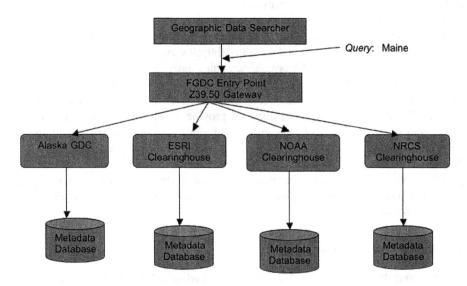

Fig. 2. FGDC Clearinghouse Metadata Model Approach

A further alternative might be to maintain distributed servers similar to the current FGDC Clearinghouse node arrangement, but provide a comprehensive centralized metadata server capability that mines metadata regularly and efficiently from all other metadata nodes and mirrors back the comprehensive collection to selected distributed full metadata sites around the globe. Determining which distributed architecture would be most efficient for serving metadata as well as the actual geographic data files with embedded IDs is a significant research question.

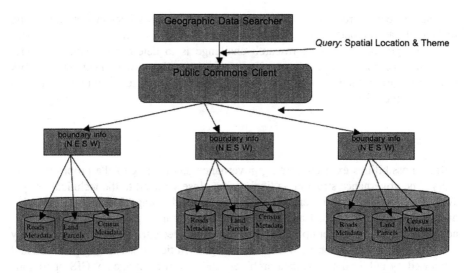

Fig. 3. Potential Public Commons Metadata Model

3.6 Peer Review and Evaluation

Metadata reported by local geographic data originators must, of course, be "truthful." Otherwise, the concept of gaining access to their data through metadata becomes dysfunctional or useless. The data, too, must be suitable for a user's purposes.

The same problem faces a number of web services today that aggregate original information from many sources, and a variety of procedures for evaluating submitted information have been developed. These range from statistical methods [23], to pledges of "neutrality" in contributing information [24], to review by founders [25], to post-publication peer review systems [7]. One promising peer-review method for assessing the reliability of reported data and for ferreting out inaccurate metadata reporting, whether purposeful or otherwise, is to use peer review methods similar to those developed by the Open-source Development Network which operates the web site www.slashdot.org. In this model, rather than hand pick or financially support editors or other "quality control evaluators," everyone in the entire community of data users becomes a potential evaluator in the quality control effort. This general methodology for quality control has worked well in online endeavors with users who are literate in the subject matter. The approach has promise as a good starting point for development of a Public Commons of Geographic Data system peer evaluation mechanism.

A further quality control issue relates to responsibility. Registration should be required for contributors in order to identify those purporting to have ownership or authority to place a specific geographic data file into the commons licensing environment. In the event of a conflict over rights in a specific file, the dispute would be primarily among claimants but administrators of the system would need to be able to be responsive to requests to remove a file until the dispute is resolved. The semi-automated license creation process includes a liability disclaimer clause and should sufficiently accommodate the liability exposure concerns of most data contributors. Thus, the primary remaining research challenge is to determine which of several alternative methods for assessing quality of data in on-line environments would work best for ensuring quality control of submissions contributed to the Public Commons of Geographic Data.

3.7 Governance

Several possibilities exist concerning governance and hosting of the proposed capability. The governance structure is likely to be closely linked to the technical design. With a design focused on centralized processing and storage facilities, the primary operations might be funded and administered by a single government agency or a non-profit organization set up for the purpose. More decentralized designs might rely more heavily on the server, network, and storage facilities already being provided and supported by public libraries, data archives, and government agency GIS operations spread across the globe. There is also the possibility that a parallel global marketplace in geographic data and services offering similar licensing, metadata, and tagging capabilities could be developed and tapped to support the ongoing operations and content expansion of the commons. The means for governing, supporting and expanding the public commons in geographic data is itself a source of numerous research questions.

4 Summary

One key fact emerges throughout all of the open-source and open-access work going on at present: in the early stages, projects have to be initiated, nurtured, and managed by a central team until a project is ready for release to the open-source/open-access community where it can effectively take on a life of its own [26, 27]. Software application models abound, including highly complex discipline specific endeavors that parallel the level of functionality of the proposed Public Commons of Geographic Data system. One such example is the Koha library automation system [28], and a number of others exist. The key to their success and, we believe, the key to the proposed Public Commons of Geographic Data system's success, is building the system to the point at which others can effectively expand upon a working model.

The intent of the Public Commons of Geographic Data is to be responsive to the tens of thousands of individuals currently creating geographic data on their desktops but who have little incentive or ability to effectively share their datasets. These individuals may want to create metadata and contribute files to a common pool only occasionally. As such, once they are familiar with the interface it should take contributors only a few minutes to create a license, complete an accurate and sufficient metadata script, and submit their geographic data file.

The significance of the conceptual model and its practical implementation is that ultimately it will provide a means to make visible a substantial body of geographic information that now exists but is effectively hidden from the view of geographic scientists and researchers, researchers in a wide range of other fields, agency analysts, nonprofit organizations, and private citizens. Looking forward, the Public Commons of Geographic Data will provide a vehicle for those who generate detailed local-level information in the future to provide access to the information they generate in a simple, "business as usual" way. The goal is to enable the sharing of locally generated geographic information to become the norm, rather than the exception that it is today, and to expand the amount and quality of locally generated data available for analysis and public use.

The broad applicability of the model is also noteworthy. That is, the word "geographic" might be appropriately dropped from the title. The concepts described may be used for the documentation, tracking and archiving of most scientific and technical datasets. Therefore, a better title for the envisioned system might be the "Public Commons of Data."

In sum, no complete analog to the Public Commons of Geographic Data system exists at present. Some proto-elements do exist. The development of the envisioned capability will require combining original research contributions (e.g., needed infrastructure specifics, simple metadata generation and ontologies, identifier embedding, etc.) with adapted processes or initiatives already in place (e.g., Creative Commons) to create seamless access to a multitude of independently developed, heterogeneous geographic data sources open to and usable by any interested citizen.

Acknowledgments

Research leading to this publication was supported by NURI grants NMA 202-99-BAA-02 and NMA 201-00-1-2003 with funds supplied by the U.S. Federal Geographic Data Committee.

References

1. National Research Council, Committee for a Study on Promoting Access to Scientific and Technical Data for the Public Interest 1999, *A Question of Balance: Private Rights and the Public Interest in Scientific and Technical Databases.* Washington, D.C. National Academy Press
2. Commission of the European Communities. 1999, *Public sector information: A key resource for Europe.* (COM(98)585 European Commission, Brussels) <http://europa.eu.int/ISPO/docs/policy/docs/COM(98)585/
3. Pira International Ltd., University of East Anglia and KnowledgeView Ltd. 2000, *Commercial exploitation of Europe's public sector information: Final report for the European Commission Directorate General for the Information Society.* Surbiton, Surrey, UK
4. Weiss, P. 2002, *Borders in Cyberspace: Conflicting Public Sector Information Policies and Their Economic Impact,* U.S. National Weather Service, http://www.weather.gov/sp/Borders_report.pdf

5. Moglen, E. 2003, Freeing the Mind: Free Software and the Death of Proprietary Culture. In *Fourth Annual Technology and Law Conference*. Portland, ME: UMaine Law School, oral recording

6. Maurer, S., P.B. Hugenholtz, and H. Onsrud. 2001, Europe's Database Experiment. *Science*, 294: 789-790, 26 Oct 2001

7. Benkler, Y. 2002, Coase's Penguin, or, Linux and The Nature of the Firm. *Yale Law Journal* 112 (3): 369-446.

8. Hill, L., G. Janee, R. Doulin, J. Frew, M. Larsgaard. 1999, Collection Metadata Solutions for Digital Library Applications. *Journal of the American Society for Information Science* 50 (13):1169-1181

9. Federal Geographic Data Committee (FGDC), 1999, Metadata Content Standards, http://www.fgdc.gov/metadata/contstan.html

10. Dublin Core Metadata Initiative (DCMI), 2003, http://www.dublincore.org/documents/dces

11. Smith, B., and D. M. Mark. Geographical Categories: An Ontological Investigation. *International Journal of Geographical Information Science* 15 (2001): 591–612.

12. Association for Geographic Information (AGI), 2003, http://www.geo.ed.ac.uk/agidict/welcome.html

13. Alexandria Digital Library Project (ADL), 2003, http://www.alexandria.ucsb.edu

14. Egenhofer, M. 2002, Toward the Semantic Geospatial Web, *ACM GIS 2002*, November 8-9, 2002, McLean, VA, 1-4.

15. Rodriguez, M. A., and M. Egenhofer, 2003, Determining Semantic Similarity Among Entity Classes from Different Ontologies. *IEEE Transactions on Knowledge and Data Engineering* 15: 442–56.

16. Bateman, J., and S. Farrar. Spatial Ontology Baseline [Preliminary]. 18 May 2004. [cited 30 Jun 2004].
    Available from http://www.sfbtr8.uni-bremen.de/project.html?project=I1

17. IBM Electronic Media Management System (EMMS), http://www-306.ibm.com/software/data/emms/features/

18. Microsoft Windows Rights Management Services (RMS) for Windows Server 2003, ttp://www.microsoft.com/windowsserver2003/technologies/rightsmgmt/default.msx

19. Freeman, N., T. Boston and A. Chapman, 1998, Integrating Internal, Intranet and Internet Access to Spatial Datasets via ERIN's Environmental Data Directory, *The 26ᵗʰ Annual Conference of AURISA*, November 23-27, 1998, Perth, Australia, np.

20. Sharad, C. N., 2003, *Public Commons for Geospatial Data: A Conceptual Model, MS Thesis*, Department of Spatial Information Science and Engineering, University of Maine

21. Huber, W., 2002, Vector Steganography: A Practical Introduction, *Directions Magzine*, http://www.directionsmag.com/article.php?article_id=195

22. Knuth, D. E., 1998, *The Art of Computer Programming, Volume 3 Searching and Sorting*, Addison Wesley Publishing, pp. 513-558

23. NASA Clickworkers Project, available at http://clickworkers.arc.nasa.gov/top

24. Wikipedia, the Free Encyclopedia, available at http://www.wikipedia.org

25. Rivlan, G., 2003, Leader of the Free World. *Wired*, November, 2003, pp. 152-154

26. Raymond, E., 2000. *Homesteading the Noosphere*. 8/00 [cited 12/26/2003], ttp://www.catb.org/~esr/writings/homesteading/homesteading/

27. Raymond, E., 2000. *The Cathedral and the Bazaar*. 9/00 [cited 12/31/2003], http://www.catb.org/~esr/writings/cathedral-bazaar/cathedral-bazaar/

28. Koha Open Source Library System, available at http://www.koha.org

Alternative Buffer Formation

Gary M. Pereira

Department of Geography, San José State University
San José, CA, USA
gpereir1@email.sjsu.edu

Abstract. Proximity buffers are used in many spatial applications in research and management. Nevertheless, they are limited in their representational validity. Since only the nearest points on the edge of an entity are used in calculating the buffer boundary, the various meanings of the 'influence' of an entity on its surrounding environment are not well-estimated. An alternative implementation of a more generalized class of buffers is described here. This method considers the contribution that the internal spatial geometry and attribute values of entities have on the buffered environment. It also considers the cumulative influence of multiple entities. The method is easily implemented by using a class of integrative spatial filters. For many applications, it is likely to yield results that are more meaningful than those obtained through proximity buffers.

1 The Limited Usefulness of Proximity Buffers

In many domains, the influence of one point on another is found to be some function of the distance between these two points (distance decay function). In Geographic Information Science, this concept is used to model the influence of polygonal forms and linear features on their environments by a calculation of influence equivalent to that used for points. The influence of a geographic entity of any size and shape on some point in its external environment is found by taking some function of the Euclidean distance from the external point (p) to the single nearest point (q) on the edge of the entity (Q). For any given external point, only one point associated with that entity is usually relevant with regard to proximity. The influence of entity Q on the external point p is a distance decay function of the Euclidean distance:

$$D(p, Q) = f(\|p - q\|) \cdot \qquad (1)$$

where q is the point on the boundary of entity Q yielding the minimum value for the Euclidean distance

$$\|p - q\| = \left((p_x - q_x)^2 + (p_y - q_y)^2 \right)^{0.5} . \qquad (2)$$

in a two-dimensional space indexed by x and y.

The locus of external points at some constant proximity to the continuous edge of an entity forms an isoline, and the area between the edge of the entity and that isoline forms a simple buffer around the entity. Defined as such, proximity buffers as indicators of spatial influence have been universally implemented in Geographic Information Systems, and they form the basis of many operations in cartographic mapping

M.J. Egenhofer, C. Freksa, and H.J. Miller (Eds.): GIScience 2004, LNCS 3234, pp. 239–250, 2004.

and geographic decision-making. The point-and-click ease with which such operations can be applied to spatial data facilitates their wide use, and they do provide useful information in many applications, particularly where the basis of influence is not well quantified. A short list of recent applications include management of riparian environments [1-3]; woodland restoration [4]; public policy [5]; estimations of cholera exposure [6]; health effects of power line radiation [7],[8]; exposure to toxic or industrial hazards [9-13]; and exposure to vehicular traffic [14]. Phenomena represented range from purely physical variables like electrical field strength [8] to purely social variables like public opposition to siting decisions [5].

It is often assumed that proximity or some function of proximity in this narrowly defined sense adequately quantifies the spatial component of real or potential influences that an entity has on its external environment or on other entities in that environment. However, a number of considerations lead us to question the general utility of this operational assumption. It is argued in this paper that the influence of an entity on any given location in its environment is not represented sufficiently by simple proximity and buffers derived from it, or even by functions of proximity that take into account the size, mass, or attributes of the entity, if the shape and internal heterogeneity of the entity is ignored. In addition, many entities do not have precise boundaries. While they can be reasonably described and modeled [15], it is unclear how fuzzy entities should be buffered using simple proximity.

A commonly available set of spatial analysis tools, the class of integrative (low-pass) spatial filters, can be used to provide robust, versatile methods of modeling the spatial components of influence and to derive more useful buffering methods. In ascertaining the influence of an entity on any point in its environment, the area within the entity itself, and not only a single point on its edge, are consulted through spatial influence functions described here. As a result, the size, shape, and possible fuzziness of entities, as well as values of the domain attributes that are responsible for the influence being generated (which may be spatially heterogeneous within each entity) are included in the assessment. The influence function may have directionality (anisotropy). All entities or portions of entities within a reasonable distance from that environmental point are consulted simultaneously, and the cumulative effects of multiple entities contributing to the influence on a point are automatically ascertained. In addition, the method can be applied over multiple spatial resolutions by using simple scaling factors and filter kernels of various sizes. Kernel-density estimation methods for the analysis of point patterns [16] are functionally similar, although they are intended for point rather than areal or field data.

For purposes of demonstration, the operations are applied to simple spatial configurations, and the results are compared to those obtained by using traditional proximity buffers. This method, termed here the receptive field approach, can be easily used in place of traditional proximity buffers to generate buffers that have greater validity in many geographic modeling applications.

2   Integrative Spatial Influence of an Entity Boundary on a Point

It is not surprising that work exploring the limitations of simple proximity has not come from the GIScience community, but from scientists working in spatial domains. Ecologists have long considered the effects of edges between different landscape types on biotic and physical characteristics of the landscape on either side of the edge. Nevertheless, there is little consensus regarding the characterization of such "edge effects" [17]. It is possible that unexamined spatial assumptions are hindering researchers from asking the right questions or noticing the most important spatial factors at work. In edge effect research, measurements are often taken along transects that extend orthogonally from edges between two landscape types, and the distance from the edge is the primary independent variable. The edge is often chosen to be free of corners or convolutions, in order to provide reliable information. The primacy of proximity to edge is unchallenged as the basis of observation, and the focus of research is always on examining the gradient changes of certain values at different proximities. Although other geometric considerations like the size and shape of the landscapes or patches that define the edge are discussed, they are seldom examined as sources of variability or reported quantitatively. The traditional focus on simple spatial proximity is consistent with the wide use of proximity buffers when edge effect results are extrapolated to landscapes.

In the context of this ecological edge effect research, J. R. Malcolm [18] pointed out that buffers ignore the cumulative influence of the continuous boundary being buffered. Malcolm argued that not only the *closest* point along a boundary exerts an influence on a given point in the nearby landscape, but that all points along the boundary whose distance to that external point is *significant* exert a joint influence. Rather than examining the distance from a point in the landscape to a single point on the edge, Malcolm suggested that a more realistic examination of the influence of that edge integrates the distances from a landscape point to the continuum of points that provide influence along the edge. Rephrasing Malcolm's argument, we can express the general integration problem as:

$$E(p, B) = \int_B g(p, i) \, db \qquad (3)$$

where E(p, B) is the total edge effect of boundary B on point p, g(p, i) is some distance decay function describing the effect of a point i along the boundary on point p, and db is the differential of arc length along boundary B. Malcolm defined the integrative summation of proximity to all relevant points along the boundary as an additive effect and demonstrated that simple buffers fail to represent them accurately. He subsequently [19] offered an analytical means of determining additive edge effects along a boundary. A modification of this method was offered by Fernández et al. [20] to allow for variability in the point edge effect along a boundary.

3 Integrative Spatial Influence of an Entity Surface on a Point

For many applications, there is a drawback of simple buffers even more significant than their neglect of the additive effects of boundaries, which Malcolm later recognized [21]. That drawback is the inability of proximity buffers to model the joint or additive influence of all points in the *interior* environment of the entity on the external environment. For many applications, the influence of an entity on some location in the landscape emerges from that entity's entire area (or volume), and not only from its perimeter. If we extent the integration of Equation 3 to two dimensions, then each point within an influential distance to a landscape point p, and which lies *within* the entity, and not only along its boundary, also exerts a 'point' edge effect. We can describe the influence of a surface A on an external point as:

$$S(p, A) = \iint_A g(p, c)\,da \qquad (4)$$

where g(p, c) is some distance decay function describing the effect of a point c within the entity's surface A on point p, and da is the differentiation of surface area over A. D(p,Q), E(p,B), and S(p,A) are visually summarized in Fig. 1.

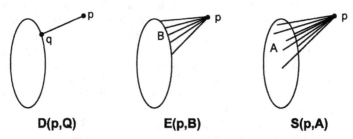

D(p,Q)   **E(p,B)**   **S(p,A)**

Fig. 1. Schematic representations of simple proximity D(p,Q), additive edge effects E(p,B), and additive surface effects S(p,A) of the entity Q, boundary B, or area A represented by the oval shape on external point p

4 Location as the Locus of a Receptive Field

The approach taken in this paper implements the spatial integration suggested in Equation 4 on a discrete spatial surface. For processing simplicity, it proceeds from the potential recipient (point p), rather than from the source (surface A), of influence. Instead of looking outward from a particular source of influence, as in traditional buffer formation, we look outward from every potential 'receptor' location in the environment in all directions to a maximum distance that depends on the type of influence being modeled. We then integrate all influences thus detected in order to derive a single cumulative influence value for each receptor location. The 'receptive field' function described below, T(p,C), examines all points within a large neighborhood C of point p. Using this method, it is possible that many points within an entity surface A are not examined in determining influence, if these points are assumed too distant from p to exert significant influence. In taking this approach, it does not matter whether the sources of influence are singular and simple or multiple and complex; the same integration process is applicable to all situations. The only requirement in using this approach is that all points not associated with influencing entities must hold the value zero for the attribute under consideration.

The spatial integration is accomplished by performing a summation of local integration approximations over the entity surface. If the landscape is rasterized so that cells within the influential entity or entities hold local integration values, the total influence on any point represented by external cell p can be determined by finding the summation of the influence of localized cells in a regular tessellation of that surface. The influence of a localized cell is attenuated by a coefficient derived from the distance decay function. This summation represented in Fig. 2 would implement the spatial integration of Equation 5 on a discrete surface:

$$T(p, C) = \sum_{i, j} \left(c_{ij} \cdot k_{ij} \right) \tag{5}$$

where c_{ij} is a cell within an envelope of influence C around p indexed in two dimensions by i and j, and k_{ij} is a coefficient describing the influence of cell c_{ij} on cell p, the result of g(p, c_{ij}). The matrix of coefficient values generated by function g is called the kernel.

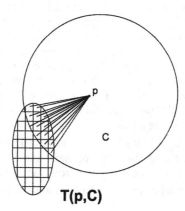

T(p,C)

Fig. 2. Schematic representation of the cumulative influence of the oval-shaped entity on external point p, using the receptive field function

Many summation filters of this general form are used in image processing and raster Geographic Information System functions. These include mean filters (in which all kernel coefficients are identical and sum to 1), Gaussian and other low-pass filters (in which coefficient values decrease with distance from the central cell), and Laplacian and other high-pass filters (in which the central coefficients are negative, peripheral coefficients generally are positive, and their summation is zero). Using serial computation, the filter is iteratively applied by centering the window on each cell of the original landscape, and the value of the corresponding cell in an output landscape is attained by the summation of the values of all cells within the matrix window multiplied by their corresponding kernel coefficients. Conceptually, the approach is massively parallel and may be implemented as such. The kernel matrix may be of any integer size, so long as it is odd in both dimensions. Generally, these filters are small: $i = j = 3, 5$, or 7. For this application, it would generally be much larger.

In creating a summation filter to use as a receptive field, the value of the coefficient k_{ij} is based on a univariate distance decay function g of the Euclidean distance between c_{ij} and p, yielding a value that is monotonic with regard to that distance.

$$k_{ij} = g \left(\left\| p - c_{ij} \right\| \right) . \tag{6}$$

In two-dimensional space, if p is indexed to the entire landscape by x and y,

$$\left\| p - c_{ij} \right\| = \left(\left| x - \frac{i-1}{2} \right|^2 + \left| y - \frac{j-1}{2} \right|^2 \right)^{0.5} . \tag{7}$$

Within a geographic information system, this can be implemented easily by using a summation filter kernel of size i by j, in which the coefficients are calculated by the application of function g. This univariate function of distance on a plane is a subset of the class of 'radial basis functions', which are often used in neural networks of arbitrary dimension for functional approximation. By locating the radial basis functions in a multidimensional space so that they form an intersecting web of receptive fields, the patterns or processes of the domain being modeled can be represented by the degrees to which combinations of field nodes fire. Similar architectures are often implemented in fuzzy systems. Here, the function is discretized, each receptive field is explicitly spatial, corresponding to each cell of the rasterized landscape, and an output 'influence' landscape is derived by the parallel application of the function to all cells of the landscape within which the influential entities are represented.

With a summation filter designed to represent a receptive field, the size of the window determines the maximum distance that any one cell may affect another. For instance, a 3 x 3 cell window allows any cell to affect any other a maximum distance of 1 unit in the horizontal, vertical, or diagonal directions. Therefore, when implementing receptive fields, we would use summation filter kernels that are much larger than kernels used in image processing applications.

The function $g \left(\| p - c_{ij} \| \right)$ may be a constant coefficient of $\| p - c_{ij} \|$, yielding a linear change of influence with distance. For many applications that involve nonlinear changes, more versatile distance decay functions may be useful. This paper investigates an adjustable sigmoidal change of influence with distance:

$$g \left(\left\| p - c_{ij} \right\| \right) = \exp \left(- a \cdot \left\| p - c_{ij} \right\|^{b} \right).$$ (8)

If p is equivalent to c_{ij}, and it is isolated from any other influencing entities, then $g(0) = 1$. In other words, the influence of any cell on itself is unity. The integrative influence on all cells within a neighborhood bounded by i and j on cell p is:

$$T \left(p, C \right) = \sum_{i,j} \left(c_{ij} \cdot \exp \left(- a \cdot \left\| p - c_{ij} \right\|^{b} \right) \right).$$ (9)

Using this formula, Gaussian functions result when a > 0 and b = 2. As a becomes larger, the curve decreases more rapidly toward zero. As b becomes larger, the top of the curve flattens. If b < 0, the top of the curve is pointed, and it is no longer asymptotic to 1. By varying a and b jointly, we may form a family of curves that effectively decay to zero at approximately the same distance but provide a versatile set of distance decay models for many applications (Fig. 3). The representation of one of these curves on a two-dimensional surface is shown in Fig. 4.

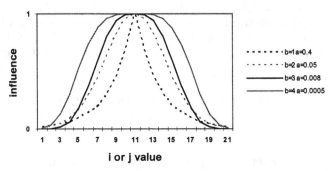

Fig. 3. A family of curves derived from Equation 9

b=3 a=0.008

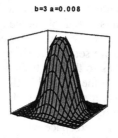

Fig. 4. Values obtained on a two-dimensional spatial surface, generated using Equation 9

4.1 Geometric Effects

In comparing buffers generated with proximity D(p,Q), a discretized implementation of edge integration E(p,B), and discrete surface integration T(p,C), different shapes are obtained for identical situations. Since buffer generation for T(p,C) is qualitatively different from traditional buffer generation, direct quantitative comparison is obtained only by estimation. We may generate equivalent buffer depths along a straight boundary segment of an entity and compare buffer behavior around the rest of the entity, as shown. Alternatively, we may generate buffers using all three methods that have equivalent areas within the buffer. The former method is used in generating the buffers shown in Fig. 5. While the depth of a simple proximity buffer is constant by definition (Fig. 5a), receptive field filters result in greatest buffer depth along concave sides of entities, and least along convex sides. This occurs both for E(p,B) and T(p,C) integration (Fig. 5b, 5c). In addition, buffer depth is constant by definition for D(p,q), but it is greater around large entities and shallower around smaller entities, for both E(p,B) and T(p,C).

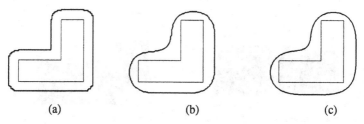

(a)	(b)	(c)

Fig. 5. A comparison of buffers (in darker shade around lighter outlined entities) generated by (a) simple proximity; (b) simulation of additive edge effects; and (c) the full application of receptive fields. A 51x51 kernel was generated using equation 9, where a=0.005 and b=3

4.2 Additive Effects of Multiple Entities and Attribute Values

While the size and shape of entities influence the formation of buffers using this receptive field approach, the cumulative effects of nearby entities can yield results that are even more strikingly different from the results of traditional buffers. Fig. 6a shows the buffering of three entities at close mutual proximity. These buffers remain identical and merge only when the entities are within twice the buffer size to one another. In contrast, receptive field buffers in Fig. 6b merge more readily when the cumulative effects of surrounding entities are quite high. If we consider the central point equidistant from all three entities, and we set the proximity buffers in Fig. 6a to be half the distance between any two entities so that the buffers do merge, the region around that equidistant point remains outside the merged buffers. It does seem reasonable to believe that, for many applications, it should be included within the buffered area, as it is in Fig. 6b. Many users of proximity buffers 'clean up' their results by removing isolated interior regions of unbuffered space when buffers merge, in a rather arbitrary manner. The use of receptive field buffers would help to alleviate this problem.

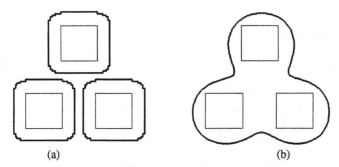

(a)   (b)

Fig. 6. A comparison of buffers (in darker shade around lighter outlined entities) generated by (a) simple proximity); and (b) the full application of receptive fields. A 51x51 kernel was generated using Equation 9, where a=0.005 and b=3

This method allows the user to vary spatial influence and the resulting buffer with varying values of the influencing attribute (Fig.7). In buffering a hazardous waste site, for example, a base value might be assigned to the entire grounds, and higher values to buildings or pits known to be of higher risk.

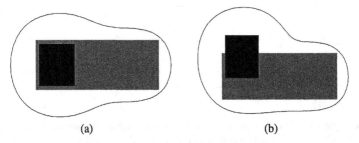

(a)   (b)

Fig. 7. Receptive field buffers generated around entities with multiple values for the influencing attribute. Value within black regions is ten times that within the gray regions. A 51x51 kernel was generated using Equation 9, where a=0.005 and b=3

4.3 Anisotropy

Anisotropy describes changes in distance decay functions with direction. Buffers that reflect anisotropy can be generated using the receptive field method:

$$E(p) = \sum_{i,j} c_{ij} \cdot \exp \left(- a_{ij} \left\| p - c_{ij} \right\|^{b_{ij}} \right). \qquad (10)$$

Here, functional variables a_{ij} and b_{ij} are coefficients of matrices A and B, which are of the same form as the receptive field matrix. This allows for full directional flexibility of in the calculation of distance decay. Many applications, like those that model potential plume extension from an areal source, would benefit from this approach.

5 Scaling the Receptive Field

Geographic scale includes both resolution (grain) and extent [22], and receptive fields should match the characteristic scales of influence in both of these important aspects. The discussion thus far has not included consideration of the actual resolution of cells used in receptive fields, although such considerations are implied in the choice of filter kernel size. For example, the entities in Fig. 5 are 60 cell units on a side, and the receptive field kernels are 51x51 cells. It is assumed in this case that nearly all of the entity would contribute to the field effect. However, kernel sizes much larger or smaller than any particular entity are valid if the spatial extent of influence has a characteristic scale different from that of any given entity.

It is often necessary to rescale receptive fields on the basis of resolution. It is easy to develop a family of equivalent receptive fields, all of the same effective extent but with different resolutions. This is done by using the kernel originally created at a particular resolution as a standard for all subsequent kernels. In calculating Euclidean distance earlier, we assumed that each cell represents one arbitrary unit. Now we must consider all other cell sizes as some multiple of this standard. The Euclidean distances are thus different, and the distance decay calculation is changed accordingly. However, we are presented with a challenge. The original receptive field is normalized to one at its center: every point has unity influence on itself. If receptive fields at finer resolutions were similarly normalized, the full integration of the field (volume under the 3D surface in Fig. 4) would result in a higher value than that of the original field.

For most applications, therefore, it is necessary to multiply each kernel coefficient by a normalizing constant, found by dividing the summation of all coefficients in the original kernel by the summation of all coefficients in the kernel derived at the finer resolution. Each of the coefficients in the new kernel must then be multiplied by this coefficient c, so that both receptive fields, at two resolutions, have equivalence. The effect of this is that we no longer have a central coefficient equal to one. We are confronted therefore with the notion that the assumption that 'a cell has unity influence on itself' must also have some characteristic scale for which it is true.

The author has been using the method described here to buffer disturbed regions of the Brazilian Amazon in order to investigate forest flammability and other edge-related changes (Figs. 8, 9). The filter used for each operation is scaled both to the resolution of the data (30 m for Landsat TM; 4 m for IKONOS) and to the nature of the edge effect being modeled.

6 Discussion

Although geographical phenomena often include influence functions analogous to electromagnetic or gravitational fields, buffers based on simple proximities, which do not capture such field effects, continue to be used extensively. Receptive field buffers yield results that are more flexibly adaptable to various models of influence. When a spatial buffer is created to help understand epidemiological or ecological data, for example, the receptive field approach is potentially far more useful than proximity buffers, which have little or no scientific justification. The drawback of the receptive field approach is that conditions and results are difficult to describe linguistically, particularly when buffers are prescribed or mandated. It is much easier to mandate a buffer of '100 meters' than to prescribe a buffer generated mathematically that takes full account of the attributes and form of the entity being buffered.

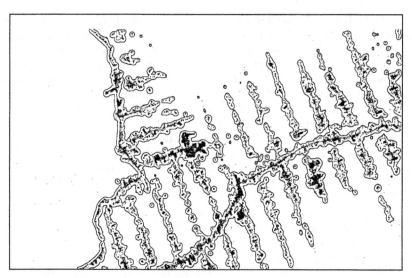

Fig. 8. A 2000 x 4500 pixel (approximately 60 x 135 km) subscene of Landsat TM data for a portion of Rondônia, Brazil, processed to reveal regions of deforestation in black, and buffered for purposes of investigating forest flammability

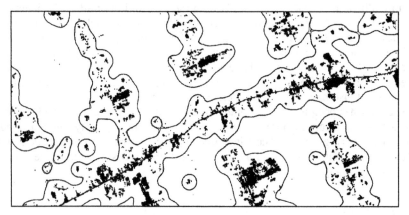

Fig. 9. A closer view of a portion of Fig. 8. Note that the filter used here does not buffer isolated pixels of deforestation, as they are not hypothesized to contribute individually to the flammability of adjacent forest

It is the opinion of this author that even mandated buffers could be more equitable and effective if they were constructed through the use of receptive fields. Such applications might include the drawing of ocean boundaries for purposes of fisheries management, or buffer zones around ecologically sensitive regions and hazardous waste sites. The wide use of GIS technologies allows for such flexibility, and it is important that we use available technologies in ways that are well suited for each case. This often requires us to go beyond the point-and-click approach, to develop the science of geographic information in response to the needs of the natural, social, and decision sciences, and to press for the improvement of functionality in commercially available systems. This paper describes a way of implementing more realistic buffer formation algorithms using the functionality already contained within most GISystems. Inefficiencies stem from the application of filters over entire landscapes, whether or not relevant entities are proximate, and from the necessity of performing vector-to-raster conversions for data that are originally in vector representations. Certainly, more efficient analytical approaches could be developed that consider only regions of a landscape containing entities, and which integrate polygon attributes on the basis of computational geometry for vector data.

References

1. Bren, L. J.: Aspects of the geometry of riparian buffer strips and its significance to forestry operations. *Forest Ecology and Management* 75 (1995) 1-10
2. Xiang, W.-N.: GIS-based riparian buffer analysis: injecting geographic information into landscape planning. *Landscape and Urban Planning* 34 (1996) 1-10
3. Lin, C. Y., Chou, W. C., and Lin, W. T.: Modeling the width and placement of riparian vegetated buffer strips: a case study on the Chi-Jia-Wang Stream, Taiwan. *Journal of Environmental Management* 66 (2002) 269-280
4. Lee, J. T., Bailey, N. and Thompson, S.: Using geographical information systems to identify and target sites for creation and restoration of native woodlands: a case study of the Chiltern Hills, UK. *Journal of Environmental Management* 64 (2002) 25-34
5. Lober, D. J.: Resolving the Siting Impasse. *Journal of the American Planning Association* 61 (1995) 482-495
6. Ali, M., Emch, M., Donnay, J.P., Yunus, M., and Sack, R.B.: Identifying environmental risk factors for epidemic cholera: a raster GIS approach. *Health & Place* 8 (2002) 201-210
7. Petridou, E., Trichopoulos, D., Kravaritis, A., Pourtsidis, A., Dessypris, N., Skalkidis, Y., Kpgevomas. M., Kalmanti, M., Koliouskas, D., Kosmidis, H., Panagiotou, J.P., Piperopoulou, F., Tzortzatou, F., and Kalapothaki, V.: Electrical Power Lines and Childhood Leukemia: A Study from Greece. *International Journal of Cancer* 73 (1997) 345-348
8. Tynes, T. and Haldorsen, T.: Electromagnetic fields and cancer in children residing near Norwegian high-voltage power lines. *American Journal of Epidemiology* 145 (1997) 219-226
9. McMaster, R. B.: Modeling community vulnerability to hazardous materials using geographic information systems. In: Peuquet D. J. and Marble, D. F. (eds.), *Introductory Readings in Geographic Information Systems*, Taylor & Francis, London (1990) 183-194
10. Finco, M. V. and Hepner, G. F.: Investigating US-Mexico Border Community Vulnerability to Industrial Hazards: A Simulation Study in Ambos Nogales. *Cartography and Geographic Information Science* 26 (1999) 243-252
11. White, E. and Aldrich, T. M.: Geographic Studies of Pediatric Cancer near Hazardous Waste Sites. *Archives of Environmental Health* 54 (1999) 390-397

12. Bolin, B., Matranga, E., Hackett, E., Sadalla, E., Pijawka, K., Brewer, D., and Sicotte, D.: Environmental equity in a sunbelt city: the spatial distribution of toxic hazards in Phoenix, Arizona. *Environmental Hazards* 2 (2000) 11-24

13. Leroyer, A., Nisse, C., Hemon, D., Gruchociak, A., Salornez, J.-L., and Haguenoer, J.-M..: Environmental Lead Exposure in a Population of Children in Northern France: Factors Affecting Lead Burden. American *Journal of Industrial Medicine* 38 (2000) 281-289

14. Reynolds, P., Elkin, E., Scalf, R., Von Behren J., and Neutra RR.: A Case-Control Pilot Study of Traffic Exposures and Early Childhood Leukemia Using a Geographic Information System. *Bioelectromagnetics Supplement* 5 (2001) S58-S68

15. Burrough, P. A. and Frank, A. U.: *Geographic Objects with Indeterminate Boundaries.* Taylor & Francis, London (1996)

16. O'Sullivan, D. and Unwin, D.: *Geographic Information Analysis.* John Wiley & Sons, Hoboken, NJ (2003)

17. Murcia, C.: Edge effects in fragmented forests: implications for conservation. *Trends in Ecology and Evolution* 10 (1995) 58-62

18. Malcolm, J. R.: Edge effects in central Amazonian forest fragments. *Ecology* 75 (1994) 2438-2445

19. Malcolm, J. R.: A Model of Conductive Heat Flow in Forest Edges and Fragmented Landscapes. *Climate Change* 39 (1998) 487-502

20. Fernández, C., Acosta, F. J., Abellá, G., López, F., and Díaz, M.: Complex edge effect fields as additive processes in patches of ecological systems. *Ecological Modelling* 149 (2001) 273-283

21. Malcolm, J. R. Extending Models of Edge Effects to Diverse Landscape Configurations, with a Test Case for the Neotropics. In: Bierregaard, Jr., R. O., Gascon, C., Lovejoy, T. E., and Mesquita, R. (eds.): *Lessons from Amazonia: The Ecology and Conservation of a Fragmented Forest*, Yale University Press, New Haven, RI (2001) 346-368

22. Pereira, G. M.: A Typology of Spatial and Temporal Scale Relations. *Geographical Analysis* 34 (2002) 21-33

Effect of Category Aggregation on Map Comparison

Robert Gilmore Pontius Jr. and Nicholas R. Malizia

Graduate School of Geography, George Perkins Marsh Institute, and
Department of International Development, Community, and Environment
Clark University
950 Main St., Worcester, MA 01610 USA
{rpontius,nmalizia}@clarku.edu

Abstract. This paper investigates the influence of category aggregation on measurement of land-use and land-cover change. To date, research concerning data aggregation has examined primarily the effects of modifying the unit of observation (i.e., the modifiable areal unit problem and the ecological inference problem); here, we examine the effects of changing the categorical definition, such as the conversion from many, detailed Anderson Level II classes to fewer, broader Anderson Level I classes. Cross-tabulation matrices are used to analyze the change between two times for aggregated and unaggregated versions of identical landscapes. A mathematical technique partitions the Total change as the sum of Net (i.e., quantity change) and Swap (i.e., location change). This paper shows that the Total and Net exhibited by maps between two points in time can be substantially reduced through land-use category aggregation, but cannot be increased. Swap, however, can be reduced or increased by the aggregation of categories. We derive five principles that dictate the effect of aggregation and illustrate the principles using both simplified examples and empirical data. The empirical data are from three Human Environment Regional Observatory sites. The principles are mathematical facts that apply to the analysis of any categorical variable.

1 Introduction

1.1 Measuring Change on a Map

Land-use and land-cover change (LUCC) analysis has become an integral component of geographic, economic, and ecological research. Changes in land-use and land-cover are either directly responsible for, or synergistically enhance, many forms of environmental change including biodiversity loss, land degradation, and climatic variation [6, 7]. Scientists study change in landscapes over time to determine its causes and effects as well as to model future landscapes. Such research directly affects conservation and development policy. This paper specifies how a decision in the early stages of a LUCC investigation regarding the definition of land-use and land-cover categories can have a profound effect on subsequent analysis and conclusions. Comparison of maps from an initial time A and a subsequent time B is the most common method to analyze LUCC. A typical first step in this comparison is the calculation of a cross-tabulation matrix. Table 1 demonstrates the format of a typical cross-tabulation matrix, where the rows represent the categories of the land-use map at time A and the columns show the categories at time B.

M.J. Egenhofer, C. Freksa, and H.J. Miller (Eds.): GIScience 2004, LNCS 3234, pp. 251–268, 2004.

Table 1. Cross-tabulation matrix to compare maps from two points in time for three categories.

		Time B				
		Category 1	Category 2	Category 3	Total Time A	Loss
Time A	Category 1	P_{11}	P_{12}	P_{13}	P_{1+}	$P_{12} + P_{13}$
	Category 2	P_{21}	P_{22}	P_{23}	P_{2+}	$P_{21} + P_{23}$
	Category 3	P_{31}	P_{32}	P_{33}	P_{3+}	$P_{31} + P_{32}$
	Total Time B	P_{+1}	P_{+2}	P_{+3}	1	
	Gain	$P_{21} + P_{31}$	$P_{12} + P_{32}$	$P_{13} + P_{23}$		

P_{ij} denotes the proportion of the map that shows a transition from category i at time A to category j at time B. Entries on the diagonal represent persistence on the map between the two points in time, thus P_{jj} identifies the proportion of the map that persists as category j. This matrix also calculates the Total amount of each category for each point in time. Entry P_{i+} sums the amount of category i at time A, while entry P_{+j} sums the amount of category j at time B. To this standard matrix we append an additional row and column to calculate the amount of Gain and Loss for each category between time A and time B. The Loss for category i is calculated by summing the off-diagonal entries for category i at time A. Thus, the amount of Loss for category i is equivalent to $P_{i+} - P_{ii}$. The Gain for category j is calculated by summing the off-diagonal entries for category j at time B, which is equivalent to $P_{+j} - P_{jj}$.

Table 2 shows how these basic statistics are further processed to yield more information that is fundamental to comparing maps of a shared categorical variable [10]. The Loss, Gain, and Total columns of Table 2 show that the sum of the Loss and the Gain for each category between time A and time B is the Total for that category. The left side of equation 1 expresses this relationship for category j.

Table 2. Map change budgets derived from the cross-tabulation matrix in Table 1.

Category	Loss	Gain	Total	Net	Swap
1	$P_{12}+P_{13}$	$P_{21}+P_{31}$	$P_{12}+P_{13}+$ $P_{21}+P_{31}$	$\|(P_{12}+P_{13})-(P_{21}+P_{31})\|$	$\mathrm{MIN}(P_{12}+P_{13}, P_{21}+P_{31}) * 2$
2	$P_{21}+P_{23}$	$P_{12}+P_{32}$	$P_{21}+P_{23}+$ $P_{12}+P_{32}$	$\|(P_{21}+P_{23})-(P_{12}+P_{32})\|$	$\mathrm{MIN}(P_{21}+P_{23}, P_{12}+P_{32}) * 2$
3	$P_{31}+P_{32}$	$P_{13}+P_{23}$	$P_{31}+P_{32}+$ $P_{13}+P_{23}$	$\|(P_{31}+P_{32})-(P_{13}+P_{23})\|$	$\mathrm{MIN}(P_{31}+P_{32}, P_{13}+P_{23}) * 2$
Map	$P_{12}+P_{13}+$ $P_{21}+P_{23}+$ $P_{31}+P_{32}$	$P_{12}+P_{13}+$ $P_{21}+P_{23}+$ $P_{31}+P_{32}$	$P_{12}+P_{13}+$ $P_{21}+P_{23}+$ $P_{31}+P_{32}$	$[\|(P_{12}+P_{13})-(P_{21}+P_{31})\| +$ $\|(P_{21}+P_{23})-(P_{12}+P_{32})\| +$ $\|(P_{31}+P_{32})-(P_{13}+P_{23})\|]/2$	$\mathrm{MIN}(P_{12}+P_{13}, P_{21}+P_{31}) +$ $\mathrm{MIN}(P_{21}+P_{23}, P_{12}+P_{32}) +$ $\mathrm{MIN}(P_{31}+P_{32}, P_{13}+P_{23})$

$$\mathrm{Loss}_j + \mathrm{Gain}_j = \mathrm{Total}_j = \mathrm{Net}_j + \mathrm{Swap}_j \qquad (1)$$

The situation is slightly different for the map-level analysis where Total equals the sum of Losses, which is also equal to the sum of Gains for the entire map. This is because a Loss of any category implies a Gain of another category. The bottom row of Table 2 demonstrates the relationship that equation 2 dictates for the map level of analysis, this is denoted with subscript M.

$$\mathrm{Loss}_M = \mathrm{Gain}_M = \mathrm{Total}_M = \mathrm{Net}_M + \mathrm{Swap}_M \qquad (2)$$

The Total column of Table 2 shows that the Total at the map level is equal to the sum of the off-diagonal entries of Table 1. The right-hand side of equations 1 and 2 show that at both the category level and at the map level, the Total can be partitioned as the sum of Net and Swap.

Net (i.e., quantity change) is the amount of uncompensated change for a category. For category j, the Net is equal to $|P_{+j} - P_{j+}|$. If P_{+j} is greater than P_{j+} then the category is Net gaining, if P_{+j} is less than P_{j+} then the category is Net losing. Table 2 shows that the Net at the map level is one half the sum of the Net for the individual categories [10].

Figure 1 demonstrates this concept, where the urban areas experience Net gain while the forest areas experience Net loss. Urban sprawl emanating from the city of Worcester in the central Massachusetts study area is overtaking the other categories on the map, most notably the eastern forest areas. In Figure 1, urban is a Net gaining category while forest is a Net losing category.

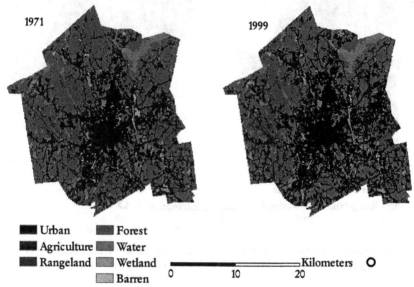

Fig. 1. These maps show the land-use change in the Massachusetts study area between 1971 and 1999. The change on these maps is predominantly Net gain of urban and Net loss of forest.

Swap (i.e., location change) is the amount of compensated change for a category. Swap for a category is equal to twice the minimum of the Gain or Loss for the category (i.e., Swap is equal to twice the minimum of P_{+j} and P_{j+} for category j). This is because the category-level Swap derives from pairing as many gaining pixels with losing pixels of the same category as possible; while the Net is the remaining un-paired pixels. Table 2 shows that Swap at the map level is equal to one half of the sum of the Swap for the individual categories [10].

Figure 2 demonstrates the concept of Swap in a map of the border between Arizona, U.S. and Sonora, Mexico. The rangeland and barren lands retain a relatively similar quantity of their respective categories on the map while the locations of those categories change. The change observed on these maps is due mainly to classification error. The maps show a considerable amount of apparent transition between the rangeland and barren categories probably because of the difficulty in distinguishing between these two land-use/cover categories using remotely sensed data. While both categories lose a large amount between time A and time B, they also gain a substantial amount that compensates for the loss, which yields a set of maps dominated by Swap.

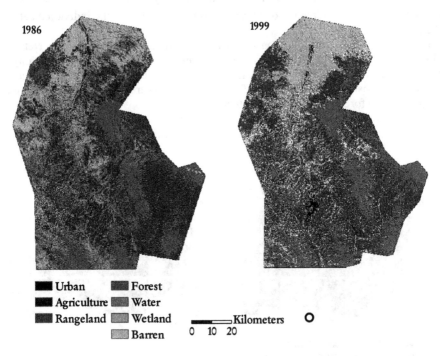

Fig. 2. These maps show the land-use change in the Arizona study area between 1986 and 1999. The change on these maps is predominantly Swap in the rangeland and barren categories.

1.2 The Problem of Aggregation

Figures 3 and 4 illustrate the potentially important effect of category aggregation on the measurement of LUCC. Both figures show maps of the landscape of Kansas' Grey County for 1985 and 2001. Figure 3 shows these maps classified at a modified Anderson Level II [1]; whereas Figure 4 shows the same maps aggregated to Anderson Level I to allow for comparison with the maps in Figures 1 and 2. The Anderson classification system was employed in the cross-site analysis to provide a standard metric with which to compare sites. Level I of this classification system can distinguish a map of 9 classes, which can be sub-divided to make a total of 37 possible categories of land uses or covers at Level II [1].

While the unaggregated maps in Figure 3 show nine categories at Level II, the aggregated maps in Figure 4 show only five categories at Level I. This is due to the amount of categorical detail that is identified within the agriculture category at

Anderson Level II. Many categories at Anderson Level II (confined feeding operations, other cropland, cool season crops, warm season crops, and two season crops) become one category in Anderson Level I (agriculture).

This change in category definition has a substantial influence on the amount of change measured on the map. At Anderson Level II, there is 61% Total in the maps between 1985 and 2001. Of that change, 48% is Swap while 13% is Net. Almost all of this change is due to the seasonal variation within the detailed agricultural categories. Analysis of the same landscape at Anderson Level I yields substantially different results. At Anderson Level I, 13% change is observed on the maps. Swap accounts for 10% of that change while Net claims the small remaining portion. This demonstrates the dramatic effect that the categorical scale can have on analysis, a problem we christen the "category aggregation problem" (CAP).

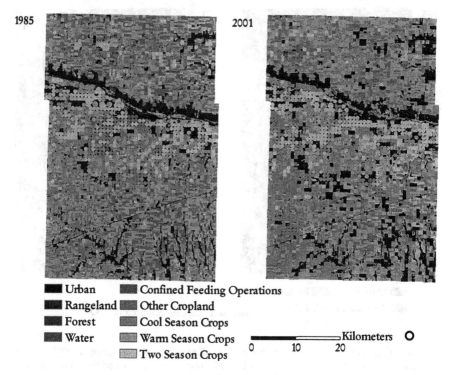

Fig. 3. These maps show the land-use change in the Kansas study area between 1985 and 2001. These maps are classified using Anderson Level II classification. Most of the change on the maps is Swap.

While some scientists might have an intuitive idea of how an aggregation scheme can change the information in a dataset, this paper provides scientists with a mathematical explanation for the effects of the manipulation of the categorical definition. It is important that scientists know exactly how category aggregation influences the measurement of difference between two maps, because category aggregation is an extremely common practice and the effect of some aggregation schemes are not intuitive. Scientists aggregate categories for many reasons. One of the most important reasons is to allow for comparison across diverse sites, times, and datasets. It is usually easiest to compare various maps when each map has the same categories; the most common way to attain this is to aggregate the data from available maps to a common set of broad categories. A second important reason for aggregation is to allow for simplification and reduction of the data. Usually the scientist wants to focus on the dynamics of the most important categories, while the available data may contain many more categories than just the most important ones. Scientists do not want the aggregation step to eliminate some potentially important information or to introduce some spurious signals. If scientists know exactly how an aggregation scheme influences the information in the maps, then they can reformat the data with confidence, ultimately in order to use a larger array of maps and statistical techniques.

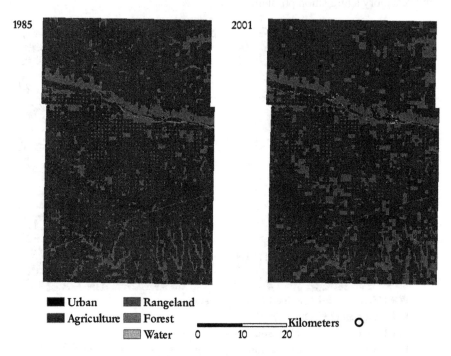

Fig. 4. These maps show the land-use change in the Kansas study area between 1985 and 2001. These maps are classified using Anderson Level I classification. Most of the change is Swap; however, there is considerably less change here than on the same maps classified using Anderson Level II.

1.3 Literature Review

Scale has been identified as a research priority within the field of GIS [11, 12]. The CAP falls under this broad issue and should be a concern for any investigation of categorical information, spatially explicit or otherwise. This problem is related to other issues of scale, including the modifiable areal unit problem and ecological inference problem.

The modifiable areal unit problem (MAUP) has been a concern in geography and related fields for over half a century [2, 14]. A detailed investigation of the problem began during the late 1970s. Openshaw [9] defined two components of the MAUP, the scale problem and the aggregation problem. The aggregation problem realizes there are numerous ways of defining the boundaries of zones, and as a result, differing conclusions are often drawn from these differing delineations [8]. The scale problem, however, is more relevant to the CAP. Openshaw [9] describes the scale problem as "the variation in results that can often be obtained when data for one set of areal units are progressively aggregated into fewer and larger units for analysis" (p. 8). The CAP can be considered a cousin of the MAUP because both problems examine the effect of aggregating from detailed information to generalized information. However, the CAP differs fundamentally from the MAUP because the MAUP modifies the unit of observation, while the CAP modifies the variable definition. Openshaw and Taylor [8] showed that researchers can obtain nearly any measurement of association between the two maps by modifying the unit of analysis. However, this is not the case for the CAP. This paper derives the mathematical principles that constrain the effect of categorical aggregation on statistical results.

Research on the ecological inference problem also has implications for the category aggregation problem. The ecological inference problem, also referred to as the ecological fallacy problem, occurs when conclusions drawn from aggregate data are applied at the individual level [4]. The ecological inference problem acknowledges that the relationships observed for groups do not always hold for individuals. Therefore, conclusions regarding a landscape established through the use of aggregated categorical land-use information may not hold true when the same landscape is analyzed using more detailed categories. In this analogy, we regard the detailed categories on the unaggregated maps as the individuals and the broader categories on the aggregated maps as the groups.

Scientists have made only limited progress on the modifiable areal unit problem and the ecological inference problem [4, 13]. For the CAP, we have been able to distill the fundamental concepts that dictate the effect of category aggregation, which the methods and results sections describe.

2 Methods

2.1 Strategy

This section of the paper demonstrates the category aggregation problem through the aggregation of various combinations of three categories that compose a simple example map. We focus on the effect of a single step of aggregating two of the three categories in order to give insight into more complex multi-category aggregations. Aggregation of several categories can be considered a step-wise process where two categories are aggregated per step. The order in which categories are combined does not affect the analysis of the final aggregated maps. Therefore, the principles gleaned from dissecting a single step of aggregation can be extrapolated to apply to any combination of any number of categories. Below, the mathematics that govern the process of a single step of aggregation are illustrated with simple example maps. These principles are then applied using empirical data in an examination of the three study landscapes mentioned above, located in Massachusetts, Arizona, and Kansas.

2.2 Examination of Illustrative Examples

Figure 5 shows maps of the example landscape at time A and time B. Through the transition on these unaggregated maps, category 1 loses while categories 2 and 3 gain. Just over 70% of the map undergoes a transition between time A and time B. Net accounts for 37% of the map change while Swap accounts for 33%. There are three possible ways to combine the categories such that the aggregated maps are composed of exactly two categories.

One possible combination is the union of categories 2 and 3. Both these categories experience a Net gain from time A to time B in the pre-aggregated maps, thus the resulting category also shows Net gain. The post-aggregation maps are shown in Figure 6. By combining categories 2 and 3, the amount of Total on the map is reduced from 70% to 50%. The Net is maintained at 37% while the Swap on the maps is reduced to 13%.

Another possible aggregation is the union of categories 1 and 2, a Net losing and Net gaining category, respectively. The resulting category, $1 \cup 2$, is Net losing (Figure 7). The combination of these categories results in the decrease of the Total observed on the map from 70% to 41%. The Net is reduced to 25% while the Swap on the maps is reduced to 16%.

The final possible aggregation with these maps is the union of categories 1 and 3. Between time A and time B category 1 loses, while category 3 gains. Category $1 \cup 3$ experiences Net loss (Figure 8). The Total on the aggregated maps is reduced to 50% and the Net decreases to 12%, while the Swap actually increases to 38% of the study area.

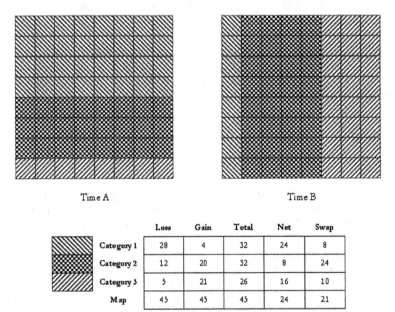

Time A   Time B

	Loss	Gain	Total	Net	Swap
Category 1	28	4	32	24	8
Category 2	12	20	32	8	24
Category 3	5	21	26	16	10
Map	45	45	45	24	21

Fig. 5. These example maps illustrate the principles of aggregation. The numbers in this budget and those in Figures 6-8 refer to map pixels, not percentages.

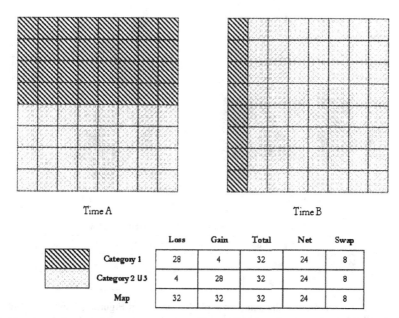

		Loss	Gain	Total	Net	Swap
	Category 1	28	4	32	24	8
	Category 2 U 3	4	28	32	24	8
	Map	32	32	32	24	8

Fig. 6. These maps represent the same example landscapes in Figure 5, however the two Net gaining categories (categories 2 and 3) have been aggregated. Net is maintained yet Swap decreases.

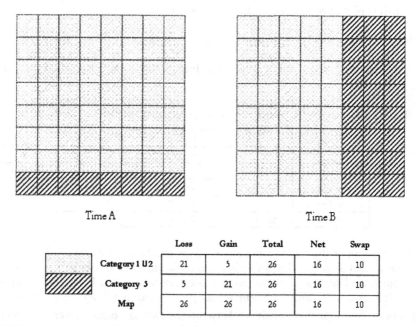

		Loss	Gain	Total	Net	Swap
	Category 1 U2	21	5	26	16	10
	Category 3	5	21	26	16	10
	Map	26	26	26	16	10

Fig. 7. These maps represent the same example landscapes in Figure 5, however a Net losing category and a Net gaining category (categories 1 and 2, respectively) have been aggregated. Both Net and Swap decrease.

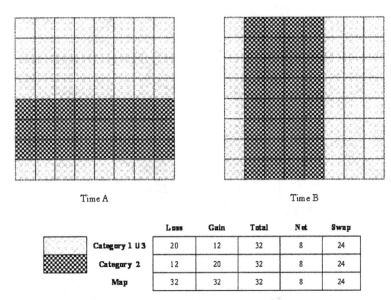

Time A   Time B

	Loss	Gain	Total	Net	Swap
Category 1 ∪ 3	20	12	32	8	24
Category 2	12	20	32	8	24
Map	32	32	32	8	24

Fig. 8. These maps represent the same example landscapes in Figure 5, however a Net losing category and a Net gaining category (categories 1 and 3, respectively) have been aggregated. Net decreases and Swap increases.

2.3 Effect of Aggregation on Change Budgets

Table 3 demonstrates how the aggregation of categories affects the amount of change on the maps. The original maps are composed of three categories, as in Table 1; however, categories 2 and 3 are aggregated to form a single category called $2 \cup 3$ in Table 3.

Table 3. Cross-tabulation matrix to examine the effect of aggregating categories 2 and 3.

		Time B		Total Time A	Loss
		Category 1	Category 2 ∪ 3		
Time A	Category 1	P_{11}	$P_{12} + P_{13}$	P_{1+}	$P_{12} + P_{13}$
	Category 2 ∪ 3	$P_{21} + P_{31}$	$P_{22} + P_{23} + P_{33} + P_{32}$	$P_{2+} + P_{3+}$	$P_{21} + P_{31}$
	Total Time B	P_{+1}	$P_{+2} + P_{+3}$	1	
	Gain	$P_{21} + P_{31}$	$P_{12} + P_{13}$		

Comparison of the cross-tabulation matrices associated with the original maps and the subsequent aggregated maps yields substantial information about the effects of the manipulation. The amount of persistence between time A and B on the unaggregated maps (i.e., P_{11}, P_{22} and P_{33}) remains as persistence in the aggregated maps. The amount of the maps that transitions between the two aggregated categories becomes persistence after aggregation. That is, entries P_{23} and P_{32}, which were off-diagonal in the cross-tabulation table for the unaggregated maps, become part of the diagonal, and therefore persistence, in the aggregated maps. This movement to the diagonal eliminates P_{23} and P_{32} from the calculation of Gain and Loss and thereby from the Total and its components of Net and Swap. Thus, Total can only be decreased or maintained through aggregation.

Therefore, the effect of aggregation is more pronounced where there are large transitions between the aggregated categories. If the aggregated categories do not transition (i.e., $P_{23} = P_{32} = 0$), then aggregation has no effect on the Total. The amount of persistence remains the same in the pre and post-aggregation maps for category 1. The amount of category 1 that transitions between time A and B also remains the same. Category 1 can transition to only category $2 \cup 3$ in the post-aggregation maps. Therefore, the sum of P_{12} and P_{13} yields a new entry that represents the amount of category 1 that transitions to the new aggregated category (Table 3). Conversely, entries P_{21} and P_{31} are summed to yield an entry representing the amount of the maps that transitions from the new aggregated category to category 1. Thus, the distribution of the Net and Swap on these aggregated maps is determined by the relative sizes of the Loss and Gain of category 1. Table 4 illustrates the effect of the aggregation on the change budget for a situation where categories 2 and 3 are aggregated.

Table 4. Map change budgets derived from the cross-tabulation matrix of Table 3.

Category	Loss	Gain	Total	Net	Swap
1	$P_{12}+P_{13}$	$P_{21}+P_{31}$	$P_{12}+P_{13}+$ $P_{21}+P_{31}$	$\lvert (P_{12}+P_{13})-(P_{21}+P_{31}) \rvert$	$MIN(P_{12}+P_{13}, P_{21}+P_{31}) * 2$
$2 \cup 3$	$P_{21}+P_{31}$	$P_{12}+P_{13}$	$P_{21}+P_{31}+$ $P_{12}+P_{13}$	$\lvert (P_{21}+P_{31})-(P_{12}+P_{13}) \rvert$	$MIN(P_{21}+P_{31}, P_{12}+P_{13}) * 2$
Map	$P_{12}+P_{13}+$ $P_{21}+P_{31}$	$P_{12}+P_{13}+$ $P_{21}+P_{31}$	$P_{12}+P_{13}+$ $P_{21}+P_{31}$	$\lvert (P_{12}+P_{13})-(P_{21}+P_{31}) \rvert$	$MIN(P_{12}+P_{13}, P_{21}+P_{31}) +$ $MIN(P_{21}+P_{31}, P_{12}+P_{13})$

2.4 Application to Kansas, Massachusetts, and Arizona Sites

To examine the effect of aggregation in practice, we use empirical data of three Human Environment Regional Observatory (HERO) study sites. The HERO Network is funded by the National Science Foundation and is composed of four sites located in Arizona, Kansas, Massachusetts, and Pennsylvania. The network addresses three core research themes: land-cover change, greenhouse-gas emissions, and the impact of these activities on climate. One explicit purpose of the HERO research is to compare land cover across its various sites, therefore we must create datasets that facilitate cross-site comparison, and hence we want to aggregate the available data to a common set of land categories. We investigate the effect of aggregation using data from Grey County in Kansas, central Massachusetts, and the Sonoran region of Arizona and Mexico.

Section 1.2 describes the Kansas data. In the central Massachusetts region, land-cover information was acquired from the Commonwealth [5], which delineates twenty land categories for the ten town region at two times: 1971 and 1999. This information is then aggregated to the Anderson level I classification, leaving seven categories. The southern Arizona data for 1986 and 1999 were classified through remote sensing to Anderson level I; however, we suspect that the method confused two categories, rangeland and barren. In this instance, we combine these two Anderson level I categories to create an aggregated classification with only six categories.

3 Results

3.1 Generalizable Effects of Aggregation

Our investigation reveals that five important principles govern the effects of category aggregation. Each principle is a mathematical fact that applies to any categorical variable and expresses the effect of the aggregation as a function of the transitions among the categories on the unaggregated maps. The post-aggregation Total, Net, and Swap are dependent upon whether the categories being aggregated transition among each other in the unaggregated maps and whether they exhibit Net loss or Net gain when transitioning from time A to time B on the unaggregated maps. All five principles are based on the fact that the Total observed on the maps is equal to the sum of the Net and Swap (equation 2). Thus, the difference in the Total due to aggregation is equal to the sum of the differences in Net and Swap as expressed in equation 3.

$$\Delta \text{Total}_M = \Delta \text{Net}_M + \Delta \text{Swap}_M \tag{3}$$

The first principle applies to all cases and is the foundation for the other principles. Principles 2 and 3 apply to cases where a losing category is aggregated with another losing category or where a gaining category is aggregated with another gaining category. Principle 2 describes the effect of aggregation on Net while principle 3 describes the effect on Swap. Principles 1 and 2 are used to prove principle 3. Alternatively, principles 4 and 5 apply to cases where a losing category is being aggregated with a gaining category. Principle 4 describes the effect of this aggregation on Net while principle 5 describes the effect on Swap. We use principles 1 and 4 to prove principle 5.

3.2 Principle 1

The first principle states "if any categories that transition between each other are aggregated, then the amount of Total on the map is reduced by the amount of their transitions." Mathematically speaking, equation 4 explains the principle where categories i and j are aggregated.

$$\Delta \text{Total}_M = -\left(P_{ij} + P_{ji}\right) \tag{4}$$

When two categories are combined, the transitions between them move to the diagonal. Thus the aggregation of categories i and j will decrease the amount of Total by $P_{ij} + P_{ji}$ because these values will be converted to persistence and be brought onto the diagonal. Consequently, the aggregation of categories cannot increase the amount of Total observed on the maps. This principle is evidenced in Tables 1 and 3. These tables show the aggregation of categories 2 and 3; as a result of this aggregation, entries P_{23} and P_{32}, which were off the diagonal in Table 1, are moved onto the diagonal in Table 3. The only instance of aggregation in which the Total is not reduced is when categories that do not transition to each other are combined, i.e., when $P_{ij} + P_{ji} = 0$.

3.3 Principle 2

The second principle states "if either a Net losing category is aggregated with a Net losing category or a Net gaining category is aggregated with a Net gaining category, then the Net on the map is maintained". Mathematically, if ($P_{+i} - P_{i+} \leq 0$ and $P_{+j} - P_{j+} \leq 0$) or ($P_{+i} - P_{i+} \geq 0$ and $P_{+j} - P_{j+} \geq 0$) then equation 5 applies.

$$\Delta Net_M = 0 \tag{5}$$

This is shown by Tables 2 and 4, where categories 2 and 3 are being aggregated and the resulting Net of category $2 \cup 3$ is equal to the sum of the individual Nets of categories 2 and 3. The proof of this is as follows. In Table 2, the Net for category 2 is equal to $|(P_{21}+P_{23}) - (P_{12}+P_{32})|$ while the Net for category 3 is equal to $|(P_{31} + P_{32}) - (P_{13} + P_{23})|$. If both categories experience a Net loss, then the absolute value signs are irrelevant, thus the Net for the unaggregated categories is equal to $(P_{21}+P_{23}) - (P_{12}+P_{32})$ for category 2 and $(P_{31} + P_{32}) - (P_{13} + P_{23})$ for category 3. The two Net losing categories are added together during the aggregation, so the values for P_{23} and P_{32} cancel. The resulting Net is $(P_{21}+P_{31}) - (P_{12}+P_{13})$, which is equal to $|(P_{21}+P_{31}) - (P_{12}+P_{13})|$, which is the Net for the aggregated category according to Table 4. Alternatively, if the two aggregated categories each demonstrate a Net gain, then a similar situation results. The Net for category 2 is $|(P_{21}+P_{23}) - (P_{12}+P_{32})|$, which would be equivalent to $(P_{12}+P_{32}) - (P_{21}+P_{23})$. Similarly, the Net for category 3 is $|(P_{31} + P_{32}) - (P_{13} + P_{23})|$, which would be equal to $(P_{13} + P_{23}) - (P_{31} + P_{32})$. During the aggregation, again cells P_{32} and P_{23} cancel and the Net for the aggregated categories is $(P_{12}+P_{13}) - (P_{21}+P_{31})$, which is equal to $|(P_{21}+P_{31}) - (P_{12}+P_{13})|$. Therefore, $Net_{2 \cup 3} = Net_2 + Net_3$ so the pre-aggregation Net equals the post-aggregation Net, thus $\Delta Net_M = 0$.

3.4   Principle 3

The third principle establishes "if either a Net losing category is combined with a Net losing category or a Net gaining category is combined with a Net gaining category, then the Swap on the map decreases by the amount of their transitions". Mathematically, if ($P_{+i} - P_{i+} \leq 0$ and $P_{+j} - P_{j+} \leq 0$) or ($P_{+i} - P_{i+} \geq 0$ and $P_{+j} - P_{j+} \geq 0$)then equation 6 results.

$$\Delta Swap_M = -\left(P_{ij} + P_{ji}\right) \tag{6}$$

The aggregation of categories i and j decreases the amount of Swap by $P_{ij} + P_{ji}$ when the categories being combined exhibit a similar direction of Net. Equation 6 results when equation 4 and equation 5 are substituted into equation 3. Thus principle 3 is a direct consequence of principles 1 and 2. This third principle applies to all cases where the second principle applies because both principles involve the aggregation of categories that have a similar direction of Net. Principles 2 and 3 apply also to instances where a category that undergoes Net on the unaggregated maps is combined with one that has zero Net, but still transitions with other categories.

3.5 Principle 4

The fourth principle states "if a Net losing category is aggregated with a Net gaining category, then Net on the map decreases by the smaller of the two Nets of the individual categories being aggregated". Mathematically, if $P_{+i} - P_{i+} \leq 0$ and $P_{+j} - P_{j+} \geq 0$, then equation 7 applies.

$$\Delta \text{Net}_M = -\text{MIN}\left(\left|P_{+i} - P_{i+}\right|, \left|P_{+j} - P_{j+}\right|\right) \tag{7}$$

To prove principle 4, we examine a general case where categories 1 and 2 are aggregated to form category $1 \cup 2$. Net for an unaggregated map is calculated in equation 8.

$$\text{Pre-Aggregation Net}_M = \frac{\sum_{j=1}^{J}\left|P_{+j} - P_{j+}\right|}{2} = \frac{\left|P_{+1} - P_{1+}\right| + \left|P_{+2} - P_{2+}\right|}{2} + \frac{\sum_{j=3}^{J}\left|P_{+j} - P_{j+}\right|}{2} \tag{8}$$

To complete the proof, we must show that equation 9 is true. Equation 9 expresses the pre-aggregation Net as the sum of the • Net and the post-aggregation Net. The • Net is located just to the right of the equal sign in equation 7, while the post-aggregation Net constitutes the remainder of the right-hand side of equation 9 in square brackets.

$$\text{Pre-Aggregtion Net}_M = \text{MIN}\left(\left|P_{+1} - P_{1+}\right|, \left|P_{+2} - P_{2+}\right|\right) + \left[\frac{\left|P_{+1\cup2} - P_{1\cup2+}\right|}{2} + \frac{\sum_{j=3}^{J}\left|P_{+j} - P_{j+}\right|}{2}\right] \tag{9}$$

Without loss of generality, call the Net losing category 1 and call the Net gaining category 2. There are two cases that need to be considered. The first case is where category 1 Net loses the same as or less than category 2 Net gains, i.e., $\left|P_{+1} - P_{1+}\right| \leq \left|P_{+2} - P_{2+}\right|$. If this case occurs, then we examine the behavior of categories 1 and 2 in equation 8. All of the Net loss of category 1 is paired with a subset of the Net gain of category 2 such that equation 10 holds. Equation 10 can then be substituted into equation 8 to yield equation 9.

$$\frac{\left|P_{+1} - P_{1+}\right| + \left|P_{+2} - P_{2+}\right|}{2} = \frac{2\times\left|P_{+1} - P_{1+}\right| + \left|P_{+1\cup2} - P_{1\cup2+}\right|}{2} = \text{MIN}\left(\left|P_{+1} - P_{1+}\right|, \left|P_{+2} - P_{2+}\right|\right) + \frac{\left|P_{+1\cup2} - P_{1\cup2+}\right|}{2} \tag{10}$$

The second case is where category 2 Net gains less than category 1 Net loses, i.e., $\left|P_{+2} - P_{2+}\right| < \left|P_{+1} - P_{1+}\right|$. If this case occurs, again we examine the behavior of categories 1 and 2 in equation 8. All the Net gain of category 2 is paired with a subset of the Net loss of category 1 such that equation 11 holds. Equation 11 can be substituted into equation 8 to yield equation 9.

$$\frac{\left|P_{+1} - P_{1+}\right| + \left|P_{+2} - P_{2+}\right|}{2} = \frac{2\times\left|P_{+2} - P_{2+}\right| + \left|P_{+1\cup2} - P_{1\cup2+}\right|}{2} = \text{MIN}\left(\left|P_{+1} - P_{1+}\right|, \left|P_{+2} - P_{2+}\right|\right) + \frac{\left|P_{+1\cup2} - P_{1\cup2+}\right|}{2} \tag{11}$$

3.6 Principle 5

The fifth principle states that "if a Net losing category is aggregated with a Net gaining category, then the Swap on the map can decrease, increase, or be maintained." Stated mathematically, if $P_{+i} - P_{i+} \leq 0$ and $P_{+j} - P_{j+} \geq 0$ then equation 12 results.

$$\Delta \text{Swap}_M = \text{MIN}\left(\left|P_{+i} - P_{i+}\right|, \left|P_{+j} - P_{j+}\right|\right) - \left(P_{ij} + P_{ji}\right) \tag{12}$$

Equation 12 is the consequence of substituting equation 4 and equation 7 into equation 3. Thus principle 5 is the direct consequence of principles 1 and 4. Principle 1 establishes the decrease in Total and principle 4 establishes the decrease in Net. Equation 3 implies the difference in Swap on the maps is equal to the difference in Total minus the difference in Net. If the decrease in Total is greater than the decrease in Net, then the Swap decreases. If decrease in Net is greater than the decrease of Total, then the Swap increases. If decrease in Total equals the decrease in Net, then the Swap is maintained.

3.7 Empirical Results

Figure 9 shows the effect of aggregation on land-use maps of study areas in Massachusetts, Arizona, and Kansas.

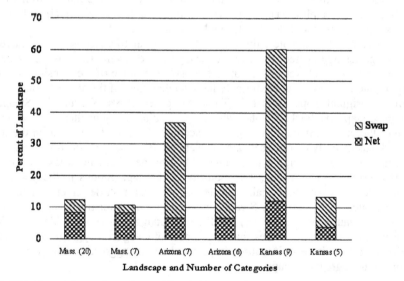

Fig. 9. This figure shows the amount of Total, Net, and Swap for each of the three study areas investigated at two different levels of categorical scale to illustrate the effect of aggregation on the maps. The height of the bar corresponds to the amount of Total on the maps.

In central Massachusetts, the twenty-category, unaggregated maps show 12% Total between 1985 and 1999; 8% is Net while the remaining 4% is Swap. After the aggregation to Anderson Level I, 11% of the maps show change where again 8% is Net and the remainder is attributable to Swap. In the Arizona maps, the rangeland and barren categories are combined, both of which are losing categories. The amount of Total on the mapped landscape decreases from 37% to 17%. The Net on the maps remains unchanged at 7% while the Swap shrinks from 30% to 11%.

4 Discussion

4.1 Interpretation of Kansas, Massachusetts, and Arizona

The empirical data demonstrate how the principles manifest on maps composed of more than three categories (Figure 9). The Arizona case provides the simplest of the empirical cases. The two most dynamic of the original Anderson Level I categories, rangeland and barren, are aggregated. Both are Net losing categories, so principles 1, 2, and 3 apply. There is a substantial amount of Swap in each of the two categories and the two categories transition between each other because of confusion in their classification. When they are aggregated, Total decreases according to principle 1, the Net is maintained according to principle 2, and Swap decreases from 30% to 11% according to principle 3.

The maps of the Kansas study site also show drastic effects of aggregation, where nine categories are aggregated to five. The aggregated categories are a mix of Net losing and Net gaining where there are substantial transitions among the categories. This leads to a dramatic reduction in the Total observed on the maps, from 61% to 13% based on principle 1. The Net shrinks from 13% to 4% by principle 4 and the Swap is reduced from 48% to 10% through principles 3 and 5.

Of the empirical examples, the maps of the central Massachusetts site have the most categories aggregated, from twenty categories to seven. There is a mix of Net gaining categories and Net losing categories; however none of the transitions among the aggregated categories are extremely large so the effect of the aggregation is small. Where the original maps show 12% Total, composed of 8% Net and 4% Swap, the aggregated maps exhibit 11% Total, where Net remains nearly unaltered at 8%. Forest is one of the original twenty categories. It accounts for 8% of the Total on the unaggregated maps; nearly all of this change is Net loss. It is not aggregated with any other categories, so its 8% change is maintained. Most of the other categories gain parts of the map surrendered by forest. The majority of the aggregations in these maps are gaining categories with other gaining categories; therefore little effect is seen on the Net as dictated by principle 2. Swap is not reduced as dramatically as in the other study sites because Swap is already small on the unaggregated maps. Thus, the Massachusetts example demonstrates that the number of categories aggregated is not necessarily important; instead, it is more important how the aggregated categories transition on the unaggregated maps.

4.2 Broader Applications

The category aggregation problem extends well beyond analysis of LUCC. It is encountered commonly in accuracy assessment. Aggregation of land categories can have a tremendous effect on the error matrices used to assess the accuracy of categorical maps. For example, Helmer et al. [3] demonstrate how this problem manifests in the accuracy assessment of a vegetation cover map. Initially, Helmer et al. classified the land cover in Puerto Rico using a 26-category classification. Their map had a classification accuracy of 79%. They reduced the total number of categories in their initial map to 19 through aggregation of categories that were being confused in the classification. As a result, the classification accuracy of their map rose to 83%. The process of aggregating categories during the production of categorical maps, especially those derived from satellite imagery, is extremely common, however it is unusual for scientists to document the effects of such manipulation in the early stages of analysis, as these authors did. Instead, most authors offer only the final classification, which makes readers blind to one of the scientist's most important decisions concerning the production of the maps. This example shows that the CAP is not limited to analysis of change. It demonstrates that researchers should investigate the effects of the problem as it applies to analysis of any categorical data.

The five principles extend to the aggregation of any categorical data because all the principles apply to any cross-tabulation analysis (Tables 1-4). While we examine the effect of this problem in terms of a spatially-explicit variable, i.e., land-use/cover, these principles apply also to the aggregation of non-spatial categorical data such as census data or labor statistics where job category definitions are analyzed by detailed sub-sectors nested within broader economic sectors. Just as the MAUP and ecological inference problem are general and transdisciplinary, the CAP is important for analysis that goes far beyond applications in geography.

5 Conclusions

A decision in the early stages of investigation regarding aggregation of categories can have a profound effect on subsequent analysis and conclusions. This paper distills five mathematical principles that dictate these effects. Principle 1 states that an aggregation can not increase the Total change. The behavior of unaggregated categories on the original maps must be considered to examine the effect of aggregation on Net and Swap. If the aggregation is of either a Net losing category with another Net losing category or a Net gaining category with another Net gaining category, then principles 2 and 3 apply. If a Net losing category is combined with a Net gaining category, then principles 4 and 5 apply. Scientists should consider the influence of category aggregation on the analysis of any categorical variable because the effects can be substantial.

Acknowledgments

The National Science Foundation supported this research through the grant "Infrastructure to Develop a Human-Environment Regional Observatory (HERO) Network" (Award ID BCS 9978052). John Harrington at Kansas State University and Cynthia Sorrenson at the University of Arizona provided land-use/cover information. Sam Ratick provided valuable insights throughout the course of this project. Mang Lung Cheuk contributed to this work through his development of the GLS macro. Clark Labs created the GIS and image processing software, Idrisi®, used in this analysis.

References

1. Anderson, J., Hardy, E., Roach, J., and Witmer, R.: A land use and land cover classification system for use with remote sensor data. USGS Professional Paper 964, Souix Falls, SD, (1976)
2. Gehlke, C. and Biehl., K.: Certain effects of grouping on the size of the correlation coefficient in census tract material. *Journal of the American Statistical Association Supplement* **29** (1934) 169-170
3. Helmer, E., Ramos, O., Lopez, T., Quinones, M., and Diaz, W.: Mapping the forest type and land cover of Puerto Rico, a component of the Caribbean biodiversity hotspot. *Caribbean Journal of Science* **38** (2002) 165-183
4. King G.: *A solution to the ecological inference problem.* Princeton University Press, Princeton, NJ (1997)
5. Massachusetts Geographic Information Systems (MASSGIS).: Land Use. (2002) http://www.state.ma.us/mgis/lus.html
6. Meyer, W. and Turner, B. eds.: *Changes in land use and land cover: A global perspective.* Cambridge University Press, Cambridge, (1994)
7. National Research Council (NRC).: *Grand challenges in the environmental sciences.* National Academy of Sciences Press, Washington, DC (2000)
8. Openshaw, S. and Taylor, P.: A million or so correlation coefficients: three experiments on the modifiable areal unit problem. In: Wrigley, N. (ed): *Statistical Applications in the Spatial Sciences.* Pion, London (1979) 127-144
9. Openshaw, S.: *The Modifiable Areal Unit Problem.* GeoBooks, Norwich (1984)
10. Pontius Jr., R., Shusas, E., and McEachern, M.: Detecting important categorical land changes while accounting for persistence. *Agriculture, Ecosystems and Environment* **101** (2004) 251-268
11. Quattrochi, D. and Goodchild, M.: (eds.): *Scale in Remote Sensing and GIS.* CRC Press, Boca Raton, FL (1997)
12. University Consortium for Geographic Information Science (UCGIS): Scale: Research White Paper (1998) http://www.ucgis.org/priorities/research/research_whi-te/1998%20Papers/scale.html
13. Wong, D. and Amrhein., C.: Research on the MAUP: old wine in a new bottle or real breakthrough? *Geographical Systems* **3** (1996) 73-76.
14. Yule, G. and Kendall, M.: *An Introduction to the Theory of Statistics.* Griffin, London (1950)

Simplifying Sets of Events
by Selecting Temporal Relations

Andrea Rodríguez[1], Nico Van de Weghe[2], and Philippe De Maeyer[2]

[1] Department of Computer Science, University of Concepción,
Center for Web Research, University of Chile, Chile
`andrea@udec.cl`
[2] Department of Geography, Ghent University, Belgium
{`nico.vandeweghe,philippe.demaeyer`}`@ugent.be`

Abstract. Reasoning about events or temporal aspects is fundamental for modeling geographic phenomena. This work concerns the analysis of events as configurations of temporal intervals. It presents two strategies to select relations that characterize configurations of temporal intervals: a strategy based on the algebraic property of composition and a strategy based on a neighboring concept in a vector representation. This type of analysis is useful for characterizing sets of events without the need of making an exhaustive specification of all temporal relations. This work complements a previous study about topological relations of regions in a 2D space and confirms the potential of using the algebraic properties of composition and the metric characteristics of intervals, even if only qualitative relations are considered.

1 Introduction

The ability to understand, represent, and manage temporal knowledge about the world is fundamental in humans and artificial agents [19, 22]. Reasons why time should be included in Geographic Information Systems (GISs) have already been discussed [13]. Such reasons are mostly associated with the fact that many applications involving the description of geographic phenomena require the treatment of dynamic aspects that are related to space and time. Examples of such applications are transportation and urban analysis, and the analysis of physical phenomena.

The temporal data model currently by far the most frequently used for understanding dynamic processes in the real world is based on the linear concept. In this concept, temporal intervals are projected onto a one-dimensional conceptual space by a one-dimensional segment. Relations between these intervals are called temporal relations. Allen [1] defined thirteen fine temporal relations, which later were extended by the sixteen coarse relations [8]. The Allen's basic thirteen relations define a relation algebra [23], which implies that the set of temporal relations is complete, each relation has a converse relation within the set, and there exists a composition operation that results in one or more relations within the set. The composition operation creates a reasoning mechanism that allows one to derive the relation between two intervals through the intervals' combination with a common third interval.

M.J. Egenhofer, C. Freksa, and H.J. Miller (Eds.): GIScience 2004, LNCS 3234, pp. 269–284, 2004.

This work concerns the temporal analysis of sets of events described by configurations of temporal intervals. In this paper we talk about events as something that occurs or happens at a particular time interval, not only at a specific point in time. Thus, we do not make a distinction between events, processes, states or actions. We consider events as everything that happens and is then gone. This work abstracts the semantics of events and focuses on temporal relations between events. In the spatial domain, the same idea is applied when one abstracts the meaning of spatial features and concentrates on spatial relations between these features. Such an abstraction has been useful for content-based retrieval where queries are expressed by sketches [5, 20]. In this work, however, sketches apply to the temporal domain. For example, if we have a database that stores the information about a diseases that occurred in a geographic area, one could be interested in searching for diseases that co-occur, diseases that precede or start to appear at the same time, and so on. In this search, the concern is not about the absolute positions of events on time, but rather on the relations between temporal intervals. This search could be extended to more than two diseases to become a search of a set of diseases, called in this paper a configuration of temporal intervals. The way the configurations of temporal intervals are expressed in a query language is out of the scope of this paper and left for future work.

The paper presents two strategies to select relations that characterize configurations of temporal intervals: a composition-based and a neighborhood-based strategy. The selection of temporal relations aims at minimizing the number of relations that are needed to characterize the temporal arrangement of events. Such temporal relations are the basis for content-based retrieval, where these relations represent query criteria or query constraints. For example, given three different events, we can have three binary temporal relations among them, without considering *equal* and *converse* relations. This study explores strategies that indicate whether or not one needs the three relations to characterize the temporal arrangement of events. The hypothesis of this work is that by minimizing the number of constraints in a query evaluation, one decreases the computational cost of content-based retrieval. In addition to applications in content-based retrieval with temporal criteria, this work can be interesting for compressing information about relations between events, or in discovering dynamic patterns by focusing on subsets of temporal relations. Associated with spatial information, where topological relations describe constraints about the spatial arrangement of objects [3, 5, 11, 20], selecting temporal relations may have an impact not only on modeling but also representing dynamic phenomena.

The composition-based strategy relies on the basic derivation property of composition operations, whereas the neighborhood-based strategy exploits the closeness of temporal intervals to select relations between intervals. In this sense, this work follows closely ideas from the work by Rodríguez *et al.* [20], which defines and compares strategies for query pre-processing of topological relations in a 2D space; however, to the best of our knowledge, there has not yet been done some research in the temporal domain in this direction. Thus, a goal of this work was to explore how well strategies that were applied to topological relations between regions in 2D space could be applied to temporal intervals.

The organization of this paper is as follows: Section 2 describes temporal relations and their composition. Section 3 presents the composition- and neighborhood-based strategies for simplifying configurations of temporal intervals. Section 4 evaluates the applicability of these strategies for content-based retrieval of configurations of temporal intervals. Conclusions and future work are presented in Section 5.

2   Temporal Relations

The representation of time by means of intervals rather than points has a history in philosophical studies of time [9, 10]. In 1983, Allen [1] defined a calculus of time intervals as a representation of temporal knowledge that could be used in artificial intelligence. Nowadays, many researchers from different disciplines still use Allen's thirteen temporal relations. Allen's Interval Algebra provides a rich formalism for expressing qualitative relations between temporal intervals. Allen's interval algebra uses the notion of binary relations between convex intervals, which are intervals without gaps. An interval I is represented as a pair $[I-; I+]$ of real numbers with $I- < I+$, denoting the left and right end points of the interval, respectively. This means that Allen only deals with pure intervals without considering point intervals.

Let the beginnings and endings of two events have three possible relations: *smaller* ($<$), *equal* ($=$), and *larger* ($>$). Then events have thirteen possible qualitative relations. The relations that are shown in Table 1 capture the qualitative aspect of event pairs as *before, meets, overlaps, starts, finishes, during,* and *equal,* in terms of constraints on the end points of the constituent temporal intervals. These basic temporal relations have their corresponding converse relations, with *equal* being self-converse. The symbolic representation of temporal relations used in this paper modifies the proposal by Kulpa [12]. It considers symbols that are based on two horizontal lines, representing the two intervals, and one or two vertical lines, representing the temporal topology.

Table 1. Topological temporal relations between two pure intervals

Relation	Symbol	Conditions	Converse
I_1 *before* I_2	⌐_	$I_1+ < I_2-$	*after* (_⌐)
I_1 *meets* I_2	⌐L	$I_1+ = I_2-$	*met_by* (_⌐)
I_1 *overlaps* I_2	⊓L	$I_2- > I_1- \wedge I_1+ < I_2+ \wedge I_1+ > I_2-$	*overlapped_by* (⊔)
I_1 *starts* I_2	⊓_	$I_1- = I_2- \wedge I_1+ < I_2+$	*started_by* (⊓)
I_1 *finishes* I_2	_⊔	$I_1+ = I_2+ \wedge I_1- > I_2-$	*finished_by* (⊓)
I_1 *during* I_2	⊓	$I_1- > I_2- \wedge I_1+ < I_2+$	*contains* (⊔)
I_1 *equal* I_2	▢	$I_1- = I_2- \wedge I_1+ = I_2+$	*equal*

Like the composition table of topological relations between regions [4], there exists a composition table for the temporal relations between intervals (Table 2). In this table, a crisp result of a composition r_{ik} and (;) r_{kj} gives a unique possible relation r_{ij}. For example, the composition I_1 is *before* I_2 ; I_2 *contains* I_3 (⌐-;⊤) has the crisp result that I_1 is *before* I_3 (⌐-). When the composition does not give any information (ni), all relations are possible.

Table 2. The composition table for the thirteen relations

$r_{ik};r_{kj}$	□	⌐-	-⌐	⊓	⊤	⊔	⊐	⌐	⌐	⊏	⊓	⊐	⊓
□	□	⌐-	-⌐	⊓	⊤	⊔	⊐	⌐	⌐	⊏	⊓	⊐	⊓
⌐-	⌐-	⌐-	ni		⌐-	⌐-		⌐-		⌐-	⌐-		⌐-
-⌐	-⌐	ni	-⌐		-⌐		-⌐		-⌐		-⌐	-⌐	-⌐
⊓	⊓	⌐-	-⌐	⊓	ni		⌐-	-⌐	⊓		⊓	⊓	
⊤	⊤				⊤						⊤		⊤
⊔	⊔	⌐-				⌐-		⌐-		⊔			⌐-
⊐	⊐		-⌐				-⌐		⊐			⊐	
⌐	⌐	⌐-			⌐-		⌐-		⌐	⌐			⌐-
⌐	⌐		-⌐	-⌐		-⌐		-⌐		⌐		⌐	⌐
⊏	⊏	⌐-	-⌐	⊓		⌐-		⌐-		⊏		⊏	
⊓	⊓		-⌐		⊤						⊓	⊐	⊤
⊐	⊐	⌐-	-⌐	⊓				⌐		⌐	⊐	⊐	⊐
⊓	⊓	⌐-			⊤		⊔		⌐		⊓		⊓

Although there is a significant analogy between spatial and temporal relations, there are certain important differences. For example, when working with temporal intervals one has to differentiate between X *meets* Y and Y *meets* X, since time has a unidirectional nature.

3 Simplifying Configurations of Temporal Intervals

A configuration of temporal intervals is considered to be a set of temporal intervals that holds particular temporal relations. In this context, a configuration can be seen as a graph G that is complete, directed and labeled, where nodes N are temporal intervals and arcs A are binary relations between these intervals (Equation 1).

$$G = (N, A)$$
$$N = \{I_1, I_2, \ldots, I_n\}$$
$$A = \{r_{ij} | \forall i, j \in [1..n], \exists I_i, I_j \in N\} \tag{1}$$

A graph G that describes a set of n intervals can be represented as a matrix of $n \times n$ elements, where these elements identify binary temporal relations r_{ij}. Elements r_{ii} along the diagonal are the identity relation (i.e., *equal*) (Figure 1).

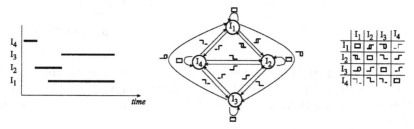

Fig. 1. Temporal configurations as a directed graph represented by an $n \times n$ matrix

The idea of simplifying configurations of temporal intervals aims at determining the subset of temporal relations that can characterize temporal configurations or events. Thus, the idea is to reduce the number of temporal relations without (significantly) reducing the amount of information conveyed by the whole set of relations. To do so, two different strategies are analyzed: (1) composition-based and (2) neighborhood-based strategies.

3.1   Composition-Based Strategy

The composition-based strategy (CBS) for analyzing configurations of temporal intervals is solely based on the composition of binary relations so that no explicit information about the duration and absolute moment on time of intervals are needed. CBS starts with a graph representing a configuration of temporal intervals and selects a minimal subgraph from which one can derive the complete and original graph without losing information about the relations between temporal intervals (i.e., relations defined in Table 1). The strategy for finding this subgraph follows the principles of logical consistency in a graph [14].

Logical consistency: Logical consistency is expressed by the composition of binary relations. Given a configuration expressed as a graph, consistency is formulated as a constraint satisfaction problem [16] over a network of binary relations [15]. To be a consistency network of binary relations, a graph must fulfill three constraints: node-consistency, arc-consistency, and path-consistency.

- Node-consistency: each node must have a self-loop arc, denoting the identity relation.
- Arc-consistency: for each directed arc there must be an arc in the reverse direction, denoting the converse binary relation.
- Path-consistency: although a variety of paths can lead from one node to another, in order to infer the path consistency of a relation it is sufficient to consider all composition paths of length two that connect the relation

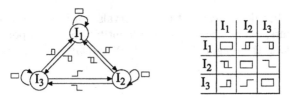

	I_1	I_2	I_3
I_1	▢	⊓	⊐
I_2	⊔	▢	⊑
I_3	⊐	⊏	▢

Fig. 2. Node-, arc-, path- consistency network of temporal relations

between nodes. Having a consistent graph, a relation must coincide with its induced relation determined by the intersection of all possible composition paths of length two (Equation 2, where ';' represents the composition operator).

$$\forall_{i,j} r_{ij} = \bigcap_{k=1}^{n} r_{ik} ; r_{kj} \tag{2}$$

Following the principle of path consistency, a relation can be completely derived if, and only if, it is the only relation that results from the intersection of all possible composition paths of length two in the graph. Consider the example in Figure 2 with three temporal intervals. The graph in the figure corresponds to a node-, arc-, and path-consistent network. The path consistency is checked with the determination of the nine induced relations (Table 3).

Table 3. All possible derivations from path-consistency

Relation	Derivation
$r_{11} = ▢$	$r_{12}; r_{21} \cap r_{11}; r_{11} \cap r_{13}; r_{31} = (⊔ \vee ⊓ \vee ▢ \vee ⊑ \vee ⊏ \vee ⊤ \vee ⊩ \vee ⊐ \vee ⊓)$ $\cap ▢ \cap (⊐ \vee ⊓ \vee ▢)$
$r_{12} = ⊓$	$r_{11}; r_{12} \cap r_{12}; r_{22} \cap r_{13}; r_{32} = ⊓ \cap ⊓ \cap (⊓ \vee ⊤ \vee ⊏)$
$r_{13} = ⊐$	$r_{11}; r_{13} \cap r_{12}; r_{23} \cap r_{13}; r_{33} = ⊐ \cap (⊓ \vee ⊤ \vee ⊐) \cap ⊐$
$r_{21} = ⊔$	$r_{21}; r_{11} \cap r_{22}; r_{21} \cap r_{23}; r_{31} = ⊔ \cap ⊔ \cap (⊔ \vee ⊩ \vee ⊑)$
$r_{22} = ▢$	$r_{21}; r_{12} \cap r_{22}; r_{22} \cap r_{23}; r_{32} = (⊔ \vee ⊓ \vee ▢ \vee ⊤ \vee ⊤ \vee ⊑ \vee ⊏ \vee ⊐ \vee ⊓)$ $\cap ▢ \cap (⊐ \vee ⊓ \vee ▢)$
$r_{23} = ⊑$	$r_{21}; r_{13} \cap r_{22}; r_{23} \cap r_{23}; r_{33} = (⌐ \vee ⊔ \vee ⊑) \cap ⊑ \cap ⊑$
$r_{31} = ⊐$	$r_{31}; r_{11} \cap r_{32}; r_{21} \cap r_{33}; r_{31} = ⊐ \cap (⊓ \vee ⊩ \vee ⊐) \cap ⊐$
$r_{32} = ⊏$	$r_{31}; r_{12} \cap r_{32}; r_{22} \cap r_{33}; r_{32} = (⌐ \vee ⊓ \vee ⊏) \cap ⊏ \cap ⊏$
$r_{33} = ▢$	$r_{31}; r_{13} \cap r_{32}; r_{23} \cap r_{33}; r_{33} = (⊐ \vee ⊐ \vee ▢) \cap (⊐ \vee ⊐ \vee ▢) \cap ▢$

Composition-based simplification: Unlike consistency checking [6], where one is interested in completing a partial graph, a strategy for simplifying configurations of temporal intervals starts from a complete and consistent graph and eliminates relations that can be consistently derived. In such a process of simplifying temporal configurations, it is important to analyze whether or not by eliminating relations we obtain a unique minimal subgraph.

The work in [20] showed that there exists one minimal subgraph that results from the application of the composition-based strategy to simplify topological configurations of regions in a 2D space. In the temporal domain with thirteen interval relations, in contrast, there may be more than one minimal subgraph. A reason for having more than one minimal subgraph is the asymmetric property of temporal relations, which has an effect on the composition of relations. Unlike the spatial domain of 2D regions, where the relation *overlaps* is a relation whose participation in a composition does not result in a crisp result, in the temporal domain each relation participates in at least one composition with a crisp result. Both *overlaps* and *overlapped_by* relations in the temporal domain participate in three different compositions with crisp results. Thus, in the temporal domain, if either *overlaps* or *overlapped_by* results from the intersection of composition paths of length two, these relations may still be used in other composition paths that derive other relations. For instance, consider a configuration with four temporal intervals and the derivable relations presented in Table 4. The relation r_{ij} is derived from the intersection of two composition paths of length two. One of these compositions involves the relation r_{ik}. In addition, r_{ik} can be derived by using a composition path that involves r_{ij}. Thus, one could use r_{ij} or r_{ik} in the composition path and get a crisp result.

Table 4. Example of alternative derivable relations

Relation	Derivation
$r_{ij} = $ ⊐	$r_{il}; r_{lj} \bigcap r_{ik}; r_{kj} = $ ⊐ ; ⌐ $\bigcap$ ⊥ ; ⌐
$r_{ik} = $ ⊓	$r_{ij}; r_{jk} \bigcap r_{il}; r_{lk} = $ ⊐ ; ⊔ $\bigcap$ ⊐ ; ⊓

A more complete example illustrates how the graph of a configuration of temporal intervals, represented as a matrix, can be reduced to a minimum number of relations. Figure 3 shows the set of temporal intervals and the corresponding matrix representation of the initial consistency network.

Table 5 shows the possible derivations from path consistency that result in crisp results. For example, the derivation of the relation between intervals I_1 and I_2 (r_{12}) is achieved by the intersection of the sets of relations that result from the compositions $r_{13}; r_{32}$, $r_{14}; r_{42}$, and $r_{15}; r_{52}$. Unlike this case, the intersection of the results of the compositions $r_{12}; r_{23}$, $r_{14}; r_{43}$, and $r_{15}; r_{53}$ is equivalent to the set $\{$ ⊥ , ⊤ , ⊐ $\}$ (i.e., any of the relations in the set is a possible result) and, therefore, this intersection does not uniquely derive the relation r_{13} (⊐). Cases when the intersections of composition paths do not result in crisp results are not included in Table 5.

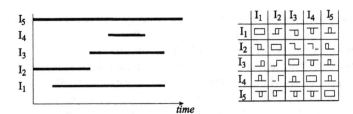

Fig. 3. Temporal configuration with the matrix representation of its consistency network

Table 5 shows that there are seven possible derivations from path consistency; however, it is not possible to eliminate all of these relations. The derivation of r_{12} requires the relation r_{14} and, vice versa, the derivation of r_{14} requires the relation r_{12}. So, we cannot eliminate both relations. The same situation occurs with relations r_{14} and r_{34}, r_{12} and r_{15}, and r_{24} and r_{34}. Note in Table 5 that to derive relation r_{35}, one just needs the composition between relations r_{31} and r_{15}, since only this composition gives a crisp result.

A minimal subgraph will eliminate the maximum number of relations that are derivable, while keeping the minimum number of relations that are needed to completely determine the original consistency network. In this example, there are two possible combinations of derivable relations obtained from node, arc- and path consistency. These combinations create minimal subgraphs with six relations (Figure 4).

Table 5. All possible derivations from path-consistency

Relation	Derivation
$r_{12} =$ ⊐	$r_{13}; r_{32} \cap r_{14}; r_{42} \cap r_{15}; r_{52} = ($ ⊏ ∨ ⊐ ∨ ⊤ $) \cap ($ ⌐ ∨ ⊐ ∨ ⊤ ∨ ⌐ ∨ ⊏ $)$
	$\cap ($ ⌐ ∨ ⊐ ∨ ⊥ ∨ ⌐ ∨ ⊐ $)$
$r_{14} =$ ⊤	$r_{12}; r_{24} \cap r_{13}; r_{34} \cap r_{15}; r_{54} = ($ ⌐ ∨ ⊐ ∨ ⌐ ∨ ⊤ ∨ ⊤ $) \cap ($ ⊤ $) \cap (ni)$
$r_{15} =$ ⊥	$r_{12}; r_{25} \cap r_{13}; r_{35} \cap r_{14}; r_{43} = ($ ⊐ ∨ ⊥ ∨ ⊐ $) \cap ($ ⊐ ∨ ⊥ ∨ ⊏ $) \cap$
	$($ ⊐ ∨ ⊐ ∨ ⊥ ∨ ⊏ ∨ ⊐ ∨ ⊤ ∨ ⊏ ∨ ⊐ ∨ ☐ $)$
$r_{24} =$ ⌐	$r_{21}; r_{14} \cap r_{23}; r_{34} \cap r_{25}; r_{54} = ($ ⌐ ∨ ⊐ ∨ ⌐ ∨ ⊤ ∨ ⊤ $) \cap ($ ⌐ $) \cap$
	$($ ⌐ ∨ ⊐ ∨ ⌐ ∨ ⊤ ∨ ⊤ $)$
$r_{34} =$ ⊤	$r_{31}; r_{14} \cap r_{32}; r_{24} \cap r_{35}; r_{54} = ($ ⌐ ∨ ⊐ ∨ ⌐ ∨ ⊤ ∨ ⊏ $) \cap (ni) \cap$
	$($ ⌐ ∨ ⊐ ∨ ⌐ ∨ ⊤ ∨ ⊤ $)$
$r_{35} =$ ⊥	$r_{31}; r_{15} \cap r_{32}; r_{25} \cap r_{34}; r_{45} = ($ ⊥ $) \cap ($ ⊐ ∨ ⊥ ∨ ⊐ $) \cap$
	$($ ⊐ ∨ ⊐ ∨ ☐ ∨ ⊏ ∨ ⊏ ∨ ⊥ ∨ ⊤ ∨ ⊐ ∨ ⊤ $)$
$r_{45} =$ ⊥	$r_{41}; r_{15} \cap r_{42}; r_{25} \cap r_{43}; r_{35} = ($ ⊥ $) \cap ($ ⌐ ∨ ⊐ ∨ ⌐ ∨ ⊥ ∨ ⊐ $) \cap ($ ⊥ $)$

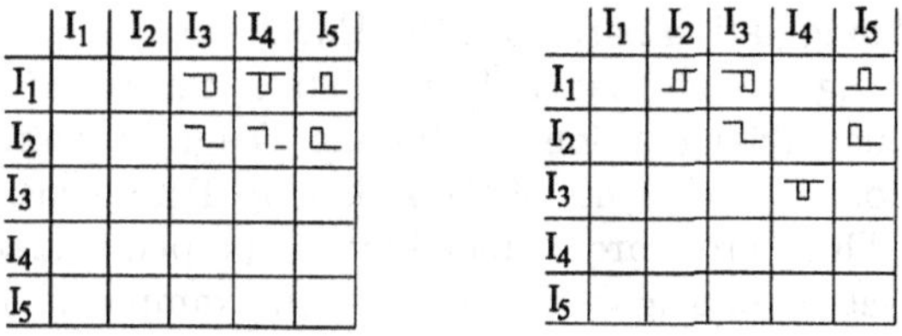

	I_1	I_2	I_3	I_4	I_5
I_1			⊐	⊓	⊔
I_2			⊐	⊓	⊔
I_3					
I_4					
I_5					

	I_1	I_2	I_3	I_4	I_5
I_1		⊓	⊐		⊔
I_2			⊐		⊔
I_3				⊓	
I_4					
I_5					

Fig. 4. Consistent minimal subgraphs derived from node-, arc-, and path-consistency

A strategy for selecting a minimal subgraph is to add heuristics in the search of minimal subgraphs. A basic heuristics is to choose the subgraph that contains more relevant relations, where relevance is a qualitative order of relations. For example, intervals that are far apart may be considered less important. Such a heuristics follows the classic principle of geography, which establishes a stronger connection between close objects [24]. Another strategy may consider to sort relations based on the information content of the relations in a context. Following the argumentation from information theory [21], information content is defined in terms of the uncertainty that the data eliminate, that is, it is negative to the likelihood or frequency of the data. For example, within a large set of intervals, the use of a less frequent relation reduces the uncertainty of which intervals hold such relation. In this work, we selected the subgraph with a larger number of *non-disjoint* relations, considering that these relations are less frequent in data sets [7].

3.2 Neighborhood-Based Strategy

The Neighborhood-Based Strategy (NBS) is a quantitative technique that concentrates on closely related temporal interval points. Making an analogy to Tobler's First Law of Geography [24] "everything is related to everything else, but nearby things are more related than distant things," this work explores a temporal closeness, that is, "everything is related to everything else, but nearby things in time are more related than distant things in time." Very important here is to say that implicitly one talks about the same location in space. It is not that obvious how applicable neighborhood in time is, especially considering the ontological object of time as temporal intervals; however, to the best of our knowledge there has been not much work that explores a neighboring concept of temporal intervals in configurations.

Neighborhood of temporal intervals: This work characterizes a temporal interval by two parameters: (1) the duration (radius) of an interval and (2) the moment when the interval happens, which is identified by the middle point of the interval [12]. These parameters are mapped onto a 2D space with the x-axis being middle points and y-axis being radii of intervals. Within this vector representation, intervals are considered connected or closely related if they are neighbors. Neighboring points are determined by using the Delaunay Triangulation [17, 18].

The Delaunay Triangulation partitions the Euclidean space, composed of a set of points, into triangles such that no four points of this set are co-circular. The dual of the Delaunay Triangulation, the Voronoi Diagram, represents a partition of space into regions where points of the Delaunay Triangulation are the nuclei of specific areas. These areas are bounded by the perpendicular bisectors of the nucleus and the set of its neighboring points. An example of the determination of neighboring temporal intervals is shown in Figure 5.

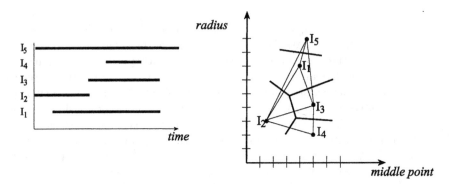

Fig. 5. Representation of temporal intervals and determination of Delaunay Triangulation

From Euler's Equation [18], which is applied for every convex polyhedron with m_n nodes and m_a arcs, one can derive that every node in the Delaunay Triangulation has a degree (i.e., number of neighboring nodes) ≥ 3. If all nodes of a Delaunay Triangulation are substituted with intervals and all arcs with binary relations, we can deduce that for a very large graph, the average number of neighbors (av_ng) of an interval is less than six (Equation 3). Thus, the average total number of arcs must be less than $3n$, that is, it grows linearly by $O(n)$, with n being the number of intervals. This upper bound of the number of relations in the final subgraph contrasts the theoretical bound of $O(n^2)$ of the initial graph.

$$av\_ng = \frac{2m_a}{m_n} \leq 6 - \frac{12}{m_n} \tag{3}$$

Neighborhood-based simplification: To simplify temporal configurations, these configurations are represented as points in a 2D space. Only the arcs that exist in the resulting graph from the Delaunay Triangulation are considered to represent relations that are part of the minimal subgraph. The graph from the Delaunay Triangulation is undirected; however, we consider the arcs of this graph as binary relations, that is, as directed arcs. For example, consider the temporal configuration and the Delaunay Triangulation in Figure 5, its corresponding minimal configuration graph represented as a matrix is shown in Figure 6.

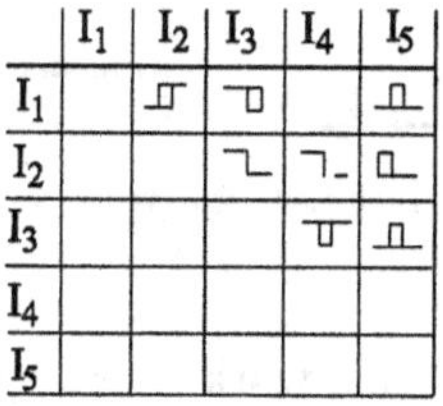

	I_1	I_2	I_3	I_4	I_5
I_1		⌐	⌐		⊓
I_2			⌐	⌐	⌐
I_3				⊓	⊓
I_4					
I_5					

Fig. 6. Minimal configuration graph that is obtained by applying the neighborhood-based simplification

Unlike the composition-based strategy for simplifying temporal configurations, given a representation of temporal configurations in a 2D space, there exists only one minimal subgraph.

4   Experimental Results of Selecting Temporal Relations

In order to evaluate how much these strategies can simplify configurations of temporal intervals, we use these strategies to simplify configurations of temporal intervals that are used as queries in a process of content-based retrieval. The idea behind this analysis is to check the effect of eliminating temporal relations of an original query when comparing the results of the retrieval process (i.e., the set of configurations that are solutions) with respect to the desired answer (i.e., the original query). Eliminations will not have a negative impact if, by considering the subset of relations between temporal intervals as constraints in a retrieval process, one obtains results whose temporal intervals satisfy the complete set of constraints of the query. In such a process, there is a basic comparison between the original configuration and the configurations retrieved from the data set.

4.1   Framework for Comparing Temporal-Interval Configurations

We distinguish two levels in the comparison of temporal relations that we have called *consistency* and *equivalence*. Temporal relations are consistent if they are the same within the set of thirteen interval relations defined by Allen (i.e., there is no contradiction in the relation between intervals); temporal relations are equivalent if they are consistent and are the same based on metric characteristics that distinguish relative size, relative overlapping, and relativity of disjointness. For example, in Figure 7 the relation between I_1 and I_2 is consistent with the relation between intervals I_3 and I_4 ($r_{12} = r_{34}$); however, the pair of intervals (I_1, I_2) is not equivalent to (I_3, I_4) in terms of the lengths of intervals and length of overlap.

In this context, consistency and equivalence are binary decisions; however, equivalence is associated with a *distance function* that depends on metric characteristics of temporal intervals. The comparison of temporal intervals is not concerned with the location of intervals on time and the intervals' absolute lengths. Absolute positions of intervals on time are typically used for *timeslice* queries,

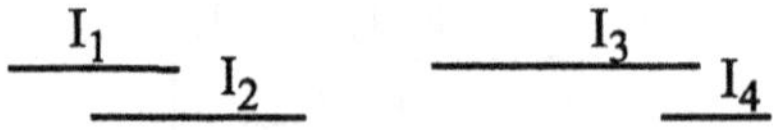

Fig. 7. Consistent but not equivalent temporal relations

which are outside the scope of this work. A relative length is used to make the distance measure less sensitive to scaling of intervals' representations. We define the distance between two pairs of intervals (I_1, I_2) and (I_3, I_4) by:

$$distance((I_1, I_2), (I_3, I_4)) = \begin{cases} |T(I_1, I_2) - T(I_3, I_4)| & \text{if } r_{12} = r_{34} \\ 1 & \text{otherwise} \end{cases} \quad \text{with}$$

$$T(I_j, I_k) = \frac{length(I_j) + length(I_k)}{2length(I_j \bigcup I_k)}$$

$$length(I_j) = I_j^+ - I_j^-$$

$$I_j \bigcup I_k \equiv \{(I^-, I^+)|I^- = min(I_j^-, I_k^-) \wedge I^+ = max(I_j^+, I_k^+)\} \quad (4)$$

This function gives values from 0 to 1, with 1 meaning that the relations are inconsistent and, therefore, non equivalent, and with 0 meaning equivalent relations. To relax the equivalence definition, one could consider that two relations are equivalent if *distance* is less than or equal to a given threshold (set in this experiment to 0.05). In this work, we say that a solution is consistent if all relations in the solution are consistent with respect to the corresponding query. Likewise, a solution is said equivalent if the relations in the solution are equivalent to the relations in the query.

4.2   Evaluation of Strategies

For the analysis of composition- versus neighborhood-based strategies, a synthetic set of temporal intervals was created that allows for experimental replication. The synthetic data set was created with all possible intervals from 0 to 10 temporal units, which results in 55 different temporal intervals. Although the data set is small, it has all possible intervals within a temporal range. These intervals, in combination, create a large set of possible configurations (e.g., for configurations with 5 intervals, there exist 3.4×10^6 possible configurations).

We started with a set of 50 configurations that were created by randomly selecting five temporal intervals (i.e., ten temporal relations or constraints) from the data set. These configurations are queries in the retrieval process and are pre-processed to obtain four different cases of simplified queries: original configurations with all constraints, configurations after eliminating two random relations, configurations derived from the composition-based strategy (CBS), and configurations derived from the neighborhood-based strategy (NBS). We used the basic strategy of random eliminations of relations to evaluate whether or not the results with subgraphs were independent of the strategies of elimination. For the comparison of CBS and NBS, and considering the multiple subgraphs obtained with CBS, we selected the minimal subgraph with more *non-disjoint* relations (i.e., *after* and *before* relations). This heuristics was defined after checking that the results of the searches based on these selected subgraphs gave, on average, better results than other subgraphs. This is in agreement with the fact that overlapping intervals are less common and, therefore, the search based on query constraints defined in terms of these relations reduces the search space.

Original and simplified configurations represent query graphs and query subgraphs, respectively. The results of the retrieval process (i.e., the sets of temporal intervals that satisfy the same relations than the relations in query subgraphs) were compared in terms of degree of consistency and degree of equivalence with respect to query graphs (i.e., graphs with all relations). The number of constraint evaluations is used as a measure of performance, which is commonly used in the evaluations of constraint satisfaction problems and makes the evaluation independent of computational resources.

In cases of using the original configurations with all constraints, the process always finds optimal configurations (i.e., configurations where all constraints are satisfied). In the case of random eliminations of constraints, we found that the number of consistent and equivalent configurations were 75% and 66%, respectively. These values less than the number of consistent and equivalent results by using CBS or NBS. A summary of results for CBS and NBS is presented in Table 6. In this table, one has to consider that each configuration may have more than one solution and that each configuration has a number of consistent and equivalent relations (100% of consistent relations implies that a solution is consistent). Figure 8 shows, in detail, the percentage of equivalent relations in the configurations of 50 random queries (i.e., percentage over 10 constraints per query) for CBS and NBS.

Table 6. Summary of results for CBS and NBS (* average with respect to a configuration that includes all relations)

Parameters	CBS	NBS
Average of consistent configurations	100%	89%
Average of equivalent configurations	73%	85%
Average of consistent relations	100%	96%
Average of equivalent relations	90%	94%
Average percentage of constraint evaluations*	76%	60%

As we increased the number of intervals (six and seven intervals) and, therefore, the number of temporal relations to fifteen or twenty-one relations, we observed a tendency to maintain the results of NBS and CBS in terms of the average percentages of relations that are consistent and equivalent in configurations. In particular, the percentages of consistent relations in configurations for NBS and CBS were 94% and 100%, respectively, and the percentages of equiva-

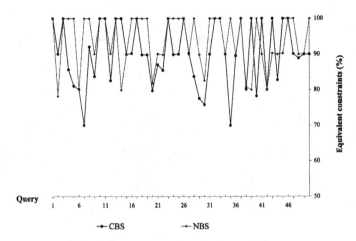

Fig. 8. Percentage of equivalent relations

lent relations were 94% and 92%, respectively. The main difference was obtained in the number of equivalent configurations for CBS, which was on average 10% less than the average of equivalent configurations with respect to the initial five intervals.

5   Conclusions and Future Work

This paper has analyzed two different strategies to define a subset of temporal relations that characterizes a set of temporal intervals. The composition-based strategy is based on the composition of temporal intervals, and the neighborhood-based strategy selects relations that connect neighboring intervals represented in a 2D space.

While the composition-based method guarantees that relations can be derived, the neighborhood-based method captures metric characteristics of temporal intervals with respect to relative length, relative overlapping, and relativity of disjointness. In most cases, such strategies were able to characterize temporal configurations. From a different perspective, finding minimal subgraphs based on the composition is a complex task that could lead to an intractable problem unless one uses heuristics to reduce the search space. The results from this work indicate the potential of using the algebraic properties of composition and the metric characteristics of intervals for reducing the number of relations considered in a process. The use of fewer relations in a search process was on average computationally less expensive than the use of the complete description of sets of temporal relations. These strategies have a potential use in applications based on the analysis of temporal relations, such as content-based retrieval and data mining.

As future work, we will continue analyzing the efficiency of algorithms to apply these strategies to a larger scale. The results of this work complement the previous results with respect to topological relations between regions in a 2D space. We expect, in the near future, to analyze spatio-temporal relations within the same framework. In particular, a continuation of this work will be to apply the strategies to the simplification of spatio-temporal configurations that are described by a combination of spatial and temporal relations [2].

Acknowledgment

We would like to thank Professor Max Egenhofer for his contribution to the previous work published in [20]. Andrea Rodríguez's research work is partially funded by Nucleus Millenium Center for Web Research, Grant P01-029-F, Mideplan, Chile. Nico Van de Weghe gratefully acknowledges the financial support of the Fund for Scientific Research - Flanders (Belgium) for his research assistantship.

References

1. J. Allen. Maintaining knowledge about temporal intervals. *Communications of the ACM*, 26(11):823–843, 1983.
2. C. Claramunt and B. Jiang. An integrated representation of spatial and temporal relations between evolving regions. *Geographical Systems*, 3(4):411–428, 2001.
3. E. Clementi, J. Sharma, and M. Egenhofer. Modelling topological spatial relations strategies for query languages. *Computing and Graphics*, 18(6):815–822, 1994.
4. M. Egenhofer. Deriving the composition of binary topological relations. *Journal of Visual Languages and Computing*, 5(2):133–149, 1994.
5. M. Egenhofer. Query preprocessing in spatial query by sketch. *Journal of Visual Computing*, 8(4):403–424, 1997.
6. M. Egenhofer and J. Sharma. Assessing the consistency of complete and incomplete topological information. *Geographical Systems*, 1:47–68, 1993.
7. J. Florence and M. Egenhofer. Distribution of topological relations in geographic databases. In *ACSM/ASPRS*, Baltimore, MD, 1996.
8. C. Freksa. Temporal reasoning based on semi intervals. *Artificial Intelligence*, 54:199–227, 1992.
9. C. Hamblin. Instants and intervals. *Studium Generale*, 24:127–134, 1971.
10. I. Humberstone. Interval semantics for tense logic: Some remarks. *Journal of Philosophical Logic*, 8:171–196, 1979.
11. H. Kriegel and T. Brinkhoff. Efficient spatial query processing. *IEEE Data Engineering Bulletin*, 16(3):10–15, 1993.
12. Z. Kulpa. Diagrammatic representation for a space of intervals. *Machine Graphics*, 6(1):5–24, 1997.
13. G. Langran. Temporal GIS design tradeoffs. *Journal of URISA*, 2(2):16–25, 1990.
14. A. Mackworth. Consistency in networks of relations. *Artificial Intelligence*, 8(1):99–118, 1977.
15. R. Maddux. *Some Algebras and Algorithms for Reasoning about Time and Space*. Technical report, Department of Mathematics, IOWA State University, Ames, IO, 1990.
16. P. Meseguer. Constraint satisfaction problems: an overview. *Artificial Intelligence Communications AICOM*, 2(1):3–17, 1989.

17. J. O'Rouke. *Computational Geometry*. Cmabridge University Press, Cambridge, MA, 1993.
18. F. Preparata and M. Shamos. *Computational Geometry: An Introduction*. Springer-Verlag, Berlin, 1985.
19. B. Ramachandran and F. MacLeod. Modelling temporal changes in GIS using an object-oriented approach. In *Proceedings of the Sixth International Symposium on Spatial Data Handling*, pages 518–537, Edinburgh, Scotland, 1994.
20. A. Rodríguez, M. Egenhofer, and A. Blaser. Query pre-processing of topological constraints: Comparing composition-based with neighborhood-based approach. In T. Hadzilacos, Y. Manolopoulos, J. Roddick, and Y. Theodoridis, editors, *Spatial and Temporal Databases. LNCS 2750*, pages 362–379, Santorini, Greece, 2003. Springer-Verlag.
21. S. Ross. *A First Course in Probability*. Macmillan, 1976.
22. O. Stock. *Spatial and Temporal Reasoning*. Kluwer Academic Publishers, Dordrecht, The Netherlands, 1997.
23. A. Tarski. On the calculus of relations. *Journal of Symbolic Logic*, 6(3):73–89, 1941.
24. W. Tobler. A computer movie simulating urban growth in the detroit region. *Economic Geography*, 46(2):234–240, 1970.

Towards a Temporal Extension
of Spatial Allocation Modeling

Takeshi Shirabe

Institute for Geoinformation
Technical University of Vienna, 1040 Vienna, Austria
shirabe@geoinfo.tuwien.ac.at

Abstract. Spatial allocation models represent decision processes of allocating discrete spatial units (e.g., land parcels and grid cells) to particular uses under the assumption that attributes of those units are unchanged over time and all decisions are made at one time. Given forecasts of spatial attributes, this assumption may be dropped, though spatial and temporal considerations will be highly interwoven and difficult to articulate in simple algebraic terms. This paper describes a systematic approach to analyzing and resolving this complexity in formulating time-dependent spatial allocation models. The implication of the paper is that geographic information systems can serve not only as devices to store and manage spatio-temporal data, but also as platforms to build spatio-temporal decision models.

1 Introduction

Consider a simple decision problem that seeks a suitable site for forestation as follows. Given a set of parcels, not previously forested, with different levels of timber productivity, select a connected subset of a required size that maximizes the total timber productivity. A problem of this kind can be modeled in terms of mathematical functions that lend themselves to an algorithmic solution. Exact models for the present example are available in the literature [27, 33]. Fig. 1 shows an optimal solution to a problem of selecting 30 cells from a ten-by-ten grid, where each cell has been arbitrarily assigned a timber productivity score.

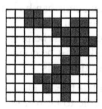

Fig. 1. An optimal site for forestation. Throughout this paper, it is assumed that two cells are adjacent if they share a cell side.

It has been found convenient to use such mathematical decision models – commonly known as mathematical programming models – along with geographic information systems (GISs) [6, 8, 24]. These two tools complement each other at least in

M.J. Egenhofer, C. Freksa, and H.J. Miller (Eds.): GIScience 2004, LNCS 3234, pp. 285–298, 2004.

two ways. First, a GIS can store and process spatial data in the form that is required by a mathematical model. This helps one to test a model with different datasets with relative ease. Second, solutions generated by a model can be displayed in a GIS for visual inspection and analyzed using GIS tools. Any information discovered will help revise the current model. These interactions are possible even with the two tools apart, but may be streamlined in one integrated platform. Such integration can be made relatively quickly, but "loosely" [24], in a manner of transferring data files between a model and a GIS. If a model is general enough, it may be embedded in a GIS as extended functionality to reach a wider range of users.

Time is becoming an important factor in GIS [3, 10, 17, 26, 35], but currently, in many GIS applications, data still tend to represent static phenomena. Decisions based on such static data tend to be static, too, as they are to be made simultaneously rather than sequentially. The shortest path problem is a good example. Path finding is a time-dependent activity in nature, since as one follows a path, time goes by. If travel time or cost associated with each street segment is fixed, however, a path finding solution can be seen as a simultaneous selection of multiple street segments. This simplified view of path finding will suffice if one travels fast or short enough to escape temporal effects. Otherwise changing traffic conditions affect one's decision of which path to take. Such dynamic network problems have recently gained growing attention in various fields, such as transportation, management, and geographic information sciences [4, 7, 10, 13, 15, 19-21, 36].

On the other hand, relatively little work has been done for dynamic allocation of space, which involves decisions of "what part of space" to use "at what time." One of a few such examples is "timber harvest scheduling," which seeks efficient harvesting patterns in space and time [5, 14, 16, 18, 22, 23, 25, 29, 31].

Now let us extend the earlier example to accommodate some elements of timber harvest scheduling. The problem is considered in a time frame divided into ten periods of equal length. Assume that the amount of timber each cell produces is given as a function of when it is forested and when it is harvested. Then, in which periods and which cells should be forested and harvested (clear-cut), while satisfying the following criteria?

1. In each period, ten or fewer cells can be forested.
2. No cell can be forested more than once.
3. All forestation activities must end before any harvest activity begins.
4. In each period, ten or fewer cells can be harvested.
5. In each period, no two adjacent cells can be harvested.
6. No cell can be harvested more than once.
7. At least half of the forested cells must be preserved.
8. All the preserved forested cells must be connected.
9. The total amount of harvested timber is to be maximized.

Although temporal dimension has been added to both data and decisions, the problem is still simple enough to be able to be formulated as a mathematical model. One such model has led to a possible plan, illustrated in Fig. 2.

The problem, however, requires a far more elaborate mathematical formulation than the previous time-independent case. Hence it does not seem easy to be generalized so that similar problems can be mechanically formulated – ultimately in a GIS on a routine basis.

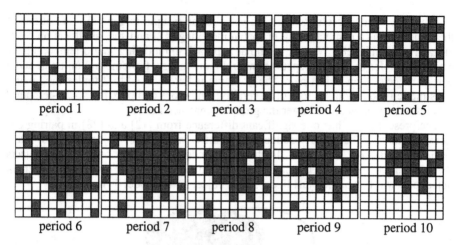

Fig. 2. Transition of forest patterns resulting from a prescribed forestation-harvest schedule. Shaded cells are forested (and not harvested).

The purpose of this paper is to make one step towards such a generalization. Here we do not attempt to give a full mathematical formulation to any particular instance, nor is a complete scheme of spatio-temporal allocation modeling presented. Rather we aim to illuminate the importance of temporal considerations in spatial allocation modeling, and propose a simple systematic view of such complex processes. This in turn should provide a sound basis for subsequent formal modeling of the problem under consideration. The rest of this paper is organized as follows: Section 2 defines what we call spatio-temporal allocation and its basic components. Section 3 analyzes criteria associated with spatio-temporal allocation in terms of those components. Section 4 addresses an exemplar problem. Section 5 concludes the paper.

2 Spatio-temporal Allocation

Spatio-temporal allocation generally deals with where and when to make certain activities. These decisions are seldom made at random, but according to some criteria. A modeler's goal is thus to put these criteria into unambiguous terms. This section describes basic ingredients of spatio-temporal allocation criteria.

2.1 Entities

It is assumed that the given study area and time frame are discretized into finite sets of *spatial units* and *temporal units*, respectively. They are not necessarily regular in size and shape. This is particularly true for spatial units which may be defined merely for administrative purposes. Census tracts and zip-code areas are such examples. Coordinating a spatial unit and a temporal unit makes a finest grain of space-time, called a *spatio-temporal unit*. One or more spatio-temporal units are to be allocated to a larger fiat entity, which we call an *object*. It depicts when and where to do a certain activity.

It has been found useful to regard a pair – or an *n*-tuple in general – of entities of the same type as a distinct entity in spatial modeling [12] and spatial allocation modeling in particular [28]. We take a similar approach to modeling spatio-temporal allocation, and refer to a pair of spatio-temporal units as a spatio-temporal unit pair and a pair of objects as an object pair. The grouping is here limited to 2-tuples (i.e., pairs) for simplicity, but without loss of generality. The more complex the spatio-temporal process to be modeled, the larger the grouping may need to be.

The present paper has one significant difference from [12] and [28] in pairing objects (not units). That is, it introduces three different forms of object pairings: intersection, subtraction, and union. Each corresponds to a subset of the paired objects (Fig. 3). An object is a set of spatio-temporal units, and so is an object pair. Then an object pair will be handled just like a single object in the model formulation.

Fig. 3. Three kinds of object pairs. The solid and the dashed ovals represent two objects to be paired. The unshaded (left) and the lightly shaded parts (right) represent two subtraction object pairs, the darkly shaded part (center) represents an intersection object pair, and all three parts collectively represents a union object pair.

The following notation is used for denoting those entities described in this section:

I: set of spatial units constituting a study area. Each spatial unit is denoted by i or j.
T: set of temporal units constituting a time frame. Each temporal unit is denoted by t or s.
O: set of objects. Each object is denoted by m or n.

In addition, the intersection of objects m and n is denoted by $m \cap n$, the subtraction of object n from object m by m / n, the subtraction of object m from object n by n / m, and the union of objects m and n by $m \cup n$.

2.2 Decision Variables

With the basic entities defined as above, a solution to a spatio-temporal allocation problem can be seen as a collection of binary decisions of whether a certain spatio-temporal unit is allocated to a certain object. Additionally, allocation decisions may involve spatio-temporal unit pairs or object pairs. There are six such cases:

– if spatio-temporal unit (i,t) is allocated to object pair $m \cap n$
– if spatio-temporal unit (i,t) is allocated to object pair m / n
– if spatio-temporal unit (i,t) is allocated to object pair n / m
– if spatio-temporal unit (i,t) is allocated to object pair $m \cup n$
– if spatio-temporal unit pair (i,t) and (j,s) are allocated to object m
– if spatio-temporal unit pair (i,t) and (j,s) are allocated to object pair $m \cup n$

At this point, we have not found any benefit of differentiating the types of object pairs in allocating spatio-temporal unit pairs. So we have arbitrarily chosen the union as the only type of object pair to which spatio-temporal unit pairs can be allocated. This "unit-pair-to-object-pair" allocation is interpreted in a way that spatio-temporal units (i,t) and (j,s) are allocated to objects m and n, respectively.

All the decisions above are expressed by combining one or more of the following 0-1 variables:

x_{itm} :   equals 1 if spatio-temporal unit (i,t) is allocated to object m, and 0 otherwise.

y_{itmjsn} : equals 1 if spatio-temporal units (i,t) and (j,s) are allocated to objects m and n, respectively, and 0 otherwise.

z_{itmjsn} : equals 1 if spatio-temporal unit (i,t) is allocated to object m, while spatio-temporal unit (j,s) is *not* allocated to object n, and 0 otherwise.

Note that i and j, t and s, and m and n may be identical.

The first type of variable is formally constrained as follows:

$$x_{itm} = \{0,1\} \qquad \forall i \in I, t \in T, m \in O \qquad (1)$$

The others are, by definition, dependent on x_{itm}'s. More specifically, $y_{itmjsn} = 1$ if and only if $x_{itm} = 1$ and $x_{jsn} = 1$, while $z_{itmjsn} = 1$ if and only if $x_{itm} = 1$ and $x_{jsn} = 0$. These relations could be formulated in quadratic terms, that is, $y_{itmjsn} = x_{itm} x_{jsn}$ and $z_{itmjsn} = x_{itm}(1 - x_{jsn})$. These formulations, however, would lead to a non-linear integer programming model, which is generally considered as a difficult class of mathematical programming model to solve. Alternatively, the following set of linear inequalities may be used to make a more tractable, linear integer programming model:

$$y_{itmjsn} \geq x_{itm} + x_{jsn} - 1 \qquad \forall i, j \in I, t, s \in T, m, n \in O \qquad (2)$$

$$y_{itmjsn} \leq x_{itm} \qquad \forall i, j \in I, t, s \in T, m, n \in O \qquad (3)$$

$$y_{itmjsn} \leq x_{jsn} \qquad \forall i, j \in I, t, s \in T, m, n \in O \qquad (4)$$

$$y_{itmjsn} \geq 0 \qquad \forall i, j \in I, t, s \in T, m, n \in O \qquad (5)$$

$$z_{itmjsn} \geq x_{itm} - x_{jsn} \qquad \forall i, j \in I, t, s \in T, m, n \in O \qquad (6)$$

$$z_{itmjsn} \leq x_{itm} \qquad \forall i, j \in I, t, s \in T, m, n \in O \qquad (7)$$

$$z_{itmjsn} \leq 1 - x_{jsn} \qquad \forall i, j \in I, t, s \in T, m, n \in O \qquad (8)$$

$$z_{itmjsn} \geq 0 \qquad \forall i, j \in I, t, s \in T, m, n \in O \qquad (9)$$

2.3 Attributes

Each spatial unit has an attribute that may change over time. Such a dynamic spatial phenomenon can be represented by a series of attributes of spatio-temporal units. Attributes can be assigned not only to individual spatio-temporal units, but to spatio-temporal unit pairs. An example of such a pairwise attribute is the degree of separation between two locations in space or time.

The following notation is used for attributes:

a_{it} :   attribute associated with spatio-temporal unit (i,t).

a_{itjs} :   attribute associated with spatio-temporal unit pair (i,t) and (j,s).

There may be cases where all spatio-temporal units, pertaining to a particular spatial or temporal unit, share an identical attribute. Then the shared attribute is denoted without the subscript corresponding to those units sharing the attribute. For example, a_t may refer to the length of the temporal unit t (shared by all spatial units), and a_{ij} to the spatial adjacency between spatial units i and j (shared by all temporal units).

3 Spatio-temporal Allocation Criteria

Our strategy for systematizing the formulation of a spatio-temporal allocation problem is to cast allocation criteria in terms of relations among attributes of objects (or object pairs). Here, object (or object pair) attribute is restricted to the kind expressible as a function of the attributes of all the spatio-temporal units (or spatio-temporal unit pairs) that constitute an object (or object pair) (Fig. 4). Certainly, it is not claimed that all allocation criteria can be articulated in this way.

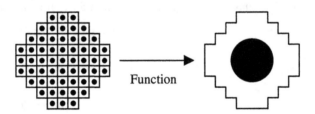

Fig. 4. Attribute of an object as a function of the attributes of all spatio-temporal units in that object. On the left-hand side, each square and its enclosed dot represent a constituent spatio-temporal unit and its attribute, respectively. On the right-hand side, the polygon and its enclosed dot represent the object and its attribute, respectively.

What functions to implement depends on both how accurate problems need to be modeled and how computationally tractable they need to be. In this paper only four functions are considered: one that sums all input attributes, one that returns the minimum of all input attributes, one that returns the maximum of all input attributes, and one that determines whether all spatio-temporal units associated with input attributes (indicating the adjacency of each unit pair) are connected. This selection is entirely for a practical reason: the functions can be formulated using linear functions only [28]. For more complex problems to be modeled, non-linear functions would need to be introduced, but at the cost of tractability. Nevertheless, the present set of functions should suffice to show the expressiveness of the proposed modeling approach.

In what follows, spatio-temporal allocation criteria are classified into spatial, temporal, and spatio-temporal according to their dominant dimension(s). Though our approach is abstract enough to deal with all these criteria in a unified manner, the distinction helps to find out how to do so.

3.1 Spatial Criteria

Some spatio-temporal allocation criteria are not concerned with when spatial units are allocated. Consider as an example that a developer plans to acquire two sets of land parcels for different uses. The developer may only be interested in the area of each set and the proximity between the two sets, no matter what order each parcel is going to be acquired. Such considerations are purely spatial, and may be easily handled by projecting objects onto a spatial plane (Fig. 5).

The relationship between an object, m, and its spatial extent (i.e., flattened object) can be represented by the following set of inequalities.

$$x_{im} \geq x_{itm} \qquad\qquad \forall i \in I, t \in T \qquad\qquad (10)$$

$$x_{im} \leq \sum_{t \in T} x_{itm} \qquad\qquad \forall i \in I \qquad\qquad (11)$$

$$x_{im} \leq 1 \qquad\qquad \forall i \in I \qquad\qquad (12)$$

where $x_{im} = 1$ if spatial unit i is allocated to object m (regardless of when), and 0 otherwise. Not surprisingly, these new variables coincide with the most fundamental decision variables of time-independent spatial allocation models [28]. It has been found that spatial criteria – often concerning size, shape, or relation – can be formulated using a small set of functions involving those variables in an open-ended manner [28].

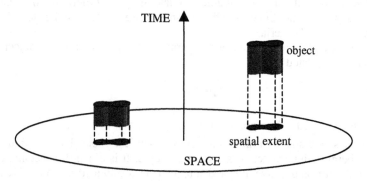

Fig. 5. Objects in space time and their spatial extents. An object can be any set of spatiotemporal units and may not be connected (much less tubular) like those in this figure.

3.2 Temporal Criteria

Contrary to spatial criteria, temporal criteria are only interested in when (any) spatial units are allocated. Examples include the duration of a construction project and the order of two such projects. Again, one can collapse objects onto the temporal axis to highlight their temporal intervals (Fig. 6).

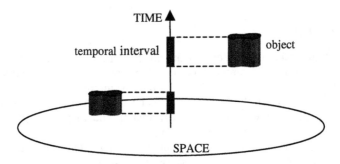

Fig. 6. Objects in space time and their temporal intervals

The relationship between an object, m, and its temporal interval can be represented by the following set of inequalities:

$$x_{tm} \geq x_{itm} \qquad \forall i \in I, t \in T \qquad (13)$$

$$x_{tm} \leq \sum_{i \in I} x_{itm} \qquad \forall t \in T \qquad (14)$$

$$x_{tm} \leq 1 \qquad \forall t \in T \qquad (15)$$

where $x_{tm} = 1$ if one or more spatial units are allocated to object m during temporal unit t, and 0 otherwise.

As with spatial criteria, it is possible to relate temporal criteria to size, shape, and relation. Due to the one-dimensionality of time, however, temporal criteria have less variation than spatial criteria.

The most straightforward kind of temporal size is the length of an object's "life" [10]. This is formulated as:

$$\sum_{t \in T} a_t x_{tm} \qquad (16)$$

where a_t is the length of temporal unit t.

Temporal shape sounds unfamiliar but is possible to conceive. Continuity – the quality of being uninterrupted – is a good example. It is a one-dimensional counterpart of contiguity and convexity, and thus can be formulated as such [27, 33, 34]. Among others are compactness and elongation, which are normally considered as spatial attributes. One useful metric for their evaluation is the difference between the time an object – any piece of it – first appears and the time it last disappears. Its actual formulation involves several linear equations, which are here encapsulated by the following expression:

$$\max_{t \in T}(e_t x_{tm}) - \min_{t \in T}(b_t x_{tm}) \tag{17}$$

where b_t and e_t indicate when temporal unit t begins and when it ends, respectively.

If the result of Eqn. 17 is greater than the result of Eqn. 16, the object experiences some breaks during its lifetime, which in turn implies how (not) compact in time the object is. The object is most compact when Eqns. 16 and 17 take the same value (i.e., there is no break).

Unlike temporal size and shape, there are a variety of distinct temporal relations. Allen [2] identified (and Freksa [11] later generalized) 13 of them that are all-inclusive and mutually exclusive. Included are "before," "equal," "meets," "over-laps," "during," "starts," "finishes," and their converse relations except for "equal." Their formulations resemble those for spatial relations [9], which are based on the overlap and adjacency of two objects, except that the arrow of time is one-way. For instance, the temporal relation "object m before object n" can be formulated as:

$$\max_{t \in T}(e_t x_{tm}) \leq \min_{t \in T}(b_t x_{tn}) \tag{18}$$

The left-hand side indicates when object m ends, while the right-hand side indicates when object n begins.

3.3 Spatio-temporal Criteria

Criteria of this type refer to both space and time. These include trivial ones such as "the object must be somewhere at *this time*" or "the object must be *here* all the time." It is easy to see that these criteria can be reduced to spatial or temporal ones by look-ing at only those spatio-temporal units associated with a particular temporal or spatial unit (Fig. 7).

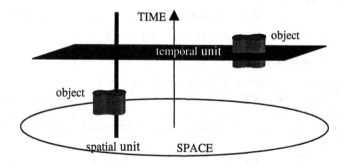

Fig. 7. The plane "slicing" an object (right) represents a temporal unit of particular interest. The line "piercing" an object (left) represents a spatial unit of particular interest.

Other spatio-temporal criteria must deal with whole objects. If objects in space time are seen in analogy to three-dimensional objects in space (as illustrated in those figures in this section), the concepts of size, shape, and relation may hold. Then many spatio-temporal criteria are modeled in the same manner as spatial or temporal criteria discussed above. Yet their interpretations differ. For example, a "total" size is not measured in volumetric terms but as an accumulation of spatial attributes (e.g. area and cost) over time. Also, the "overlap" relation refers to two objects that share a part occurring at the same time and at the same location. Similarly, "contiguity" is shaped by temporally adjacent spatio-temporal unit pairs or spatially adjacent ones.

More interesting, though challenging, spatio-temporal criteria concern the way objects change – perhaps in size, shape, and relation – with time. In a context of urban planning, for example, a county's park system may be expected to grow at a certain rate. In forestry, a timber harvest schedule often prohibits clear-cutting adjacent harvest units within a short period of time. Elsewhere, a marketing strategy may program to seed initial sales districts at different locations and gradually expand them until they completely merge. These criteria may seem rather chaotic, but possible to model under the present scheme. Still, change-related criteria have been largely unexplored.

4 Example

To illustrate how the present scheme is applied, we return to the forestation-harvest problem introduced in Section 1 as an example. In doing so, it is assumed that the spatial units (ten-by-ten grid cells), the temporal units (ten periods), and their associated variables are denoted by the notation introduced in Section 2, unless any confusion occurs.

The first step is to identify the objects to be modeled. Since forestation and harvest are different activities, it seems reasonable to assign one object to each. Then let m and n be the forestation object and the harvest object, respectively. The next step is to catalogue relevant attributes, which include:

b_t :   time when period t begins.

e_t :   time when period t ends.

a_{ij} :   spatial adjacency between cells i and j, which takes on 1 if they are adjacent, and 0 otherwise.

t_{itjs} :   amount of timber to be produced if cell i is forested during period t and cell j is harvested during period s. If $i \neq j$ or $t \geq s$, $t_{itjs} = 0$.

Then each criterion is addressed below.

1.   In each period, ten or fewer cells can be forested.
     This is a trivial spatio-temporal criterion, which can be reduced to a spatial criterion concerning size. It is done by taking into account only those spatio-temporal units associated with each temporal unit. The criterion takes the following formulation:

$$\sum_{i \in I} x_{itm} \leq 10 \qquad\qquad \forall t \in T \qquad\qquad (19)$$

     The left-hand side counts the number of cells harvested during each period.

2.   No cell can be forested more than once.
     This is similar to the previous criterion, except that it is analyzed from the perspective of each spatial unit instead of each temporal unit. The criterion is then formulated as follows:

$$\sum_{t \in T} x_{itm} \leq 1 \qquad\qquad \forall i \in I \qquad\qquad (20)$$

The left-hand side counts the number of harvesting periods for each cell.

3. All forestation activities must end before any harvest activity begins.

This is a temporal criterion concerning the relation "before" between two objects. This relation can be clarified by projecting both objects onto the temporal axis, and formulated such that:

$$\max_{t \in T}(e_t x_{tm}) \leq \min_{t \in T}(b_t x_{tm}) \qquad\qquad (21)$$

The left-hand side indicates the last time any cell is forested, while the right-hand side indicates the first time any cell is harvested.

4. In each period, ten or fewer cells can be harvested.

This is the same as criterion 1, except that this is for object n. Its formulation is:

$$\sum_{i \in I} x_{itn} \leq 10 \qquad\qquad \forall t \in T \qquad\qquad (22)$$

5. In each period, no two adjacent cells can be harvested.

This spatio-temporal criterion can be reduced to a spatial criterion concerning the shape "fragmentation" [30]. This shape results from excluding any adjacent spatio-temporal unit pair from an object for each temporal unit, as expressed by:

$$\sum_{i,j \in I} a_{ij} y_{itnjtn} = 0 \qquad\qquad \forall t \in T \qquad\qquad (23)$$

The left-hand side indicates the number of adjacent pairs of cells harvested during each period.

6. No cell can be harvested more than once.

This is the same as criterion 2, except that this is for object n. Its formulation is:

$$\sum_{t \in T} x_{itn} \leq 1 \qquad\qquad \forall i \in I \qquad\qquad (24)$$

7. At least half of the forested cells must be preserved.

This is a simple spatio-temporal criterion concerning size. The size here refers to the number of spatio-temporal units of an object. The criterion compares two such sizes in the following form:

$$\sum_{t \in T} \sum_{i \in I} x_{itm} = 2 \sum_{t \in T} \sum_{i \in I} x_{itn} \qquad\qquad (25)$$

The left-hand side counts the number of cells in the forestation object, while the right-hand side twice counts the number of cells in the harvest object.

8. All the preserved forested cells must be connected.

This is a spatial criterion concerning the shape "contiguity." Unlike others, no notation has been given to the contiguity (or connectedness) function. So, let us here only paraphrase the criterion in a way that fits the present scheme:

The spatial extent of object pair m/n, projected on a spatial plane, is to be connected in terms of the cell-adjacency defined by a_{ij}.

Object pair m/n is a part of the forestation object that does not overlap the harvest object.

9. The total amount of harvested timber is to be maximized.
    This is another spatio-temporal criterion regarding size. The size here is the sum
    of attributes of spatio-temporal unit pairs allocated to an object pair. The use of
    an object pair is necessary since timber production results from both forestation
    and harvest activities. Considering its objective of maximization, this criterion is
    formulated as:

$$\text{Maximize} \qquad \sum_{t,s \in Ti, j \in I} \sum t_{itjs}\, y_{itmjsn} \qquad\qquad (26)$$

Each term in the summation function indicates the amount of timber gained from
a pair of forestation and harvest activities.

There can be more than one possible formulation for each criterion. Those presented here are not necessarily most efficient (but fairly intuitive). For example, criterion 2 can be formulated in a more tractable form such that:

$$\max_{i \in I, t \in T}(e_t x_{itm}) \le \min_{i \in I, t \in T}(b_t x_{itn}) \qquad\qquad (27)$$

Thus, for any system to implement the present scheme as a modeling language, an
internal engine would be useful for transforming specified criteria into simpler
equivalents.

5 Conclusions

A rudimentary strategy has been presented for systematizing the formulation of spatio-temporal allocation problems. These problems often involve a number of complex
criteria that might not be required by those without temporal considerations. To cope
with this complexity in establishing spatio-temporal allocation criteria, we limit entities in a model to a few kinds – spatial units, temporal units, spatio-temporal units,
objects, and their pairings – and employ a small set of elementary functions involving
(attributes of) those entities – sum, max, min, and connectedness – encapsulating
details of actual algebraic formulations. These building blocks can then be combined
to specify particular criteria in an open-ended manner.

It must be admitted, however, that very complex criteria will not be able to be articulated in this way. Take, for example, the adjacency criterion described above (criterion 5). Considering its aims to prevent large openings, it would be more practical to
restate the criterion such that "in each period, no contiguous set of cells that exceeds a
certain limit can be harvested" (adapted from [18, 22]). The present method is unable
to deal with this version of adjacency constraint, unless every impermissible combination of cells is groped for an n-tuple and recognized as an entity of a model. The enumeration of all such n-tuples, however, could be as hard as the original problem. As
such, future research should identify the types of problems that the present approach
would (and would not) address.

The scheme described here seems possible to be implemented in GIS. Spatio-
temporal units and their attributes can be stored in the form of maps or tables, while
objects can be manipulated and analyzed – "flattened," "pierced," or "sliced" –
graphically. These features would help GIS to be more like "systems to think with"

[6, 32]. Such systems still would need to await more powerful mathematical programming optimizers or heuristics than what current standards offer.

Acknowledgments

The author would like to thank Simon McBride and the anonymous reviewers for their valuable comments on the manuscript.

References

1. Ahuja, R.K., Orlin, J.B., Pallottino, and S., Scutella, M.G.: Minimum Time and Minimum Cost Path Problems in Street Networks with Traffic Lights. Transportation Science 36(3) (2002) 326-336
2. Allen, J.F.: Maintaining Knowledge about Temporal Intervals. Communication of the ACM 26(11) (1983) 832-843
3. Al-Taha, K.K. and Barrera, R.: Temporal Data and GIS: An Overview. Proceedings of GIS/LIS '90 (1990) 244-254
4. Chabini, I: Discrete Dynamic Shortest Path Problems in Transportation Applications: Complexity and Algorithms with Optimal Run Time. Transportation Research Record 1645 (1998) 170-175
5. Dahlin, B. and Sallnäs, O.: Harvest Scheduling under Adjacency Constraints – A Case Study from the Swedish Sub-alpine Region. Scandinavian Journal of Forest Research 8 (1993) 281-290
6. Densham, P.J.: Integrating GIS and Spatial Modeling: Visual Interactive Modeling and Location Selection. Geographical Systems 1(3) (1994) 203-219
7. Desaulniers, G. and Villeneuve D.: The Shortest Path Problem with Time Windows and Linear Waiting Costs. Transportation Science 34(3) (2000) 312-319
8. Diamond, J.T. and Wright, J.R.: Design of an Integrated Spatial Information System for Multiobjective Land-use Planning. Environment and Planning B 15(2) (1988) 205-214
9. Egenhofer, M. and Franzosa, R.: Point-Set Topological Spatial Relations. International Journal of Geographical Information Systems 5(2) (1991) 161-174
10. Frank, A.U., Raper, J. and Cheylan, J.P.: Life and Motion of Socio-Economic Units, Taylor & Francis, London (2001)
11. Freksa, C.: Conceptual neighborhood and its role in temporal and spatial reasoning. In: Singh, M. and Travé-Massuyès, L. (eds.): Decision Support Systems and Qualitative Reasoning, North-Holland, Amsterdam (1991) 181-187
12. Goodchild, M.F.: Towards an Enumeration and Classification of GIS Functions. Proceedings, International Geographic Information Systems (IGIS) Symposium: The Research Agenda, NASA, Washington, DC, Vol. 2 (1988) 67-77
13. Jason, B.N.: Dynamic Traffic Assignment for Urban Road Networks. Transportation Research B 25 (2/3) (1991) 143-161
14. Johnson, K.N. and Scheurman, H.L.: Techniques for Prescribing Optimal Timber Harvest and Investment under Different Objectives: Discussion and Synthesis. Forest Science Monograph 18 (1977)
15. Kaufman, D.E., Nonis, J. and Smith, R.L.: A Mixed Integer Linear Programming Model for Dynamic Route Guidance. Transportation Research B 32(6) (1998) 431-440
16. Kirby, M.W., Hager, W.A. and Wong, P.: Simultaneous Planning of Wildland Management and Transportation Alternatives. In: Kallio, M., Andersson, A.E., Seppala, R., and Morgan, A. (eds.): Systems Analysis in Forestry and Forest Industries, Vol. 21. North-Holland/TIMS Studies in Management. Elsevier Scientific Publishing Co., New York, (1986) 371-387
17. Langran, G.: Time in Geographic Information Systems, Taylor & Francis, London (1992)

18. McDill, M.E., Rebain, S., and Braze, J.: Harvest Scheduling with Area-Based Adjacency Constraints. Forest Science 48(4) (2002) 631-642
19. Merchant, D.K. and Nemhauser, G.L.: A Model and an Algorithm for the Dynamic Traffic Assignment Problems. Transportation Science 12(3) (1978) 183-207
20. Miller, H.J.: Measuring Space-Time Accessibility Benefits within Transportation Networks: Basic Theory and Computational Methods. Geographical Analysis 31(2) (1999) 187-212
21. Miller, H.J. and Wu, Y.-H.: GIS Software for Measuring Space-Time Accessibility in Transportation Planning and Analysis. GeoInformatica 4(2) (2000) 141-159
22. Murray, A.T.: Spatial Restrictions in Harvest Scheduling. Forest Science 45(1) (1999) 45-52
23. Nelson, J. and Brodie, J.D.: Comparison of a Random Search Algorithm and Mixed Integer Programming for Solving Area-based Forest Plans. Canadian Journal of Forest Research 20 (1990) 934-942.
24. Nyerges, T.: Coupling GIS and Spatial Analytic Models. Proceedings, 5th International Spatial Data Handling Symposium, Charleston, SC (1992) 534-543
25. O'Hare, A.J., Faaland, B.H., and Bare, B.B.: Spatially Constrained Timber Harvest Scheduling. Canadian Journal of Forest Research 19 (1989) 715-724
26. Peuquet, D.J.: Time in GIS and Geographical Databases. In: Longley, P.A., Goodchild, M.F., Maguire, D.J., and Rhind, D.W. (eds.): Geographical Information Systems, Vol. 1, Principles and Technical Issues, 2nd edition. John Wiley & Sons, New York, (1999) 91-103
27. Shirabe, T.: Modeling Topological Properties of a Raster Region for Spatial Optimization. Proceedings, 11th International Symposium on Spatial Data Handling, Leicester, UK (2004)
28. Shirabe, T. and Tomlin, C.D.: Decomposing Integer Programming Models for Spatial Allocation. In: Egenhofer, M.J. and Mark, D.M. (eds.): Proceedings, GIScience 2002, Lecture Notes in Computer Science, Vol. 2478 (2002) 300-312
29. Snyder, S. and ReVelle, C.: Temporal and Spatial Harvesting of Irregular Systems of Parcels. Canadian Journal of Forest Research 26 (1996) 1079-1088
30. Taylor, P.J.: Distances within Shapes: An Introduction to a Family of Finite Frequency Distributions. Discussion paper, No. 16, Department of Geography, University of Iowa, Iowa City, IA (1970) 1-20
31. Weintraub, A., Barahona, F., and Epstein, R.: A Column Generation Algorithm for Solving General Forest Planning Problems with Adjacency Constraints. Forest Science 40(1) (1994) 142-161
32. Weber, E.S.: Systems to Think with: A Response to "A Vision for Decision Support Systems." Journal of Management Information Systems 2 (4) (1986) 85-97
33. Williams, J.C.: A Zero-One Programming Model for Contiguous Land Acquisition, Geographical Analysis 34(4) (2002) 330-349
34. Williams, J.C.: Convex Land Acquisition with Zero-One Programming, Environment and Planning B 30(2) (2003) 255-270
35. Yuan, M.: Modeling Semantical, Temporal, and Spatial Information in Geographic Information Systems. In: Craglia, M. and Couclelis, H. (eds.): Geographic Information Research: Bridging the Atlantic, Taylor & Francis, London (1996) 334-347
36. Ziliaskopoulos, A.K. and Mahmassani, H.S.: A Note on Least Time Path Computation Considering Delays and Prohibitions for Intersection. Transportation Research B 30(5) (1996) 359-367

Formalizing User Actions for Ontologies

Keanhuat Soon and Werner Kuhn

Institute for Geoinformatics, University of Muenster
Robert-Koch-Str. 26-28, D-48149 Muenster, Germany
{kh.soon,kuhn}@uni-muenster.de

Abstract. We demonstrate an approach to the formalization of user actions[1] for the purpose of producing task-oriented ontologies. Verbs and nouns are extracted from a document that depicts user actions during a surface water monitoring process. The conceptual structure of the user actions is formalized through a combination of Formal Concept Analysis and Entailment theory. The subconcept-superconcept relations between user actions in a concept lattice are refined through the entailments of troponymy, proper inclusion, and backward presupposition. The approach is intended to strengthen the consideration of user tasks in ontology engineering for geographic information applications. We aim to enhance the context specification in semantic similarity assessments with *action lattices*.

1 Introduction

Semantic interoperability [4] has received attention in the geospatial information community as well as in Artificial Intelligence (AI) and Knowledge Engineering. The idea of an ontology as an explicit specification of a conceptualization [14] has been proposed for realizing it. An ontology describes the contents of a particular domain or task setting, so that different contents can be matched and integrated during decision-making. So far, most ontologies for geospatial information and most ontology languages are limited to entities, leaving out the conceptualization of actions.

1.1 Goal

We propose a method using ideas from Formal Concept Analysis (FCA) and Entailment theory to construct a conceptual structure of user actions for ontologies. An *action lattice* is built from the relations found between the user actions through FCA [11]. Complementing the *action lattice*, Fellbaum's [9] notion of *entailment* between two actions is adopted. The subconcept-superconcept relations between two user actions are thus supplemented by *entailment* relations.

A case study is introduced to present the construction of an action structure from the user perspective. An official document, which contains the user actions, is selected as the main knowledge source. This document describes all the user actions that are to be performed during surface water monitoring in the context of the Euro-

---

[1] Actions in our sense are often called *tasks* in the human-computer interaction literature [18].

M.J. Egenhofer, C. Freksa, and H.J. Miller (Eds.): GIScience 2004, LNCS 3234, pp. 299–312, 2004.

pean Water Framework Directive (WFD). The user actions are extracted from the verb and noun phrases in the document.

The overall goal of this ongoing research is to bring task-related knowledge to ontology engineering [13], resulting in the development of ontologies that support user actions [21]. With the developed *action lattice*, we aim to enhance the context specification in semantic similarity assessment [27]. We call the result of our knowledge extraction and formalization process an "action structure," as it does not yet have the form of an ontology expressed in a standard language (such as OWL, the web ontology language).

1.2 Theoretical Motivation

The motivation for this work is to enhance the definition of context specification to be used in semantic similarity assessment [27]. The definition of context specification in [27] rests on a string matching between user intentions and functions afforded by the entities. The user is prompted to specify the intended use of the entities from a list of functions, which the entities afford. For example, after the intended use is specified as *play* for *sport facilities*, entities like *athletic field, ball park, tennis court, sport arena, swimming pool* and *stadium*, all of which afford the function *play*, are included in the context specification. However, the verb *play* has different senses: *perform* (e.g., drama, concert, and show), *meet* (e.g., Manchester United plays Newcastle United this weekend), and *act* (e.g., Albert played Vespucci). If the user is interested in *sport facilities* with the function *play* in the sense of *perform*, the result is not a good match. A more precise result would be *stadium* and *sport arena* which facilitate not only open-air sport but entertainments[2]. So far, there has been no further consideration of the semantics of intended use.

Our study extends the context specification of [27] by developing a user action ontology. This ontology will serve as the action vocabulary for the user to specify intended use on the one hand, and facilitate semantic similarity measurements between the intended use and the functions afforded by the entities on the other hand. With the possibility that each verb can have more than one sense, the action ontology supports users in selecting an intended use that suits their needs.

In forming the user action ontology, Formal Concept Analysis (FCA) is used to formalize the actions. In FCA, the formal objects of subconcepts share the formal attributes of the superconcepts. For example, in the *action lattice* of Figure 1, the formal attribute *play* is shared by the formal objects *stadium, sport arena* and *tennis court*. However, the formal attribute *perform* is only shared by *stadium* and *sport arena* (FCA will be discussed in more detail in section 4).

When the user is searching for entities that afford the action *play*, the context specification will involve the entities *tennis court, stadium* and *sport arena*. The ac-

---

[2] According to WordNet 2.0,

athletic field – a piece of land prepared for playing a game,

swimming pool – pool that provides a facility for swimming,

tennis court – the court on which tennis is played,

stadium – a large structure for open-air sports or entertainments,

arena – a large field for open-air sports or entertainments,

ball park – a facility in which ball games are played.

tion lattice allows the user to refine the concept of *play* with the action *perform*. The entities satisfying this interest are *stadium* and *sport arena* only. This approach not only produces more specific results, it also increases the accuracy of the context specification.

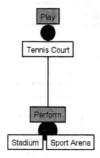

Fig. 1. A simple action lattice

1.3 Practical Motivation

The European Water Framework Directive (http://europa.eu.int/comm/environment/water/water-framework/index_en.html) foresees three types of monitoring programs: surveillance monitoring, operational monitoring and investigative monitoring. Surveillance monitoring is the earliest stage of the monitoring program. In order to establish the surveillance monitoring network, monitoring stations are included based on the significance of the flow rate and the volume of the water within the river basin. The monitoring stations which are found at risk after surveillance monitoring are required to perform operational monitoring. At monitoring stations where further investigation is needed to find the causes of accidental pollution, investigative monitoring is carried out [32].

In managing the data used in water monitoring, the *Guidance Document on Implementing the GIS Elements of the WFD* has defined the *Surface Monitoring Station* as subsumed by the *Monitoring Station* ([33], p. 43) (Figure 2).

According to Figure 2, the surface monitoring station has the following attributes that can be used to identify the functions it performs:

- Drinking – Y/N if the station monitors drinking water
- Investigative – Y/N if the station is an investigative station
- Operational – Y/N if the station is an operational station
- Habitat – Y/N if the station is a habitat monitoring station
- Surveillance – Y/N if the station is surveillance station
- Reference – Y/N if the station is a reference station
- Depth – Depth in meters.

The given attributes of the surface water monitoring station have no semantic description. Each attribute indicates a function of the surface water monitoring station (e.g., drinking, surveillance, etc). When a user is looking for stations doing surveillance monitoring, is the monitoring station for operational monitoring of interest? This situation occurs when the user plans to establish a particular monitoring network (e.g., surveillance monitoring network) from the available monitoring stations. Figure 3 illustrates the case where three surface water monitoring stations have different attribute values. In the European Water Framework Directive context, every operational monitoring station also has the function for surveillance. When the user invokes the above query, however, the operational station C would not be found.

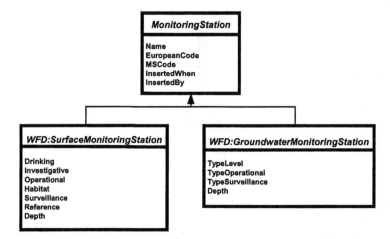

Fig. 2. Monitoring Station, Surface Monitoring Station and Ground Water Monitoring Station in the WFD implementation guide

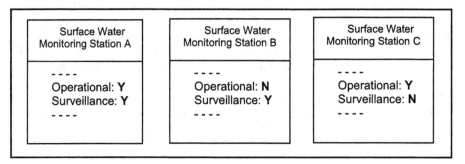

Fig. 3. Three stations with different attribute values

One may try to use constraints on the database to solve this problem. When the value of operational attribute is "Y" (yes), the database would set "Y" for the surveillance attribute. However, this is a surface treatment for the deeper issue of action-related dependencies between the attributes in a database [31].

The current attributes for the monitoring station, with the lack of semantic relations between the functions, would not give the exact answer to the user. Our study constructs an action ontology in surface water monitoring. Based on the affordances from the *action lattice*, the user can get their intended results efficiently.

In the following section, some background for this study is discussed, along with the reasons user actions need to be described in more detail. In section 3, the idea to extract task knowledge from a typical mandate document is explored. The formalization of the *action lattice* is explained in section 4 and section 5 describes the use of the developed *action lattice*. Section 6 draws conclusions and outlines future work.

2 Background

This section gives a description of notions used throughout this study. Current achievements in developing action-centered ontologies are also discussed.

2.1 Ontologies

In the context of information extraction and retrieval, different kinds of ontologies can be distinguished [15]:

- *Top-level ontologies* describe very general concepts like space and time, not depending on a particular domain,
- *Domain ontologies and task ontologies* describe the vocabulary related to a generic domain or kind of task, detailing the terms used in the top-level ontology,
- *Application ontologies* describe the concepts that depend on the particular domain and task within a specific activity.

Several investigations have been conducted to bring actions (tasks) to bear on ontologies. Among them are Chandrasekaran *et al.* [6] and Mizoguchi *et al.* [23] in the fields of AI and Knowledge Engineering. For the geospatial domain, Kuhn [21] and Raubal and Kuhn [26] have attempted to support human actions in ontologies for transportation. Acknowledging the importance of human actions in the geographic domain, a research workshop was held in 2002, bringing together experts from different disciplines to share the knowledge and work on this issue [1]. Camara [5], one of the workshop participants, has proposed that action-driven spatial ontologies are formed via category theory, for the case of emergency action plans.

2.2 Affordances

The idea of affordances goes back at least to Gibson [12] and has been discussed in the geographic domain since the '90s [20, 17, 28]. Under this notion, objects are characterized by the actions they afford. In surface water monitoring context, the monitoring stations are used to monitor the surface water bodies. According to the affordances notion, the monitoring stations afford the action *monitor* to the users.

Modeling differences between monitoring stations by affordances requires afforded actions below the *monitor* level. The *monitor* action needs to be decomposed into several sub-actions. For example, the quality level should be *identified* in the surface water monitoring context. The action to *identify* a quality level is somehow refining the *monitor* action (as will be explained in more detail in section 4).

Refining the *monitor* action makes the distinction between the station types more apparent. Different from surveillance monitoring stations, the quality level of operational monitoring stations must be *identified* beforehand. The quality level is then used to categorize whether the particular monitoring station is at risk and operational monitoring is needed. Each operational monitoring station not only affords the action *monitor*, but also the *identify* action lacking for surveillance monitoring stations.

The affordance notion inspires the approach used in this study, where the user actions decide on the (affording) objects to be included. Introducing the notion into FCA, afforded-actions-as-attributes characterize the objects and imply other user actions through entailments.

3   Case Study

In order to capture user actions in the surface water monitoring domain, the official document from WFD [32] has been used. The document describes the framework for community (Member States) action in the field of water policy. Only one particular section in the appendix of the document, describing the monitoring domain, has been selected for the study.

Table 1 represents the extracted verbs from the text with corresponding affording objects (direct objects of the extracted verbs) in alphabetical order. The verbs whose subjects are not referring to actors that perform the monitoring activity are filtered out. In addition, the noun phrases (affording objects) that are synonymous are grouped into one meaning. For example, parameters and quality elements are generalized to quality elements.

Table 1. Extracted verbs with corresponding affording objects

Verbs (Actions)	Affording Objects
ascertain	*magnitude of accidental pollution*
assess	*water bodies status changes*
ensure	*appropriate monitoring points*
establish	*monitoring program*
identify	*appropriate taxonomic level*
monitor	*discharged substances*
provide	*maps*
select	*quality elements*

Instead of the manual extraction adopted here, automated retrieval [29] could be used to retrieve this kind of information from textual descriptions. Manual extraction is a time consuming method, but it achieves better results. Furthermore, the length of the document used in this study is moderate. The time used to extract and revise the information would not be significantly different with automated approaches.

4   Formalization

Formal Concept Analysis [11] has received attention in various domains such as database re-engineering [16] and software engineering [30]. In the geographic domain, FCA has been adopted by Kavouras and Kokla [19] in coping with the problem of integrating different land use categorizations.

FCA has been selected for this study due to its capacity to produce simple lattice representations of entity *and* action structure. The diagrammatic representation of a formal context gives readers a clear picture of the relations between the formal concepts [25]. In addition, the mathematical lattice can be interpreted as a classification system. Within the cross-tabulation of FCA, formal attributes are classified according to their relationships with formal objects. The following section will explain the basic notions of FCA. For detailed descriptions refer to [11].

4.1 Formal Concept Analysis

In FCA, a *formal context* is a triple (G, M, I) if G and M are sets and $I \subseteq G \times M$ is a binary relation between G and M. We call the elements in G *formal objects*, the elements in M *formal attributes* and I, the *incidence* of the context (G, M, I), is written as gIm or $(g, m) \in I$ if $g \in G$ and $m \in M$. This *incidence relation* of the context can be read as "the object g *has* the attribute m".

A pair (A, B) is a *formal concept* (denoted by c) of the formal context (G, M, I) iff

$$A \subseteq G, B \subseteq M, A' = B, \text{ and } B' = A \tag{1}$$

where A is called the *extent* (denoted by $Ext(c)$) and B the *intent* (denoted by $Int(c)$) of the formal concept $c := (A, B)$. The set of all common formal attributes of a set of formal objects, $A \subseteq G$ is denoted by A'

$$A' := \{ m \in M \mid (g, m) \in I \text{ for all } g \in A \} \tag{2}$$

and, the set of all common formal objects of a set of formal attribute, $B \subseteq M$ is denoted by B'

$$B' := \{ g \in G \mid (g, m) \in I \text{ for all } m \in B \} \tag{3}$$

The set of all formal concepts of (G, M, I) is denoted by $ß(G, M, I)$. The ordered set of all formal concepts forms a mathematical lattice and is denoted by $\underline{ß}(G, M, I)$. The formal subconcept-superconcept relation is the most important feature in $\underline{ß}(G, M, I)$. The formal concept is ordered by the formal subconcept-superconcept relation and is defined as

$$(A_1, B_1) \le (A_2, B_2) : \Leftrightarrow A_1 \subseteq A_2 (\Leftrightarrow B_2 \subseteq B_1) \tag{4}$$

The formal concept (A_1, B_1) is a subconcept of the formal concept (A_2, B_2) (denoted by $(A_1, B_1) \le (A_2, B_2)$) if $A_1 \subseteq A_2$, which is equivalent to $B_2 \subseteq B_1$. In other words, c_1 is subconcept of c_2 if the extent of $c_1 \subseteq$ extent of c_2, and the intent of $c_2 \subseteq$ intent of c_1.

We illustrate the context of *woman, man* and *adult* with attributes *female, grown-up, male, mother* and *parent*. In FCA, *woman, man* and *adult* in our context are the formal objects while *female, grown-up, male, mother* and *parent* are the formal attributes. We classify the formal attributes according to the related formal objects within the cross tabulation. The mathematical lattice can be constructed with open source tools, such as the Concept Explorer (http://sourceforge.net/projects/conexp) (Figure 4).

The set of all common formal attributes (A') of a set of formal objects is ({*grown-up*}) which is commonly shared by the extent (A'', ({*adult, woman, man*}) (indicated with bold rectangle in the table). For a set of all common formal objects (B') of a set of formal attributes is ({*woman*}) which is shared by the intent (B'', ({*grown-up, female, mother, parent*}) (indicated with shading). Following this, the formal concepts of the context are formed as ({*woman*}, {*female, grown-up, mother, parent*}), ({*man*}, {*male, grown-up*}), ({*adult*}, {*grown-up*}) and ({*adult, woman, man*}, {*grown-up*}).

There is a subconcept-superconcept relation between the formal concepts of the context. For example, the formal concept ({*man*}, {*grown-up, male*}) is a subconcept of the formal concept ({*adult, woman, man*}, {*grown-up*}). The formal object/formal attribute of ({*man*}, {*grown-up, male*}) is less/more than ({*adult, woman, man*}, {*grown-up*}).

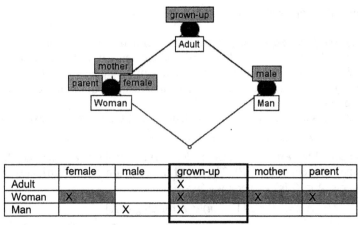

	female	male	grown-up	mother	parent
Adult			X		
Woman	X		X	X	X
Man		X	X		

Fig. 4. The mathematical lattice and cross tabulation

The notion of *implication between attributes* is used in FCA. Based on this notion, the formal context is not stated explicitly, but we can infer it with respect to the implications between the attributes ([11], p. 79). For example, the formal attribute *mother*, from the previous example, *implies* the formal attribute *parent*. Someone is a *mother* implies that she is also a *parent*.

4.2 Formal Concepts of User Actions

Using the notion of *implication between attributes*, a cross tabulation is formed between affording objects and corresponding verbal adjectives (verbs with suffix "-able") as attributes (Table 2). The cells, which are shaded, imply the following attributes at the same rows. For instance, the quality elements are *selectable* implies a monitoring program is *establishable*, appropriate monitoring points are *ensurable*, discharged substances are *monitorable*, appropriate taxonomic level is *identifiable* and water body status changes are *assessable* (at the third row of Table 2). Similarly, {*selectable*} → {*establishable, ensurable, monitorable, identifiable, assessable*}. In other words, when a user is able to *select* the quality elements, the actions *establish, ensure, monitor, identify* and *assess* are also afforded to the user.

The idea of forming Table 2, by using verbal adjectives as attributes, is motivated from [8]. The purpose of adding the suffix "-able" to verbs is to turn them into attributes. Moreover, the verbal adjectives convey the meaning "capable of" or "able to", which coincide with the notion of affordances. For example, *quality elements are selectable* means that *quality elements are able to be selected* or quality elements *afford* the action select. Subsequently, the extent (A'') and intent (A') of the formal concept of cross tabulation is represented in Table 3 with respect to the corresponding concepts.

For each monitoring program, member states are required to produce maps showing the surface water monitoring network in a river basin management plan. Accordingly, the results of monitoring (i.e., the identified levels for the selected quality elements) have to be shown in maps. The formal object *maps* thus appears in all the extent (A'') of Table 3.

Table 2. Cross tabulation of affording objects with attributes

	providable	ascertainable	selectable	assessable	identifiable	monitorable	ensurable	establishable
maps	x	x	x	x	x	x	x	x
magnitude of accidental pollution		x		x		x	x	x
quality elements			x	x	x	x	x	x
water body status changes				x		x	x	x
appropriate taxonomic level					x	x	x	x
discharged substances						x	x	x
appropriate monitoring points							x	x
monitoring program								x

Table 3. Extent (A'') and intent (A') with corresponding concepts

(A'', A')	Concepts
({maps}, {providable, ascertainable, selectable, assessable, identifiable, monitorable, ensurable, establishable})	provide maps
({magnitude of accidental pollution, maps}, {ascertainable, assessable, monitorable, ensurable, establishable})	ascertain magnitude of accidental pollution
({quality elements, maps}, {selectable, assessable, identifiable, monitorable, ensurable, establishable})	select quality elements
({water body status changes, quality elements, magnitude of accidental pollution, maps}, {assessable, monitorable, ensurable, establishable})	assess water body status changes
({appropriate taxonomic level, quality elements, maps}, {identifiable, monitorable, ensurable, establishable})	identify appropriate taxonomic level
({discharged substances, appropriate taxonomic level, water body status changes, quality elements, magnitude of accidental pollution, maps}, {monitorable, ensurable, establishable})	monitor discharged substances
({appropriate monitoring points, discharged substances, appropriate taxonomic level, water body status changes, quality elements, magnitude of accidental pollution, maps}, {ensurable, establishable})	ensure appropriate monitoring points
({monitoring program, appropriate monitoring points, discharged substances, appropriate taxonomic level, water body status changes, quality elements, magnitude of accidental pollution, maps}, {establishable})	establish monitoring program

The subconcept-superconcept relations of these concepts are then depicted as an *action lattice* in Figure 5. Through this *action lattice*, the monitoring domain is now analyzed and modeled in terms of afforded actions [7].

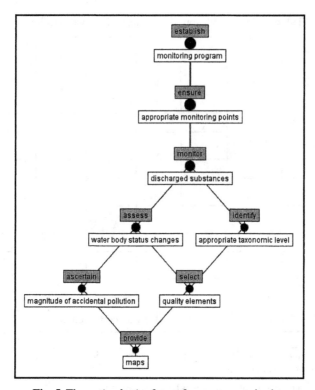

Fig. 5. The *action lattice* for surface water monitoring

However, the representation in Figure 5 does not provide further details of the relations found between these actions. It only presents a general relation (that of subconcept-superconcept) between two actions with affording objects. For example, *assess water body status changes* is a subconcept of *monitor discharged substances*, but we are not informed whether these actions could happen at the same time or separately or whether they have any other logical connections. In order to investigate the meaning in additional relations, the notion of *entailment* from Fellbaum [9] is applied to the description of the semantic relations between two actions.

4.3 Entailment: Semantic Relations Between Actions

The semantic relations between verbs (actions) differ from the relations between nouns, pronouns, adjectives and other parts of speech [2, 22, 10]. According to Fellbaum [9], the semantic relations between actions can be divided into four different relationships with *entailments*:

- Troponymy – two actions co-occur (temporally co-extensive),
- Proper Inclusion – two actions occur in the same time interval but not for the entire duration,
- Backward Presupposition – the second action precedes the first action,
- Causation – the first action causes the second action.

Troponymy. Troponymy is a semantic relation between actions where the first action co-occurs with the second action, that is, the two actions happen in the same time duration (start and end at the same time point). In the example given by Fellbaum [9], *limp* entails *walk*, when the action *limp* happens, the action *walk* co-occurs in a certain manner (in this case, slowly).

Proposition 1. Assess is troponymy of *monitor*.

Rationale 1. The *assess*ment of water body status changes implicitly contains the meaning to *monitor* water bodies. When the *assess* action is started, the *monitor* meaning composed in the *assess* action intrinsically is also started, and when the *assess* action is stopped, the *monitor* meaning in the *assess* action is terminated as well. When someone is *monitoring* something formally (in a certain manner), he is also *assessing* at the same time.

Proper Inclusion. Proper inclusion means two actions could happen at the same time but not for the entire duration. For example, *snore* entails *sleep* by proper inclusion. *Snore* does not have any similar meaning to *sleep*, but when someone is *snoring* he is *sleeping* as well. When someone is *sleeping*, someone may not be *snoring*, or not for the entire period of sleep, and *sleeping* can continue when someone stops *snoring*.

Proposition 2. Identify is properly included in *monitor*.

Rationale 2. Identify is performed during *monitor*ing. Someone needs to *identify* the appropriate level of the quality elements during the *monitoring. Identify* may not happen in the entire *monitoring* process.

Backward Presupposition. Backward Presupposition is the semantic relation in which the second action precedes the first action. For instance, *succeed* entails *try* by backward presupposition. Someone who *succeeds* has *tried* before.

Proposition 3. Select backwardly presupposes *identify*.

Rationale 3. If someone *selects* the quality elements of a water body, they have already *identified* the appropriate level of the quality elements and then made a decision to *select*.

The following examples also follow this relation in the case study:

- *ensure* backwardly presupposes *establish*,
- *select* backwardly presupposes *assess*,
- *provide* backwardly presupposes *select*,
- *monitor* backwardly presupposes *ensure*,
- *ascertain* backwardly presupposes *assess*,
- *provide* backwardly presupposes *ascertain*.

Causation. This semantic relation is describing the causality between two actions. The example given by Fellbaum [9] is *give* entails *have*. When someone is *giving*, it causes others to *have* something (although this seems somewhat less than a logical necessity). No relation has been found between the actions in our case study that represents this semantic relation.

Following the investigation of the semantic relations between actions, Table 4 is formed, where the actions are grouped with respect to the types of semantic relations. Table 4 shows that besides the subconcept-superconcept relation between *assess* and

monitor, for instance, these two actions also comprise the semantic relation of troponymy, that is, when the *assess* action happens, the *monitor* action also happens in
the same time duration.

Table 4. Semantic relations of actions

Type of Relation	Relation
Troponymy	*assess – monitor*
Proper Inclusion	*identify – monitor*
Backward	*ensure – establish, monitor – ensure, select – identify, select – as-*
Presupposition	*sess, provide – select, ascertain – assess, provide – ascertain*
Causation	None

5   Application

The developed *action lattice* facilitates the selection of monitoring stations for the
surface water monitoring network. The user can specify the intended use from the
action lattice. During the establishment of the operational monitoring network, the
user needs to find out monitoring stations at risk. The stations that at risk are the stations with their quality levels (e.g., good, medium, bad) *identified*. Based on this notion, users can specify the function of *identify* from the *action lattice* to get the intended monitoring stations for operational monitoring. The intended result of this
search is the monitoring stations that afford the function of *identify*.

The monitoring stations with at least one of these functions: *monitor, select, identify, assess, ascertain, provide*, are the user's intended result of the query "is the operational monitoring station of interest for surveillance monitoring" (section 1.3). In
establishing the surveillance monitoring network, the monitoring stations that afford
monitor would be appropriate (the surveillance monitoring is the earliest monitoring
phase where detailed information, like the quality level of the station, is not needed).
The monitoring stations of the intended result have been selected owing to the afforded function of *monitor*.

6   Conclusions and Future Work

Conceptual hierarchies are crucial for the development of any knowledge-based system and in particular for ontologies [24]. The strength of ontologies depends on the
conceptual structures they are composed of [34]. This paper presents a study in constructing conceptual hierarchies for user actions in a geospatial application domain.
Such a structure captures the problem solving process and describes user tasks at the
conceptual level.

We do not claim that our approach will achieve a semantic similarity evaluation
between entities. It enhances the context specification, since, prior to a semantic similarity measurement, entities are identified that match the user's interest more specifically.

In future research, we will investigate how mappings between actions can support
the reuse of information sources (data and services) for related tasks. Besides applying Formal Concept Analysis and entailments, we will consider the notion of Information Flow [3] for semantic integration between action concepts.

Acknowledgments

Support for this work of Muenster's Semantic Interoperability Lab (MUSIL) (http://musil.uni-muenster.de) is partially provided by the European Commission, through the ACE-GIS project (IST-2001-37724). We thank to three anonymous reviewers for their invaluable comments and suggestions.

References

1. ACTOR: Research Workshop. Action-oriented Approaches in Geographic Information Science (2002) http://www.spatial.maine.edu/~actor2002/index.html
2. Ballmer, T. and Brennenstuhl, W.: Speech Act Classification. A Study in the Lexical Analysis of English Speech Activity Verbs. Springer-Verlag, Berlin (1981)
3. Barwise, J. and Seligman, J.: Information Flow. The Logic of Distributed Systems. Cambridge Tracts in Theoretical Computer Science 44. Cambridge University Press, US (1997)
4. Bishr, Y.: Semantic Aspects of Interoperable GIS. ITC Publication No. 56, Enschede, Netherlands (1997)
5. Câmara, G., de Carvalho, M. T. M., Fonseca, F., Casanova, M. A., Paiva, J., and Gibotti, F.: From Workflows to Action-Driven Ontologies. Towards a Framework for Interoperability of Geographical Processes. In: Research Workshop. Action-Oriented Approaches in Geographic Information Science, Holden, ME (2002)
   http://www.spatial.maine.edu/~actor2002/participants/camara.pdf
6. Chandrasekaran, B., Josephson, J. R. and Benjamins, V. R: Ontology of Tasks and Methods. In: Knowledge Acquisition Workshop, Banff, Canada (1998)
   http://www.cis.ohio-state.edu/~chandra/Ontology-of-Tasks-Methods.pdf
7. Chandrasekaran, B., Johnson, T. and Smith J. W.: Task Structure Analysis for Knowledge Modeling. In: Communications of the ACM **33** (1992) 124-136
8. Cimiano, P., Hotho, A., Stumme, G. and Tane, J.: Conceptual Knowledge Processing with Formal Concept Analysis and Ontologies. In: Eklund, P. (ed.): Proceedings of Second International Conference on Formal Concept Analysis, ICFCA 2004, Sydney Australia. LNAI 2961 Springer, Berlin (2004) 189-207
9. Fellbaum, C.: English Verbs as a Semantic Net. International Journal of Lexicography **3** (1990) 270-301
10. Fellbaum, C.: On the Semantics of Troponymy. In: Green, R, Bean, C. A. and Myaeng, S. H. (eds.): The Semantics of Relationships. An Interdisciplinary Perspective. Kluwer Academic Publishers, Netherlands (2002) 23-34
11. Ganter, B. and Wille, R.: Formal Concept Analysis. Mathematical Foundations. Springer-Verlag, Berlin Heidelberg New York (1999)
12. Gibson, J. J.: The Ecological Approach to Visual Perception. Lawrence Erlbaum, Hillsdale, NJ (1986)
13. Gomez-Perez, A., Fernandez-Lopez, M. and Corcho, O.: Ontological Engineering. With Examples from the Areas of Knowledge Management, E-Commerce and the Semantic Web. Springer, London (2004)
14. Gruber, T. R.: Towards Principles for the Design of Ontologies Used for Knowledge Sharing. In: Guarino, N and Poli, R. (eds.): Formal Ontology in Conceptual Analysis and Knowledge Representation. Kluwer Academic Publishers, Dordrecht (1993)
15. Guarino, N.: Semantic Matching. Formal Ontological Distinctions for Information Organization, Extraction and Integration (1997)
   http://citeseer.nj.nec.com/guarino97semantic.html

16. Hernández, C., Prieto, F., Laguna, M. A., and Crespo Y.: Formal Concept Analysis Support for Conceptual Abstraction in Database Reengineering. In: Piattini, M. and Calero, C. (eds.): Proceedings of Database Maintenance and Reengineering Workshop (DBMR '2002), Montreal, Canada (2002)

17. Jordan, T., Raubal, M., Gartrell, B., and Egenhofer, M.: An Affordance-Based Model of Place in GIS. In: Poiker, T.K and Chrisman, N. (eds.): 8th International Symposium on Spatial Data Handling, SDH'98, Vancouver, Canada (1998) 98-109

18. Kaptelinin, V., Nardi, B. and Macaulav, C.: The Activity Checklist. A Tool for Representing the "Space" of Context. Interactions 6(4). (1999) 27-39

19. Kavouras, M. and Kokla, M.: A Method for the Formalization and Integration of Geographical Categorizations. International Journal of Geographical Information Science 16(5). (2002) 439-453

20. Kuhn, W.: Formalizing Spatial Affordances, Formal Models of Commonsense Worlds (Special Meeting, NCGIA Research Initiative 21). National Center for Geographic Information and Analysis (NCGIA), San Marcos, TX. (1996)
http://www.geog.buffalo.edu/ncgia/i21/papers/kuhn.html

21. Kuhn, W.: Ontologies in Support of Activities in Geographical Space. International Journal of Geographical Information Science 15(7). (2001) 613-631

22. Levin, B.: English Verbs Classes and Alternations. A Preliminary Investigation. University of Chicago Press, Chicago, IL (1993)

23. Mizoguchi, R., Sinitsa, K. and Ikeda, M.: Task Ontology Design for Intelligent Educational/Training Systems. In: Workshop on Architectures and Methods for Designing Cost-Effective and Reusable ITSs, Montreal. (1996)
http://www.ei.sanken.osaka-u.ac.jp/.index.html

24. Pinto, H. S. and Martins, J. P.: Ontologies. How Can They be Built? Journal of Knowledge and Information Systems 6(4). (2004) 441–464

25. Priss, U. E.: The Formalization of WordNet by Method of Relational Concept Analysis. In: Fellbaum, C. (ed.): WordNet. An Electronic Lexical Database. MIT Press (1998) 179–196

26. Raubal, M. and Kuhn, W.: Ontology-Based Task Simulation. Spatial Cognition and Computation 4(1). (2004) 15-37

27. Rodríguez, M. A. and Egenhofer, M.: Comparing Geospatial Entity Classes. An Asymmetric and Context-dependent Similarity Measure. International Journal of Geographical Information Science 18(3). (2004) 229–256

28. Ruether, C., Kuhn, W., and Bishr, Y.: An Algebraic Description of a Common Ontology for ATKIS and GDF. In: Proceedings of 3rd AGILE Conference on Geographic Information Science, Helsinki/Espoo, Finland (2000)
http://agile.isegi.unl.pt/Conference/Helsinki2000/Abstract/Agile2000_58.pdf

29. Sanderson, M. and Croft, B.: Deriving Concept Hierarchies from Text. In: Proceedings of the 22nd Annual International ACM SIGIR Conference on Research and Development in Information Retrieval. ACM Press, New York (1999) 206-213

30. Snelting, G. and Tip, F.: Reengineering Class Hierarchies Using Concept Analysis. In: Osterweil, L. J. and Scherlis, W. L. (eds.): Proceedings of the ACM SIGSOFT Symposium on the Foundations Software Engineering. ACM Press, New York (1998) 99-110

31. Ullman, J. D.: Principles of Database Systems. 2nd edition. Computer Science Press, Rockwell, MD (1982)

32. Water Framework Directive, WFD: Official Journal of the European Communities (2000)
http://europa.eu.int/eur-lex/pri/en/oj/dat/2000/l_327/l_32720001222en00010072.pdf

33. Water Framework Directive, WFD: Guidance Document on Implementing the GIS Elements of the WFD. Water Framework Directive Common Implementation Strategy. European Communities (2002)

34. Welty, C. and Guarino, N.: Supporting Ontological Analysis of Taxonomic Relationships. Data and Knowledge Engineering 39(1). (2001) 51–74

Landmarks in the Communication of Route Directions

Elisabeth Weissensteiner and Stephan Winter

Department of Geomatics, The University of Melbourne
Parkville, Victoria 3010, Australia
winter@unimelb.edu.au

Abstract. We investigate the understanding of landmarks using a model of embedding procedures that sees affordances established on three levels. On the
first level there are landmark experience and direct communication process as
distinct affordance structures. On the second level the initial landmark experience has become part of the speech act thus establishing direct wayfinding
communication. On the third level the communication situation of the speech
act incorporating the landmark experience is changed into indirect communication thus transforming into a narrative. We apply the leveled model of narrative
structure to human-generated route directions. We demonstrate how the initial
experience–the landmark–is incorporated into communication structures, how it
turns into a narrative in order to secure understanding and how understanding is
guided via narrative structures. With this approach from Literature we contribute to the problem of in-depth natural language understanding. Ultimately our
interest is in automatic generation of route directions. We show that simply referring to landmarks in route directions is insufficient for successful communication.

1 Introduction

Landmarks are essential parts of wayfinding directions and any communication about
space. In the research literature we find a thorough body of publications about the role
of landmarks for wayfinding from a cognitive point of view. The focus is on landmark
selection and on the interaction between subject (the wayfinder) and object (the landmark) in the orientation process. In contrast, this paper investigates for the first time
how landmarks are incorporated into spoken and written narratives. Route directions
given by humans as well as by services have to be considered as narratives if the
communication situation lacks the direct interaction of two individuals. Hence, understanding the structure of narratives is a prerequisite of designing navigation services
that include landmarks in their route instructions. Specifically we will show that it is
insufficient for successful communication to simply name landmarks in route instructions. As Mark and Gould state, "Many of the properties and principles related to the
understanding of fictional narrative apply to verbal wayfinding directions as well"
(Mark and Gould 1995, p. 388). However, their interest is in cross-linguistic aspects,
whereas our goal is a better understanding of the communication process itself. Our
case study is a human-generated route instruction, which we use to exemplify a leveled model of narrative structure. A sound theory of the communication process will
enable navigation services to more effectively deliver the message.

M.J. Egenhofer, C. Freksa, and H.J. Miller (Eds.): GIScience 2004, LNCS 3234, pp. 313–326, 2004.
© Springer-Verlag Berlin Heidelberg 2004

This paper develops an analytical approach to the *structural position* of landmarks in narrative wayfinding communications. In this way, it is the first time that a structural approach to the communication of landmarks has been undertaken. We believe that our investigations will help clarify the interlacement of physical world, oral and mediated narrative. Thus preliminary steps are taken to understand how wayfinding texts function and how they might be improved. Further work will be necessary to formalize and apply this model to accommodate the use of landmarks in the automatic generation of wayfinding texts.

As we show, landmarks are subject to an embedding structure in speech acts and narratives. Landmarks themselves are products of an elemental cognitive relationship. An object becomes a landmark through the directional perceiving process of a subjective mind, resulting in a spontaneous action of the individual. This action can be communicated from one individual to another, thus incorporating the previous action into a direct communication process. Furthermore each communication partner may take account of this communication about the original action either via oral or other communication media, such as writing, video, photography, or digital interaction. Submitting the original subject-object relationship to a translation process, transforming it from one structural level to another, evolves a complex structure of communication. Thus three levels of structural complexity emerge: the initial relationship between the subjective mind and the object, the direct verbal communication between two subjects, and the taking account of that communication process by means of other media. By studying those three levels it will be possible to investigate the requirements of preserving the wayfinding properties of the original cognitive process between individual and object–the landmark.

This paper presents previous work to introduce the terminology used, based on literature from psychology, linguistics, narratology, and natural language processing. Then it develops a hierarchical model of increasing verbal explicitness, starting from a holistic experience situation, passing into a speech act, and finally into a written narrative. This analytical model is applied to a human-generated wayfinding narrative, which serves as a demonstration as well as a test of the proposed method. In the last section the model is summarized and discussed in relation to navigation services and automatically generated wayfinding narratives, and an outlook to further questions is given.

2 Previous Work

This section introduces the terminology for this paper by collecting concepts from different disciplines, such as psychology, linguistics, narratology, and natural language processing. It also puts Location-Based Services and landmarks in context.

The concept of *landmarks* is central. In the literature relating to this paper landmarks are not formally defined. Instead, landmarks are seen as points of reference, or, more focused, as features that are relatively better known and define the location of other points (Presson and Montello 1988). Thus, properties that can make features prominent show distinct characteristics in relation to properties of features in the neighborhood. The red color of a building lets it be perceived as distinct in a neighborhood only as long as the other buildings have a different color (Nothegger et al. 2004). Consequently it needs a perceiving subject to use (and qualify) a distinct feature as a landmark. In the context of route directions, it is the experience of relevant distinct features along a route which make these distinct features landmarks. Several authors classify landmarks in route directions according to their relation to the route (Habel 1988; Lovelace et al. 1999; Michon and Denis 2001).

Landmarks support the building of a mental representation of an advance model of a route (Siegel and White 1975; Hirtle and Heidorn 1993; Tversky 1993). People use landmarks preferably at decision points of routes (Michon and Denis 2001), and communicate route directions by using landmarks (Daniel and Denis 1998; Denis et al. 1999; Lovelace et al. 1999). Lynch (1960) characterizes landmarks by their singularity or contrast to the background. This is a concept from Gestalt theory (Wertheimer 1925), a reference that supports the understanding that being a landmark is a relative property. In this paper we focus on the structures that arise through narrative communication incorporating landmarks.

Analysis of the research literature reveals a strong interest in understanding spatial perception and the related individual cognitive process (e.g., Kaplan and Kaplan 1982; Golledge 1999). Moreover linguistics investigates how spatial orientation appears in language (Jackendoff 1983). This means that either interviews are recorded of how people explain a certain way (Klein 1979; Klein 1982; Couclelis 1996; Denis et al. 1999) or concepts of location in different languages are investigated by comparing verbal expressions with spatial content (Talmy 1983; Herskovits 1986; Mark and Gould 1995). In addition there is ongoing research in modeling natural language for computerized devices, not least are navigation services (Dale et al. 2003). As communication cannot be exhaustively understood through isolated words and sentences alone, linguistic research concentrates on narrative structures (Duchan et al. 1995). In addition, the question of how to make use of accumulated data in favor of clients has led to several research initiatives employing the narrative as a device (Kim et al. 2002; Miles-Board et al. 2003; Rutledge et al. 2003).

At the same time there has been research conducted on the construction and function of narratives with a growing focus on the relation between cognitive science and narratology, and on media analysis (Ryan 1993; Herman 2003; Ryan 2003) and what it needs to model narratives (Herman 2000). Furthermore during the seventies a strong theory was developed to understand the process of understanding reading (Iser 1978). Iser builds on the speech act theory (Searle 1969; Austin 1975) and on structuralism (for a comprehensive overview see Bal 1997; Jahn 2003), on the basis of the philosophy of Roman Ingarden. Iser thus developed a theoretical approach on how reading functions, or why a reader cannot help being drawn into the experience of reading. In linguistics an approach to narrative structures can be found in the theory of deixis outlining how coherence is established within a text (Duchan et al. 1995).

In regards to wayfinding directions the process of understanding seems to be crucial. In wayfinding communications understanding has to be non-ambiguous, otherwise wayfinding might fail. Thus there is demand for research focusing on the seemingly passive recipient of uttered words. The ways computers understand natural language is distinct from how humans understand natural language. However, looking into how computers understand natural language shows us how far we have formalized what we know about human language understanding. The classical model of information communication is based on a source, a channel, and a destination, and assumes the information successfully transferred if a signal is received and decoded completely at the destination (Shannon and Weaver 1949). For explaining the understanding process of language this model comes too short. Research in natural language understanding by AI identified four levels of processing for understanding: the syntactic level (dealing with grammatical structures), the semantic level (dealing with the meaning of words and their ambiguity), the use of context (in the local and global discourse), and knowledge about the world (e.g., Winograd 1972; Allen 1987). Progress has been made on these levels in many directions, based on logical deduction and semantical networks, finally reaching story understanding to a degree that simple question about the story can be answered by the computer (e.g., Shapiro and Rapaport 1995). However, McCarthy demonstrates with an example that logical deduction is not sufficient to reach *deep* or genuine story understanding, and develops a research program for open challenges at all four levels of language understanding (McCarthy 1990). In the same direction argues Mueller in a recent literature review, stating that "the problems raised by McCarthy remain unsolved" and "it is time to return to the harder problem of [deep] story understanding" (Mueller 1998). With our approach, based on the theory of aesthetic response from literature, we aim to contribute to the problem of in-depth natural language understanding. With this approach we start from how humans understand language, in the expectation that this will turn out to be helpful for building computer interfaces that communicate more successfully to human beings. It is open for discussion whether this contribution can also help build computers that better understand language.

3 A Model of Narrative Understanding

Not always do we try to comprehend route descriptions through direct conversation. Probably we try more often to acquire understanding from written words, pictures or other devices. This means the same expectation of understanding applies even though there is no partner to be questioned. On one hand we have to make sense of what we read, hear, or see in order to find our way. On the other hand we cannot refrain from making sense when we perceive a narrative. But what we understand intuitively and what we should understand to find the right way does not necessarily coincide.

Hence there must be structures generating, guiding and shaping the process of understanding. Therefore the question of landmarks in texts involves both the initial cognitive process and the way in which landmarks are incorporated in the bipolar and multilayered structure of language use.

Three questions should be asked:

- What is the initial cognitive process like?
- How is it transferred into a speech act?
- How is a speech act incorporating landmarks transferred into a narrative?

To deal with these questions the mentioned disciplines offer important approaches. Cognitive science will help clarify the first question. The latter questions need speech act theory and narratology to provide the tools of understanding how spoken and mediated communication are related. Furthermore a theory of the process of making sense is needed to understand why we do not understand when understanding seems to be taken for granted. The approach of this paper is to point out that it is vital to analyze the *receiving part* of a communication process in order to generate comprehensibility of mediated communication of whatever kind in the future.

Landmarks can only function in the very sense of the word through the cognitive act of a subject. To mark an object among others has to be done by an individual. It cannot be the quality of the object itself. It needs a perceiving and reflecting mind. Although its quality to stand out from its surroundings is partly due to its physical qualities, complex procedures of perceiving and understanding are needed to make use of it. Therefore, a landmark is an object within a relation to a subject.

The special connection between the subject and the object was identified and analyzed by Gibson. He called it affordance (Gibson 1979). Affordance is neither the quality of the object alone nor the mere perceiving action on the subject's side. The term captures the connection. There have to be structures of the object that afford the subject to do something in regard to it. Doing means both the physical and mental activity of the subject. Gibson concentrates on affordance of objects within a defined and stable perceiving situation. But he also states that affordance can be found in more complex situations like events and communication. Therefore, the term affordance will help understand more about the relationship of objects and subjects within communication.

It is not only the subject engaged in various affordances. The subject may talk about her experiences to another subject. Then the above outlined relationships between subject and object are embedded in a *speech act* (Wittgenstein 1963; Searle 1969; Austin 1975). According to Austin the speech act consists of three parts, which have to be in place simultaneously. There has to be a physical utterance, which can be understood as an encoded message. This is the *locutionary act*. The second part ensures that the recipient of the message knows what he is supposed to understand. This part is called the *illocutionary act*. And the third part tries to make the recipient do something. This is the *perlocutionary act*. All three parts happen at the same time and are not necessarily verbal. The illocutionary act needs agreement between the communicating partners about *accepted procedures*. The perlocutionary act constructs the verbal material according to the accepted procedures in order to initiate an action on the recipient's side. The action may be both physical and mental.

Finally the speech act, referring to the event of a certain wayfinding process, can be accounted for without a direct conversation partner using other media. This means an additional encoding process. Thus narrative procedures come into place. The narrative can be understood as the encoded structure of a speech act. According to Iser (1978), the accepted procedures of the illocutionary act have to be made explicit to produce a narrative as there is no externally defined speech act situation. Instead there is an agent accounting for what has happened.

The elements of a narrative have to be arranged consecutively. Hence understanding develops through a step-by-step process. As there are distinct elements positioned beside each other, *gaps* appear, which function as initiators of the text's reception (Iser 1978). The reader will try to fill those gaps according to the basic logic of cognitive models (Lakoff 1987). Spontaneous interrelation through the reading process is ensured; the text executes affordance.

Thus both speech acts and narratives offer structures to spontaneously involve the recipient. Therefore, it can be said that they exert affordance: they make people do something meaningful.

Hence communicating landmarks is a translating process of the basic relation between a subject and an object. The process results in a completed orientation procedure through a narrative. It makes the recipient experience the initial affordance of the object and perform a similar orientation procedure without having previous knowledge of the landmarks. This can be performed in two steps of increasing complexity according to our understanding of the embedding structure of communication.

4 A Leveled Model of Narrative Structure

When we talk of landmarks we think of an object and a perceiving subject. Apparently a basic relationship has to be formed between the two in order to become a landmark. This relationship can be understood in the terms of Gibson's affordance. There is something on the object that fits into the physical and mental structure of the subject, into his or her subjective cognition, which triggers an action on the subject's side.

This means three constitutive structural elements can be discerned: an object, a subject, and the action triggered by their interplay. It means furthermore that the term landmark apparently conveys an experience, and not a single reference to an object in the world.

Hence, when we want to communicate landmarks we have to communicate experiences. When we want to comprehend an experience of somebody else we usually try to put ourselves into his situation. This metaphor says clearly that we try to have the same feelings, thoughts, and physical perceptions: we are ready to do something.

As stated above, communication may happen on different levels of complexity (Figure 1). At the beginning there is the initial encounter of Ego and an object. According to the wayfinding properties of this relationship the object can be called landmark, thus depicting the affordance. This is a nonverbal process containing only the perceiving subjective person and the object with its physical properties. We call this an experience.

Then, this initial experience may be incorporated into a speech act with two or more partners who are physically present, sharing the same (not verbalized) situation. This means that the whole experience has to be put into signs and organized sequentially. The speech act being partly organized nonverbally, based on accepted procedures, provides the affordance to have a corporeal experience.

Finally the initial experience can be accounted for without a shared situation and even by means of other media. Now the situational specificities have to be accounted for by an agent. The accepted procedures have to be built-in and the agent orchestrates them. Thus the orchestration of the accepted procedures, together with the initial experience, produces the needed affordance to activate the recipient's mind.

In all three stages experiences are negotiated. At the beginning there is the initial encounter of person and object. Then the initial experience becomes part of another experience–the speech act. Thereafter the experience of a speech act merges with the initial experience and becomes an experience itself–a narrative. Thus a basic experience can apparently only be communicated by translating it into another experience, either in one (speech act) or two stages (narrative). Note that this understanding of the narrative is distinct from the widespread understanding of the narrative as written speech acts. Our approach to define narrative is structural, and acknowledges the higher complexity of the communication process.

What does this mean for the landmark? The three constituents, consisting of object, subjectivity, and action, have to appear in the speech act as well as in the narrative. They have to appear in their interrelation. Neither the object nor the subject, not even both of them, is sufficient. It is not sufficient to mention three entities. It has to be the affordance that constitutes the experience and therefore has to be incorporated in verbal structures.

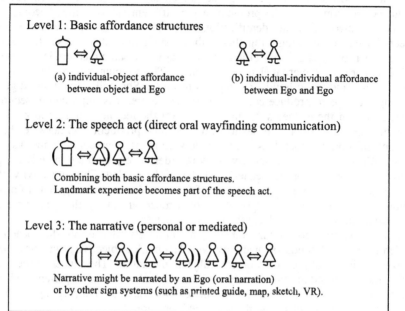

Level 1: Basic affordance structures

(a) individual-object affordance
between object and Ego

(b) individual-individual affordance
between Ego and Ego

Level 2: The speech act (direct oral wayfinding communication)

Combining both basic affordance structures.
Landmark experience becomes part of the speech act.

Level 3: The narrative (personal or mediated)

Narrative might be narrated by an Ego (oral narration)
or by other sign systems (such as printed guide, map, sketch, VR).

Fig. 1. The three levels of the presented model (see text).

The subjectivity of the landmark experience will be found in strategies to make sure that the partner who makes the utterances has really had this very experience with the landmark. He will do everything (with words) to assure that he is the one to have had that experience and to know it. This has to be understood as a mere strategic measure to express subjectivity, not as a question of personal reliability. The subjectivity of the landmark experience and the subjectivity of the uttering partner of the conversation merge into one, even if the person giving directions has not himself experienced the landmark but merely knows about it. He usually will make it explicit if he has not experienced it himself, which is a strategy in itself to transfer reliability. At the same time the subjectivity of the real person making statements about directions and objects within a speech act gives rise to doubts about whether his references to reality will be reliable. Therefore the direction giving person will additionally try to assure his personal integrity.

Two opposed goals have to be pursued to accomplish the task of transferring affordance of an initial landmark experience into a speech act: the subjectivity of the direction giving person has to be pointed out positively in order to transfer affordance, and at the same time the subjectivity has to be shown under control in order to ensure confidence of the recipient. The accepted procedures will control the sender's subjectivity by ensuring the most possible compliance between the communication partners. On this basis the perlocutionary act can be realized by pointing out the involvement of the sender and, therefore, being justified to expect the same experiences from the recipient. Insofar Grice's famous maxims of quality and relevance for verbal communication are also satisfied (Grice 1975).

In the written narrative affordance of the initial landmark experience will appear as the event that will be told about. As the narrative has to incorporate both illocutionary act and perlocutionary act the problem of the different subjectivities becomes even more complicated. Now the identification of the initial subjectivity with the sender subjectivity has to be brought together with measures to control the sender's subjectivity in order to be reliable. All three will appear in the same medium in consecutive units. There are no nonverbal elements any more. The recipient will read the units one after the other and try to build a continuous meaning. She will actively build gestalts by filling the gaps to produce connectedness. Therefore, it is important to secure the connectability of the elements. It must be easy for the recipient to introduce his own mental images. Nevertheless it has to be a controlled production. Otherwise she will make herself a picture that does not coincide with reality elements (landmarks). This means controlling subjective activity. Now in order to rely on the transferred experiences, the subjectivity on the reader's side has to be guided. The subjectivity of the sender thus moves into the gap structure of the text. The text strategy has to provide possibilities for the reader to activate her own repertoire but at the same time limit those possibilities in order to have the desired experience.

The assuring subjectivity that transfers the initial affordance into the speech act–the identification of sender subjectivity with landmark subjectivity–will now be the focalized depiction of the initial object, subject, and action. The sender of the speech act who has identified his subjectivity with the original landmark experience will now appear as the perspective on the event. The text will have a *focalizing agent* who decides in what way the object and the action will be depicted.

As a narrative always has a narrator, even though he will not always function as a focalizer (Bal 1997), the question of reliability arises again. A narrative *per se* has subjectivity being an utterance whose origin cannot be identified. There are no accepted procedures to support the communicative process outside the written text. The narrative has to produce this situation by itself. The elements that have secured the compliance of the speech act partners now appear as the repertoire to deal with. *Narrative strategies* will be employed to organize the *repertoire* in such a way that the reader may believe what she understands has meaning.

5 A Wayfinding Narrative

This section will show how the above developed analysis model can be applied to a route instruction. On the one hand the abstract model can be explained with a specific example; on the other hand the model is tested for applicability.

The following route instruction is human-generated, and was given to one of the authors in a real-world wayfinding situation.

I am looking forward to your visit to my home, at Abcstrasse 33 in Münster. Here is how you will get there by bus from the train station.

*After you leave the Münster **train station** (take the main exit, "Westausgang," not "Ostausgang"), continue across the **square** leaving the **big glass house** (containing the bike parking) on your left, and cross the **main road** at the **traffic lights**.*

*On the other side of the road, turn left in front of the first row of houses, and walk toward the last of the **bus stations** on that side (approximately 150 meters). The sign should list bus number 10, which you take (direction Mecklenbeck).*

*Enter in front of the bus (it runs every 20 minutes) and buy a ticket from the driver for 0.65 Euro (try to have small change ready!). Stay on this bus for approximately 15 minutes. It takes you through the **center of town** (watch the **Dome** on your right after 5 minutes) and along **Aasee** (the lake) to the bus station Abcstrasse (the second after a sharp left turn, away from the lake).*

*Exit and walk in the direction back to where the bus came from for 200 meters. Just across the **small intersection** with Bischopstrasse you find Abcstrasse Nr. 33, the second entrance in a **row of red brick buildings** on your right.*

Have a safe trip!

The emphasized nouns are landmarks, some of them prototypical ones (such as *train station, Dome*), others marginal ones (such as *small intersection, row of red brick buildings*). For centrality in categories see Rosch (1978) or Lakoff (1987). Without having (or trying to give) a formal definition of landmarks, we do not claim that the identified landmarks in the text above are complete. The landmarks can be identified because they are objects and they trigger some kind of bodily awareness in the recipient. As mentioned above the landmark consists of an object, a subject, and the affordance enacted between the two. This means qualities of the object and qualities of the subject fit together in a way that the perceiving process and the physical action cannot be separated. Both perceiving process and physical action are part of the spatial learning process of the individual. Whenever these basic experiences are addressed the individual will produce them in her mind (Gibson 1979; Lakoff 1987). The individual will produce the sensation, not an intellectual reference.

At the same time other expressions refer to special characteristics that have had to be agreed upon before. That a train station is a building with exits, that one may travel by train at all, that a house can be made of glass, what kind of material glass is, that a square consists of intersecting streets, what traffic lights are, that buses have stations which can be found by signs, what a cathedral might look like and what is meant by a lake–by mentioning those nouns the text assumes, and thus makes explicit, the compliance between the sender of the information and his recipient. These nouns function as gaps and are elements of the repertoire at the same time. As gaps they invite the recipient to make a picture herself, as elements of the repertoire they make sure that what is said is possible. They evoke mental models, presumably prototypes, thus engaging the recipient spontaneously. Nevertheless the produced pictures will be unspecific.

In order to convert the initial affordance–the landmark and the individual experience connected to it–into a written language without an actual speech act situation, the *I-you* structure of the underlying speech act is transferred into the focalizing agent thus turning the initial experience into the event of a narrative.

The *you* provides the perspective of the narrative units. The performed actions belong to the realm of basic logic of the recipient (such as *leave, follow, cross, watch*). The *you* in the speech act justifies the invitational quality of the utterance by referring to the corresponding *I. You* cannot be understood without an *I*. That means that even when there is no actual dialog situation the recipient has to infer one spontaneously thus getting assurance that the sender knows the experiences he is telling about. The *you* as the focalizing agent of the narrative has to recall the *I* in the sense of a fore-ground-background relationship to substitute the lack of an actual *I*. This means furthermore that this focalizing process (and with it the transference of the speech act into a narrative) produces gaps. The recipient being a real person furnishes the gap with her personal experiences of wayfinding. Thus subjectivity on the reader's side comes into play. To ensure the reliability of the communication the filling of the gaps is controlled by mentioning location qualities. These indicate special knowledge of the area (such as *exit "Westausgang", exit "Ostausgang", the bike parking*, and location names such as *Aasee, Bischopstrasse*). This narrative structure coincides with the actual speech act situation of a wayfinding dialog: there the subjectivity of the speech act has to be controlled in order to ensure the reliability of the communication. The recipient of wayfinding directions wants permission to believe what she is hearing.

As the narrative consists of the events (in our case landmarks) and the focalizing agent (in direct communication: the sender making an utterance) there has to be an entity that arranges all these elements. The narrator sets the stage. He has to be understood as an agent, as a strategy of the narrative how to make the recipient experience and understand. Although the narrator may not be identifiable as a personality of any kind he/she is one of the basic elements of a narration.

In our case the narrator adopts personality by saying "I" and thus connecting to the basic *I-you* structure of the wayfinding speech act and thus ensuring the reliability of the following story (*I am looking forward to your visit..., Have a safe trip*). If there is no specification of the narrator the recipient will fill specifications in. This means the narrator can act as a gap, and the *I* can be understood as a controlling strategy. The recipient is required to fill in a communication experience when addressed with *you*. To have the desired effect she has to produce the desired experiences in order to be able to identify the landmarks correctly and further to find the right way. The narrator does not necessarily have to identify himself. The narrator can be less personal and even withdraw into other strategies of assuring as in computerized wayfinding directions or may take on all kinds of strategies as the history of fiction tells us (Jahn 2003).

The goal of the whole procedure of conveying landmarks (and furthermore of complete wayfinding directions) is that the recipient may develop a picture of the intended way and eventually find and use this way. This means that the recipient has to be enabled to produce a gestalt which coincides with the intended way. She has to connect the pictures she has been triggered to produce to a gestalt as complete as possible with regard to the goal of the narrative. Relating to landmarks, this means that landmarks have to be selected and placed in an order where the affordance of one landmark can be closely related to the affordance of the next (see examples such as *after you leave...continue across...leaving on your left side...on the other side of the road...it takes you through...exit and walk...across...*). Thus those gaps that appear through arranging the landmark stories consecutively can be controlled. The recipient will provide images to connect them. To prevent her from losing the right way in the very sense of the word, connecting the different narrative units has to be made easy, thus reproducing the basic logic of *source-path-goal* (Lakoff 1987) of the initial landmark experience in the narrative structure.

To summarize we can state that two main goals have to be performed in order to translate the basic landmark experience: subjectivity has to be made explicit to make the affordance traceable and at the same time subjectivity has to be controlled to ensure reliability. This happens at two different levels according to the structures of communication, either the speech act or the narrative. Wayfinding with the help of landmarks via narrative has to provide gaps for the recipient to fill in her personal experiences. This ensures the translation of affordance and at the same time has to control these gaps to make sure that the intended picture may be produced.

6 Conclusions

In this paper we investigated the structure of understanding landmarks in verbal wayfinding directions. The initial landmark experience can be understood as a semantic structure that can be explained in cognitive terms. After the transformation into speech acts, the initial experience transfers into a syntactic structure, which produces what we call an event worth telling about. The speech act partners provide for the semantics. In the narrative the speech act transfers into an event as well, and a new syntactic structure evolves. Now the narrative structures establish the new semantics. The embedding procedures always function through establishing affordances (Figure 1).

For a human-generated wayfinding narrative we demonstrated how the initial experience–the landmark–is incorporated into communication structures, how it then turns into a narrative in order to secure understanding, and how understanding is guided via narrative structures. However, our final interest is in computer-generated route instructions, as provided by navigation services. Is the presented and discussed model of narrative structure applicable to the communication of navigation services to their clients? And if so, what are the issues to be learned from the model in this context? Finally, if deficiencies can be identified, can we develop computable models filling these gaps, such as narrative agents? Investigating these questions is part of our ongoing work, and beyond the scope of this paper. However, some preliminary observations give evidence of the relevance of these questions.

Location-Based Services, especially navigation services, are direct competitors to people generating route instructions: they produce and communicate route instructions on demand, either in advance (offline route planners) or sequentially, adapting to the location of the client (online navigation services). Their means of communication may vary, for instance using visual means, like maps or symbols, or verbal means, like written or spoken text. However, their communication has to be of the complexity of the third level in Figure 1. This is because the immediate personal communication situation is missing, and the non-verbal part of spoken communication, responding to a specific client and a specific situation, cannot be produced automatically and has to be explicated. In this sense, an oral instruction, for instance, by a car navigation system, is structurally a narrative put through speech synthesis.

Compared to the complex communication task, current navigation services deliver relatively plain narratives. They are based only on the geometry of routes (Winter 2004), and lack landmarks. The major obstacle for not including landmarks is the lack of landmark information in spatial databases, or, in cases where services confuse landmarks with pre-selected points of interest, a lack of understanding what properties make a good landmark in a specific wayfinding situation. This situation will change in the future since recent research addresses the automatic selection of landmarks from databases to enrich route instructions (Elias 2003; Nothegger et al. 2004). As soon as landmark information becomes available the problem of integration in text arises. In this regard our research is timely.

The above discussion has made it clear that it is not sufficient just to name landmarks in a computer-generated narrative. Landmarks should generate an experience, and this process should be controlled by the service. Hence, a narrating agent is required that is equipped with formal models of communication experience or the narrative structure of landmarks. A formalization of our findings is needed and left for future work.

Landmarks are just one element of orientation in the complex interplay of mind and object within the discussed narrative structure. Other elements of the narrative can be studied in the same way, which forms another line of ongoing and future work.

Acknowledgments

This work was supported by an ECR grant of The University of Melbourne. The authors would like to thank the anonymous reviewers for their helpful comments, and also Robert Dale, Macquarie University, for discussions of our ideas.

References

Allen, J., 1987: Natural Language Understanding. The Benjamin/Cummings Publishing Company, Inc., Menlo Park, CA.

Austin, J.L., 1975: How to Do Things with Words. Clarendon Press, Oxford.

Bal, M., 1997: Narratology: Introduction to the Theory of Narrative. University of Toronto Press, Toronto.

Couclelis, H., 1996: Verbal Directions for Way-Finding: Space, Cognition, and Language. In: Portugali, J. (Ed.), The Construction of Cognitive Maps, Kluwer, Dordrecht, pp. 133-153.

Dale, R.; Geldof, S.; Prost, J.-P., 2003: CORAL: Using Natural Language Generation for Navigational Assistance. . In: Oudshoorn, M.J. (Ed.), Proceedings of the 26th Australasian Computer Science Conference. Conferences in Research and Practice in Information Technology, 16: 35-44.

Daniel, M.-P.; Denis, M., 1998: Spatial Descriptions as Navigational Aids: A Cognitive Analysis of Route Directions. Kognitionswissenschaft, 7 (1): 45-52.

Denis, M.; Pazzaglia, F.; Cornoldi, C.; Bertolo, L., 1999: Spatial Discourse and Navigation: An Analysis of Route Directions in the City of Venice. Applied Cognitive Psychology, 13: 145-174.

Duchan, J.F.; Bruder, G.A.; Hewitt, L.E. (Eds.), 1995: Deixis in Narrative: A Cognitive Science Perspective. Lawrence Erlbaum Associates, Hillsdale, NJ.

Elias, B., 2003: Extracting Landmarks with Data Mining Methods. In: Kuhn, W.; Worboys, M.F.; Timpf, S. (Eds.), Spatial Information Theory. Lecture Notes in Computer Science, 2825. Springer, Berlin, pp. 398-412.

Gibson, J.J., 1979: The Ecological Approach to Visual Perception. Houghton Mifflin Company, Boston, MA.

Golledge, R.G. (Ed.), 1999: Wayfinding Behavior: Cognitive Mapping and Other Spatial Processes. The Johns Hopkins University Press, Baltimore, MA.

Grice, P., 1975: Logic and Conversation. Syntax and Semantics, 3: 41-58.

Habel, C., 1988: Prozedurale Aspekte der Wegplanung und Wegbeschreibung. In: Schnelle, H.; Rickheit, G. (Eds.), Sprache in Mensch und Computer, Westdeutscher Verlag, Opladen, pp. 107-133.

Herman, D., 2000: Narratology as a Cognitive Science. Image and Narrative, online journal (http://www.imageandnarrative.be), Issue 1.

Herman, D. (Ed.), 2003: Narrative Theory and the Cognitive Sciences. CSLI Lecture Notes, 158. CSLI Publications, Stanford.

Herskovits, A., 1986: Language and Spatial Cognition. Cambridge University Press, Cambridge.

Hirtle, S.C.; Heidorn, P.B., 1993: The Structure of Cognitive Maps: Representations and Processes. In: Gärling, T.; Golledge, R.G. (Eds.), Behavior and Environment: Psychological and Geographical Approaches, North Holland, Amsterdam, pp. 1-29.

Iser, W., 1978: The Act of Reading: A Theory of Aesthetic Response. John Hopkins University Press, Baltimore, MD.

Jackendoff, R., 1983: Semantics and Cognition. The MIT Press, Cambridge, MA.

Jahn, M., 2003: Narratology: A Guide to the Theory of Narrative. English Department, University of Cologne, http://www.uni-koeln.de/~ame02/pppn.htm. Last accessed 8.3.2004.

Kaplan, S.; Kaplan, R., 1982: Cognition and Environment: Functioning in an Uncertain World. Praeger, New York, NY.

Kim, S.; Alani, H.; Hall, W.; Lewis, P.H.; Millard, D.E.; Shadbolt, N.R.; Weal, M.J., 2002: Artequakt: Generating Tailored Biographies from Automatically Annotated Fragments from the Web, Proceedings of Workshop on Semantic Authoring, Annotation & Knowledge Markup (SAAKM'02), Lyon, France, pp. 1-6.

Klein, W., 1979: Wegauskünfte. Zeitschrift für Literaturwissenschaft und Linguistik, 33: 9-57.

Klein, W., 1982: Local Deixis in Route Directions. In: Jarvella, R.J.; Klein, W. (Eds.), Speech, Place, and Action, John Wiley & Sons, Chichester, pp. 161-182.

Lakoff, G., 1987: Women, Fire, and Dangerous Things - What Categories Reveal about the Mind. The University of Chicago Press, Chicago, IL.

Lovelace, K.L.; Hegarty, M.; Montello, D.R., 1999: Elements of Good Route Directions in Familiar and Unfamiliar Environments. In: Freksa, C.; Mark, D.M. (Eds.), Spatial Information Theory. Lecture Notes in Computer Science, 1661. Springer, Berlin, pp. 65-82.

Lynch, K., 1960: The Image of the City. MIT Press, Cambridge, MA.

Mark, D.M.; Gould, M.D., 1995: Wayfinding Directions as Discourse: Verbal Directions in English and Spanish. In: Duchan, J.F.; Bruder, G.A.; Hewitt, L.E. (Eds.), Deixis in Narrative, Lawrence Erlbaum Asscociates, Hillsdale, NJ, pp. 387-405.

McCarthy, J., 1990: An Example for Natural Language Understanding and the AI Problems It Raises. In: McCarthy, J. (Ed.), Formalizing Common Sense, Ablex, Norwood, NJ, pp. 70-76.

Michon, P.-E.; Denis, M., 2001: When and Why are Visual Landmarks Used in Giving Directions? In: Montello, D.R. (Ed.), Spatial Information Theory. Lecture Notes in Computer Science, 2205. Springer, Berlin, pp. 292-305.

Miles-Board, T.; Deveril; Lansdale, J.; Carr, L.; Hall, W., 2003: Decentering the Dancing Text: From Dance Intertext to Hypertext, Proceedings of the Fourteenth ACM Conference on Hypertext and Hypermedia. ACM Press, pp. 108-119.

Mueller, E.T., 1998: Natural Language Processing with ThoughtTreasure. Signiform, New York, NY.

Nothegger, C.; Winter, S.; Raubal, M., 2004: Selection of Salient Features for Route Directions. Spatial Cognition and Computation, 4 (2): 113-136.

Presson, C.C.; Montello, D.R., 1988: Points of Reference in Spatial Cognition: Stalking the Elusive Landmark. British Journal of Developmental Psychology, 6: 378-381.

Rosch, E., 1978: Principles of Categorization. In: Rosch, E.; Lloyd, B.B. (Eds.), Cognition and Categorization, Lawrence Erlbaum Associates, Hillsdale, NJ, pp. 27-48.

Rutledge, L.; Alberink, M.; Brussee, R.; Pokraev, S.; van Dieten, W.; Veenstra, M., 2003: Finding the Story - Broader Applicability of Semantics and Discourse for Hypermedia Generation, Proceedings of the Fourteenth ACM Conference on Hypertext and Hypermedia. ACM Press, pp. 67-76.

Ryan, M.-L., 1993: Narrative in Real Time: Chronicle, Mimesis and Plot in the Baseball Broadcast. Narrative, 1 (2): 138-155.

Ryan, M.-L., 2003: On Defining Narrative Media. Image and Narrative, online journal (http://www.imageandnarrative.be), Issue 6.

Searle, J.R., 1969: Speech Acts. Cambridge University Press, Cambridge.

Shannon, C.E.; Weaver, W., 1949: The Mathematical Theory of Communication. The University of Illinois Press, Urbana, IL.

Shapiro, S.C.; Rapaport, W.J., 1995: An Introduction to a Computational Reader of Narratives. In: Duchan, J.F.; Bruder, G.A.; Hewitt, L.E. (Eds.), Deixis in Narrative, Lawrence Erlbaum Asscociated, Inc, Hillsdale, NJ, pp. 79-105.

Siegel, A.W.; White, S.H., 1975: The Development of Spatial Representations of Large-Scale Environments. In: Reese, H.W. (Ed.), Advances in Child Development and Behavior, 10. Academic Press, New York, NY, pp. 9-55.

Talmy, L., 1983: How Language Structures Space. In: Pick, H.; Acredolo, L. (Eds.), Spatial Orientation: Theory, Research, and Application, Plenum Press, New York, NY, pp. 225-282.

Tversky, B., 1993: Cognitive Maps, Cognitive Collages, and Spatial Mental Models. In: Frank, A.U.; Campari, I. (Eds.), Spatial Information Theory. Lecture Notes in Computer Science, 716. Springer, Heidelberg, pp. 14-24.

Wertheimer, M., 1925: Über Gestalttheorie. Philosophische Zeitschrift für Forschung und Aussprache, 1: 39-60.

Winograd, T., 1972: Understanding Natural Language. Academic Press, New York, NY.

Winter, S., 2004: Communication about Space. Transactions in GIS, 8 (3): 291-296.

Wittgenstein, L., 1963: Philosophical Investigations. Basil Blackwell, Oxford.

From Objects to Events: GEM, the Geospatial Event Model

Michael Worboys and Kathleen Hornsby

National Center for Geographic Information and Analysis
University of Maine, Orono, ME 04469-5711, USA
{worboys,khornsby}@spatial.maine.edu

Abstract. This paper discusses the construction of a modeling approach for dynamic geospatial domains based on the concepts of object and event. The paper shows how such a model extends traditional object-based geospatial models. The focus of the research is the introduction of events into the object-based paradigm, and consequent work on the classification of object-event and event-event relationships. The specific geospatial nature of this model is captured in the concept of a geosetting. The paper also introduces an extension of UML diagrams to incorporate events and their relationship to each other, and to objects. The paper briefly considers an example to show the working of some of the modeling constructs, and concludes with a discussion of further research needed on event aggregation and event-based query languages.

1   Introduction

This paper reports on introductory work on conceptual models for information systems concerned with dynamic, geospatial domains. Here, the term *conceptual model* refers to a structured representation of part of the world that is to be captured by the information system. Although the representation is structured, it is not yet at the level for direct translation into a form accepted by the database system. In Guarino's terms [1], the paper constructs an *ontology*; "a specific vocabulary used to describe a certain reality, plus a set of explicit assumptions regarding the intended meaning of the vocabulary words." Moreover, the ontology is *upper-level*, as the constructs are general enough to be used in many scenarios involving dynamic, spatial entities. Throughout, we shall use the term *model* to describe this structured collection of upper-level constructs.

From an ontological perspective, an initial distinction can be made between entities existing in the world either as *continuants* or *occurrents*. Continuant entities endure through some extended (although possibly very short) interval of time (e.g., houses, roads, cities, and people). Occurrent entities happen and are then gone (e.g., a house repair job, road construction project, urban expansion, a person's life). There is a difference between a city, whose characteristics are recorded by census and surveyed once each decade, say, and the processes of urban growth and decline, migration, and expansion, that constitute the city in flux. From an information system perspective, continuants and occurrents that have a unique identity in the system are

M.J. Egenhofer, C. Freksa, and H.J. Miller (Eds.): GIScience 2004, LNCS 3234, pp. 327–343, 2004.
© Springer-Verlag Berlin Heidelberg 2004

referred to as *objects* and *events*, respectively. Events and processes have distinct, although rather elusively definable, shades of meaning, but essentially speak to the same underlying idea. The event of constructing a new ramp from Stillwater Avenue to the I95 highway is clearly related to the process of ramp construction. In this paper we focus on and use the term *event*.

Objects and events are both needed to model fully a dynamic system. While continuant entities endure through time, they will usually change some or all of their attributes. Traditionally, such time-varying continuants are represented in information systems as temporally indexed collections of objects or collections of objects, called *snapshots*. The shortcomings of such a representation are that the events that underlie changes are not explicitly represented; indeed the changes are themselves not explicitly represented [2]. Events are needed to capture the mechanisms of change.

Objects and events are fundamentally different, but, as we will see in this paper, from an information modeling perspective, can be treated in many ways as structurally similar. The approach of this paper is to use as much as possible of the methods of object-orientation, designed for modeling static entities, to event modeling. Therefore, we assume a background of basic object-oriented modeling. Our constructions will be framed in a UML-like formalism [3].

This paper contributes to the development of a general approach to modeling dynamic geospatial phenomena for information system development. Current modeling approaches are limited in that they are capable in only a limited way of expressing dynamic aspects of the world. This work is connected to work [4] concerned with more formal aspects of event specification using process algebras, and to the construction of temporal and event-oriented ontologies (for example, [5, 6]). In [4] a pure event model is developed, in which all entities are modeled as occurrents. In this work we adopt a hybrid approach, allowing three main categories of entities: objects, events, and settings.

The motivation behind this research is that a dynamic model should be capable of formal verification, and should be translatable into a logical model – the next stage of system development. As an example of the kind of capability that this work provides is the ability to specify in what ways events may be aggregated (this is the content of Section 7). An important direction in which this work is leading is to the construction of a query language in which configurations of objects, events, and their relationships, can be framed.

2   Geospatial Settings and the Situation Function

The distinguishing characteristic of a geospatial entity, whether object or event, is its *setting*. Each geospatial object or event is *situated* in a setting. A setting may be either spatial, temporal, or a combination of both. An object or event, however, cannot be situated in more than one setting at the same time. Here, settings do not just refer to point locations, so we allow the possibility of an object or event being situated over an extended location (such as a region, or a time period). Geospatial entities also have the property that their settings are at an appropriate scale, neither too small (e.g., not at the quantum scale) nor too large (not at the cosmic scale).

A setting may be:

1. Purely *spatial* (e.g., point, line, region, or composition thereof). Spatial geosettings have been extensively studied in the GIScience literature [7]. Spatial settings may be zero, one, two, or three-dimensional, and are often explicitly or implicitly assumed to be embedded in a space, such as a Euclidean space, a metric space, or a topological space.

2. Purely *temporal* (e.g., instant, interval, period). Temporal settings have also been well researched in the artificial intelligence and temporal database literature [8, 9, 10). Temporal settings may be zero or one-dimensional, and are explicitly or implicitly assumed to be embedded in a space, such as a linearly ordered set, a partially ordered set, a discrete or continuous set, or a cyclical structure.

3. Mixed *spatio-temporal*. By our constraint that an object or event cannot be situated in more than one setting at the same time, we do not allow the full Cartesian product of space and time here. Formally, spatio-temporal settings are functions from a temporal setting to a spatial setting. Spatio-temporal settings are called *trajectories* [11], *histories* [12,13], or *geospatial lifelines* [14].

Just as with objects and events, settings may be abstracted into classes. The classes of all purely spatial and purely temporal settings are denoted **SpatialSetting-Class** and **TemporalSettingClass**, respectively. The class of all spatio-temporal settings is denoted **STSettingClass.** Settings may have attributes (e.g., the duration of a time interval, the area of a region). Settings may be organized into subsumption hierarchies (e.g., **Region** subsumes **SimplyConnectedRegion**). Settings also have spatial, temporal, or spatio-temporal parts, depending upon context, and may be composed into composite entities (e.g., **Regions** as a composition of settings in class **Region**). However, in the strict sense, settings are not to be considered as objects in the object-oriented sense. A setting does not have an identifier that remains the same when its attributes change. For example, a time interval [3,6], becomes a different interval, when its end-point changes from 6 to 7. In programming language terminology, settings are *literals* or *constants*.

2.1 The **Situate** Function

A function **Situate** maps each geospatial object or event to its situation or location in a setting. We will see below that **Situate** acts differently, depending on whether the situated entity is an object or an event. However, each geospatial object and event has a unique setting, specified by the action of the **Situate** function. If **GOClass, GEClass**, and **SettingClass,** denote the classes of geospatial objects, events and settings, respectively, then we have the following definition:

$$\text{Situate: GOClass} \cup \text{GEClass} \rightarrow \text{SettingClass}$$

2.2 Relationships Between Settings

This section briefly reviews work on setting – setting relationships, in the case where the settings are spatial, temporal, or spatio-temporal. The first remark is that these three kinds of settings do not mix; we do not allow, for example, a spatial setting to stand in relation to a temporal setting.

Spatial settings: Subsumption hierarchies of zero, one and two-dimensional spatial settings are discussed in [7]. Part-whole relationships are classically handled by treating a spatial setting as a set of points, and using the subset relationship. Non-classical, mereological approaches that do not assume sets of points stem from the work of Brentano [15]. Spatial relationships between spatial settings have been investigated by many authors [16, 17].

Temporal settings: These have been extensively investigated by the artificial intelligence and temporal database communities. There are three basic classes: `TemporalInstant`, `TemporalInterval` and `TemporalPeriod` (a composition of `TemporalInterval`). Reasoning about temporal relationships between temporal settings that are intervals is handled by Allen's interval calculus [8] for the linear case and for cycles by Hornsby *et al.* [18].

Spatio-temporal settings: As discussed above, members of `STSettingClass` can considered as functions from temporal to spatial settings, and inherit in this way the properties of their components. Formally,

$$\texttt{STSettingClass} =_{\text{def}} [\texttt{TemporalSettingClass} \rightarrow$$
$$\texttt{SpatialSettingClass}]$$

Thus each member of `STSettingClass` assigns to each element of a temporal setting (e.g., each time instant in a temporal interval) an element of a spatial setting (e.g. a point in a spatial region).

Spatio-temporal relations between two spatio-temporal settings may be defined with reference to relations between the domains and codomains of the functions representing them. For example, we can define *disjunction* relations as follows. Let $\texttt{STSet}_1 : \texttt{TS}_1 \rightarrow \texttt{SS}_1$ and $\texttt{STSet}_2 : \texttt{TS}_2 \rightarrow \texttt{SS}_2$ be two instances of `STSetting-Class`.

$$\texttt{TDisjoint}(\texttt{STSet}_1, \texttt{STSet}_2) =_{\text{def}} \texttt{TDisjoint}(\texttt{TS}_1, \texttt{TS}_2)$$

$$\texttt{SDisjoint}(\texttt{STSet}_1, \texttt{STSet}_2) =_{\text{def}} \texttt{SDisjoint}(\texttt{Image}(\texttt{STSet}_1),$$
$$\texttt{Image}(\texttt{STSet}_2))$$

$$\texttt{STDisjoint}(\texttt{STSet}_1, \texttt{STSet}_2) =_{\text{def}} \forall t \in \texttt{TS}_1 \cap \texttt{TS}_2.$$

$$\texttt{STSet}_1(t) \cap \texttt{STSet}_2(t) = \varnothing$$

The first of these equations defines two spatio-temporal settings to be temporally disjoint if and only if their temporal domains are temporally disjoint (note the overloading of the `TDisjoint` relation). The second defines two spatio-temporal settings to be spatially disjoint if and only if their spatial images are spatially disjoint (note the overloading of the `SDisjoint` relation). The third defines two spatio-temporal settings to be spatio-temporally disjoint if and only if at each temporal element that they share, their spatial settings are spatially disjoint.

These definitions are not independent. In particular, we have:

$$\texttt{TDisjoint}(\texttt{STSet}_1, \texttt{STSet}_2) \Rightarrow \texttt{STDisjoint}(\texttt{STSet}_1, \texttt{STSet}_2)$$

$$\texttt{SDisjoint}(\texttt{STSet}_1, \texttt{STSet}_2) \Rightarrow \texttt{STDisjoint}(\texttt{STSet}_1, \texttt{STSet}_2)$$

These definitions provide examples of possible relations between settings. There are many others that could be defined similarly.

3 Static Geospatial Object Model

The general conceptual model of static geospatial entities has been thoroughly investigated, both in the object [19, 20] and field settings [21]. In this paper, a geospatial object is defined as an object that is situated in a setting (Figure 1). As the diagram shows, geospatial objects can be instances or classes, organized into composition and generalization hierarchies, and related to each other in object-object relationships.

Object situation. Time is abstracted from the static geospatial object model, and **the settings in which geospatial objects are situated are purely spatial.** Purely spatial settings are described in the previous section. We have the following functional restriction of **Situate** to geospatial objects.

$$\texttt{Situate: GOClass} \rightarrow \texttt{SpatialSettingClass}$$

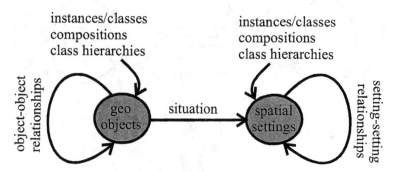

Fig. 1. Underlying geospatial object model.

A *temporal snapshot* of a collection of geospatial objects is an association of a timestamp to the collection. Object evolution can be modeled as a sequence of temporal snapshots. Within each snapshot, each geospatial object has a purely spatial setting. Temporal snapshot sequences can be represented formally as a function from a purely temporal setting, such as a time interval, to a collection of geospatial objects, as modeled in the previous section. This kind of representation has been discussed by Al-Taha and Barrera [22], Worboys [23], Claramunt [24], Medak [25], and Hornsby and Egenhofer [26].

4 The Geospatial Event Model (GEM)

The principal contribution of this paper and the new aspect of the research is the introduction of geospatial events into the model. Figure 2 shows the scope of the geospatial event model (GEM), which encompasses geospatial objects, events and their geosettings. The gray region in the figure indicates the additional linkages and relations that need to be considered with the addition of events. The GEM, therefore, extends the geospatial object model and consists of the following:

1. Geospatial object instances and classes, their attributes, subsumption and composition hierarchies, and object-object relationships.
2. Geospatial event instances and classes, their attributes, subsumption and composition hierarchies, and event-event relationships.
3. Settings in which geospatial objects and geospatial events are situated, their instances and classes, attributes, subsumption and composition hierarchies.
4. The situation function between geospatial objects and events, and their settings.
5. Geospatial object-event participation relationships, and converse geospatial event-object involvement relationships.

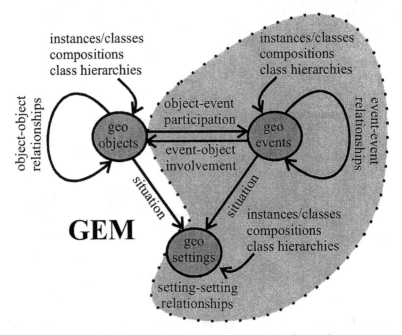

Fig. 2. The GEM model: objects, events and their interaction.

4.1 Underlying Model of Events

There are several close structural similarities between objects and events. The following lists some of the parallels:

Classification. It is fundamental to the object-oriented modeling literature that object instances with similar properties and behaviors may be grouped into classes. In the same way, collections of event instances (e.g. the car overturning on Third Avenue at 3.00pm, the traffic signal failure at Main/Fifth at 3.30pm, and the gridlock on the highway at 5.00pm) can be abstracted into classes (e.g., `TrafficEvent`). So, events may be modeled as instances or classes.

Attribution. As with objects, events can have *attributes* that describe their properties and qualities. So the event class `TrafficEvent` might have the attribute `severity`, and a specific instance of the class, such as the car overturning on Third Avenue at 3.00pm might have a specific severity level "moderately severe".

Subsumption or *generalization.* Classes of both objects and events may be arranged in a hierarchy, based on the relation of subsumption. So, for example, just as object class `Building` subsumes object class `Church`, so event class `TrafficEvent` subsumes event class `TrafficAccident`.

Composition. Both objects and events may be aggregated into composite entities, and decomposed into parts. However, there is a major distinction between objects and events. Objects may have spatial parts, but events may have temporal or spatio-temporal parts. So, for example, just as object class `Building` may be composed of a collection of objects in classes `Roof`, `Window`, `Door`, and so on, so event class `TrafficAccident` may be composed of events in classes `TrafficAccidentReport`, and `EmergencyResponse`. The issue of composition is discussed at greater length below.

Event situation. As with geospatial objects, what makes an event geospatial is its situation in an appropriate setting. The static geospatial object model is timeless **and the settings in which geospatial events are situated are spatio-temporal.** The nature of `STSettingClass` has been discussed above. We have the following functional restriction of `Situate` to geospatial events.

$$\text{Situate: GEClass} \rightarrow \text{STSettingClass}$$

Note. We are glossing over some issues here. As discussed in Casati and Varzi [27], the ways in which an object occupies a region of space and an event occupies a region of spacetime are rather different. In some approaches, a geospatial event might be better thought of as occupying a purely temporal location. Indeed, the distinction has often been made between the spatial parts of a geospatial object, such as the rooms of a house, and the temporal parts of geospatial event, such as the stages of urban decay. Even here there is some looseness, as strictly the adjectives "spatial" and "temporal" should really apply to the settings.

4.2 Geospatial Event-Event Relationships

While there is considerable research in the general data modeling and GIScience literatures on categories of geospatial object-object relationships, there is less work on event-event relationships. Many geospatial event relationships may reduced to relationships between their settings. For example, the issue of whether our trip will take us through the thunderstorm is actually a question of spatio-temporal overlap between the geospatial settings of "trip" and "thunderstorm". However, there are cases where the relationship is directly between events themselves. This section provides a brief categorization of such relationships.

General kinds of event-event relationships [28] include, in order of decreasing positive effect:

- *Initiation*: The occurrence of event *A* starts event *B* in progress. E.g., the lights turning green initiates the progress of the vehicle along the road segment.

- *Perpetuation/facilitation*: The occurrence of event *A* plays a positive role in the initiation or continuation of event *B*. E.g., opening the door allows the procession to continue; the opening of a second toll booth facilitated traffic flow in the evening rush hour.

- *Hindrance/blocking*: The occurrence of event *A* plays a positive role in the weakening, temporary stoppage, or termination of event *B*. E.g., the closing of a second toll booth hindered traffic flow in the evening rush hour.

- *Termination*: The occurrence of event *A* allows/forces event *B*, already initiated, to terminate. E.g., running out of fuel terminates the progress of the vehicle along the road segment.

Even though settings have been excluded from this discussion, it is clear that setting-setting relationships will provide constraints on the existence of the above event-event relationship types. For example, the spatio-temporal setting of a traffic light changing to red must be closely related to that of the vehicle journey for the behavior of the lights to impact the journey.

4.3 Geospatial Object-Event Relationships: Participation and Involvement

Objects and events are closely bound up with each other. Without the occurrence of events (e.g., object creation), objects will not exist. Conversely, without objects events will have a vacuity; a traffic jam cannot exist without traffic. This section explores the kinds of relationships can obtain between objects and events, and conversely between events and objects? Again, following Grenon and Smith [28], these relationships are characterized as *participation* and *involvement* relationships. So, objects participate in events and events involve objects. As with event-event relationships, we are not concerned in this section with relationships between the settings of the objects and events.

A fundamental example of an object-event relationship is the *agentive* relationship, where an object acts to produce a particular event (e.g., a person opens the door). This leads to a subcategory of event, the category of *actions*, in which at least one agent is involved. The dual of the agentive relationship is the *patientive* object-event relationship. Characterizations of objects based on their role with respect to events include the following: *perpetrator objects (initiators, perpetuators, terminators)*, *influencing objects (facilitators and hindrances)*, and *mediator objects*. The semantics for these categories is similar to the event-event cases above. A new case is *mediation*, where the mediator plays a positive but indirect role in a process involving other participating objects. For example, the building mediates the meeting between John and Mary.

At the object or event class level, in any participation relation, it may be important to know whether participation in an event, or class of events, is mandatory for a particular object, or class of objects. We have already seen in the participation of traffic in a traffic jam, a relationship where the object class traffic is mandatory for an event of class traffic jam to occur. On the other hand, the participation of faulty traffic signals is not mandatory.

The classification of types of involvement of events with objects, following Grenon and Smith, includes:

- *Creation*: An event that results in the creation of an object. For example, a bridge-building event may result in a new bridge.

- *Sustaining in being*: An event that results in the continuation in existence of an object. For example, a bridge-painting event may result in the continued life of the bridge.

- *Reinforcement/degradation*: An event that has positive/negative effects on the existence of an object. For example, plowing snow from a road keeps the road open to traffic; a storm event may result in the loss of some functionality (load-bearing ability) of the bridge.

- *Destruction*: An event that results in the destruction of an object. For example, an explosion event may result in the loss of a bridge.

- *Splitting/merger*: An event that creates/destroys a boundary between objects. For example, the splitting/merging of East and West Germany.

As with participation events, we can categorize involvements as optional or mandatory on the involving events. For example, changing ownership is mandatory upon selling a land parcel, but painting the bridge may only be optional on the continuation of the bridge.

5 GEM as a Modeling Approach

This analysis offers an approach to modeling information systems that deal with dynamic aspects of the geospatial world. This section considers some aspects of this approach, in particular presenting a diagrammatic notation for representing entities and relationships of the GEM model. We use a notation that extends Unified Modeling Language (UML) constructs [3]. For previous work on spatial visual languages that are UML-based see, for example, Bédard [29].

From the discussion above, geospatial object and event classes, `GOClass` and `GEClass`, respectively have the following structure to their definitions:

```
GOClass
     identifier
            GOID
     setting
            Situation:      SpatialSettingClass
     attributes
            Attribute₁:    OClass₁
            ...
            Attributeₘ:    OClassₘ
     object relationships
            ORel₁                      OClass₁
            ...
            ORelₙ                      OClassₙ
     event participation relationships
            PartRel₁                   EClass₁
            ...
            PartRelₚ                   EClassₚ


GEClass
     identifier
            GEID
     setting
            Situation:      SpatialSettingClass
     attributes
            Attribute₁:    OClass₁
            ...
            Attributeₘ:    OClassₘ
     event relationships
            ERel₁                      EClass₁
            ...
            ERelₙ                      EClassₙ
     object involvement relationships
            InvRel₁                    OClass₁
            ...
            InvRelₚ                    OClassₚ
```

These signatures require an extension of UML's class diagrams where no distinction is typically made concerning the type of class (i.e., object *vs.* event class). A GOClass diagram is a rectangular icon (Figure 3a) while GEClasses are distinguished from GOClasses through use of a rounded rectangle icon (Figure 3b). Each of these class constructs has four main components, a class name, setting, attributes, and operations. Settings, as presented in section 3, refer to the spatial, temporal, or spatio-temporal situations that hold for a given class. Spatial settings are either zero, one, or two-dimensional. Temporal settings are either zero or one-dimensional and spatio-temporal settings are described by space-time trajectories, histories, or lifelines. GOClasses are associated with settings that are either spatial or spatio-temporal while spatio-temporal settings hold for GEClasses. Both GOClass and GEClass diagrams represent settings with a specification in the form, settingName: settingClass. Attributes and operations for GOClass and GEClass are represented in a similar way as UML class diagrams. Any number of attributes and operations may be specified for either `GOClass` or `GEClass`. The signature for attributes is `attributeName: Oclass` while operations are described through an operation name with an optional return class specified.

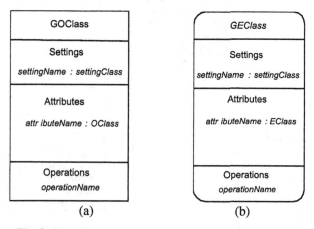

Fig. 3. Class diagrams for (a) `GOClass` and (b) `GEClass`.

6   Case Study: Object and Event Detection

We apply the GEM to a case study that illustrates how entities and events may be modeled for a specific domain. The example scenario is an airport, where entities include airport terminals, plane hangars, various other buildings, and planes. At any given time, an airport also contains pedestrians, passenger and visitor vehicles, and other miscellaneous object entities (e.g., runways, roads, traffic signs, etc.). These entities and the relationships that hold between them can be modeled using subclasses of `GOClass`. For example, Figure 4 shows entities `Plane`, `Runway`, `Hangar`, and `People` GOClasses are associated with an `Airport GOClass` through a composition relation that models the case where these classes are *part-of* an `Airport` (aggregations are considered in more detail in the next section). Multiplicities that describe the constraints of the relationships (e.g., 1: 1..*, read as one to one or more), are attached to the class diagrams. These multiplicities are useful for detailing, for example, whether an object class has a mandatory relationship with another object class (e.g., an airport has at least one runway, and perhaps more).

The GEM model also incorporates events that are related to these object classes, for example, `PlaneTakeOff` and `PassengerBoarding` events (Figure 5). Relating object and event classes allows an investigation of the kinds of possible relationships between object and events, (i.e., object-event participation relationships and event-object involvement relationships). The diagrammatic representation of the model depicts these dual relations through a unidirectional arrow that may be read in either direction. In the same way that certain object classes are mandatory for particu-

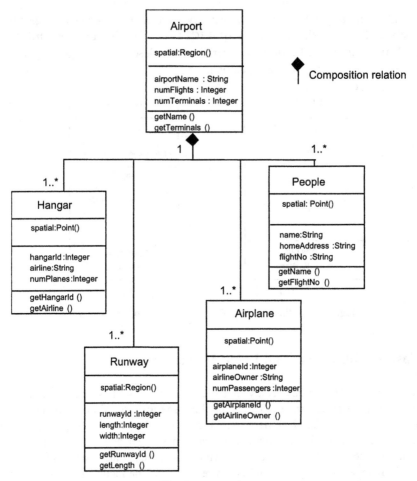

Fig. 4. Airport classes.

lar relationships, certain object-event relationships have mandatory components, for example, a `PlaneTakeOff` event cannot occur without a plane. This event, however, does not always have to occur on a runway, as other scenarios such as a takeoff from a lake or road (i.e., not an airport) may be possible. Thus evidence from the objects in a given domain ontology (e.g., an airport ontology with runway and plane classes) along with other object indicators (e.g, plane on the runway), contribute to the inferences of particular events.

It is possible to label object-event and event-event relations, for example, a `Runway` object class serves as a *facilitator* for `PlaneTakeOff` events and `PassengerBoarding` events *initiate* `PlaneTakeOff` events. (Note that cardinality constraints on the latter relationship allow the possibility of the plane taking off without any passengers).

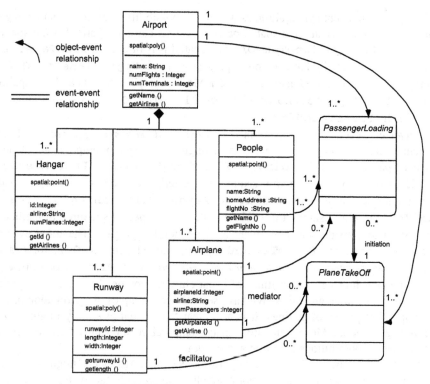

Fig. 5. GEM diagram of airport objects and events.

7 Conclusions and Further Work

The motivation for this work is to provide the foundations of a general approach to modeling dynamic geospatial domains. This model is based on the three basic entity types: geospatial object, geospatial event, and geospatial setting. The paper has discussed some of the details of the attributes and relationships between these basic entity types. The constructions of the model were represented diagrammatically by means of extensions to UML. The previous section briefly showed, by means of a case study scenario, how some aspects of the model might be applied. In this conclusion, we examine two further areas of work: extensions of the model to handle aggregate entities and the related issue of granularity, and how the model facilitates the development of query language capability in which events may be explicitly represented. Other ongoing work not described here relates to the translation from this conceptual model to the information system design level.

7.1 Aggregation and Granularity

Constructions for object aggregation (composition, grouping, etc.) are a well-understood aspect of the object-oriented approach. This section briefly points to further directions in event aggregation, as well as the possibility of event-object aggregations. Just as objects can have spatial parts, so events can have parts that are spatial (e.g., event `Runway6PlaneTakeOff` is a spatial part of event `PlaneTakeOff`.) Events can more usually have temporal parts (e.g., event `NightPlaneTakeOff` is a temporal part of event `PlaneTakeOff`.). More generally, events can have spatio-temporal parts ((e.g., event `Runway6NightPlaneTakeOff` is a spatio-temporal part of event `PlaneTakeOff`.)

Conversely, events can be combined into *temporal sequence aggregations*. Such sequences can be compared in order to distinguish any spatio-temporal anomalies. For example, if: `PlaneLanding < PlaneTaxiToGate < PassengerDeplaning` is the typical temporal sequence of events that models a plane's arrival, then a sequence of `PlaneLanding < PassengerDeplaning < PlaneTaxiToGate` would signal a possibly unusual or unexpected situation. Sometimes, it is useful to consider the new event as a temporal aggregate. The three events, `PlaneDepartGate`, `PlaneTaxiOnRunway`, and `PlaneTakeOff`, for example, can be aggregated to form a *temporal sequence composite* `PlaneDeparture`.

The relationships between settings of event clusters provide clues as to the appropriate type of event aggregation. We have seen above examples of aggregation based on temporal sequence of settings. It is also clear that spatial proximity might signal useful aggregations. More generally, spatio-temporal relationships between event settings lead to possibilities for event aggregation.

An interesting question is whether hybrid object-event aggregates might be useful modeling constructs in this model. For example, aggregates of the events related to a traffic accident and the vehicles involved. This aggregation would be a closer coupling than the event-object involvement and participation relationships discussed in an earlier section.

An important consideration for modeling geospatial objects and events relates to the *granularity* of events. Granularity refers to the amount of detail [5, 14] necessary for a modeling task. Events, like objects, can be treated over multiple granularities. Shifting perspectives from single events to composite events, for example, involves a change in granularity. If, for example, an event, (e.g., a deer is on the runway) occurs during an `PlaneDeparture` event, that granularity may be too coarse to capture whether the event was actually an obstruction event (i.e., the plane was blocked from departing) or that the event occurred during the passenger loading phase of the plane's departure and did not cause a problem for takeoff. Sequences of events may be aggregated based on changes of temporal, spatial, or spatio-temporal granularities.

These, and other issues related to object-event aggregation and granularity, are the subject of ongoing research.

7.2 Query Languages for Dynamic Systems

The GEM model offers a foundation for querying dynamic systems. The addition of GEClasses, dual object-event relationships, as well as event-event relationships provides a richer basis and more expressive power for querying. Some of the general kinds of queries supported by this model include:

- What are all the events related to object X?
- What are the objects that are related to event Y?
- Can event Y happen without object X?
- What are all the events that are related to event Y?
- What events serve as initiator events for event Y?
- How many objects serve as event-initiating or facilitating objects?
- What is the spatio-temporal setting for event Y?

These queries all involve some dynamic aspect that would not be captured by a strictly object-oriented model. For example, "What events are necessary for a passenger to board a plane?" and "What events could hinder a passenger from getting their checked luggage?" Semantics common in dynamic scenarios, such as initiating, facilitating, and blocking, for example, suggest new predicates for event-based queries. In addition, temporal sequences of events and aggregated events offer even further opportunities for querying. This includes queries, such as "What passenger-related events can hinder a `PlaneDeparture` event?" and "What luggage-related events occur between when a passenger checks in at the airline counter and the flight leaves?"

This section has highlighted two key areas of research that relate to event modeling and need further development. It seems clear that the geographic information systems of the future will need the capability to treat dynamic geospatial domains. This paper is a contribution to the development of techniques for conceptual modeling of dynamic geospatial phenomena. We have shown how it is possible to extend object-based approaches to dynamic entities, and seen how this will lead to more powerful modeling representations and query languages.

Acknowledgments

Mike Worboys' work is supported by the National Science Foundation under NSF grant numbers EPS-9983432 and BCS-0327615, the Ordnance Survey of Great Britain, and the UK EPSRC. Kathleen Hornsby's work is supported by the National Geospatial-Intelligence Agency under grant number, NMA201-00-1-200 and the National Institute of Environmental Health Sciences, under grant number 1 R 01 ES09816-01.

References

1. Guarino, N. (1998). Formal ontology and information systems. In N. Guarino, editor, Formal Ontology in Information Systems, Proceedings of FOIS'98, pp. 3-17, Amsterdam: IOS Press.
2. Chrisman, N. (1998). Beyond the snapshot: changing the approach to change, error, and process. in: M. J. Egenhofer and R. G. Golledge (Eds.), Spatial and Temporal Reasoning in Geographic Information Systems. Spatial Information Systems, Oxford University Press, New York, NY, pp. 85-93.

3. Booch, G., J. Rumbaugh, I. Jacobson (1999). The Unified Modeling Language User Guide, Addison-Wesley Longman.
4. Worboys, M.F. (2004). Event-oriented approaches to geographic phenomena. Accepted for publication in *International Journal of Geographic Information Systems*.
5. Hobbs, J. (2002). A DAML Ontology of Time. http://www.cs.rochester.edu/~ferguson/daml/daml-timenov2002.
6. Pan, F. and Hobbs, J.R. (2004). Time in OWL-S. *Proceedings of AAAI 2004*, pp. 29-36.
7. Worboys, M.F. (1995). Geographic Information Systems: A Computing Perspective, Taylor & Francis, London.
8. Allen, J. (1983). Maintaining knowledge about temporal intervals. *Communications of the ACM* 26(11): 832-843.
9. R. Snodgrass, editor (1995). *The TSQL2 Temporal Query Language*. Kluwer Academic Publishers.
10. R. Snodgrass (2000). *Developing time-oriented database applications in SQL*. San Francisco, CA: Morgan Kaufmann Publishers.
11. Partsinevelos, P., Stefanidis, A., Agouris, P. (2001). Automated Spatiotemporal Scaling for Video Generalization, *IEEE International Conference on Image Processing (ICIP) 2001*, Vol. 1, pp. 177-180, Thessaloniki, Greece.
12. Griffiths, T., Fernandes, A. Paton,, N., Jeong, S.-H., Djafri, N., Mason, K.T., Huang, B., and Worboys, M. (2002). TRIPOD: A Spatio-Historical Object Database System, in Ladner, R., Shaw, K. and Abdelguerfi, L. (eds.) *Mining Spatio-Temporal Information Systems*, Amsterdam: Kluwer Academic Publishers, pp. 127-146.
13. Galton, A. (2004). Fields and objects in space, time, and space-time, *Journal of Spatial Cognition and Computation*, 4(1):39-68, 2004.
14. Hornsby, K. and M. Egenhofer (2002). Modeling moving objects over multiple granularities, Special issue on Spatial and Temporal Granularity, *Annals of Mathematics and Artificial Intelligence* 36: 177-194.
15. Simons, P. M. (1987). *Parts. A Study in Ontology*, Oxford: Clarendon.
16. Egenhofer, M.J. and Franzosa, R.D. (1991). Point-set topological spatial relations, *International Journal of Geographical Information Systems*, 5(2): 161-174.
17. Cui, Z., A.G. Cohn and D.A. Randell (1993). Qualitative and topological relationships in spatial databases, In D.J. Abel and B.C. Ooi (eds.) *Advances in Spatial Databases - Third International Symposium, SSD '93, Singapore*, Lecture Notes in Computer Science 692, Berlin: Springer-Verlag 296-315.
18. Hornsby, K., M. Egenhofer, and P. Hayes (1999). Modeling cyclic change. in: P. Chen, D. Embley, J. Kouloumdjian, S. Liddle, and J. Roddick (Ed.), *Advances in Conceptual Modeling, ER'99 Workshop on Evolution and Change in Data Management*, Paris, France, pp. 98-109.
19. Egenhofer, M. and A. Frank (1992). Object-Oriented Modeling for GIS. *Journal of the Urban and Regional Information Systems Association* 4 (2): 3-19.
20. Worboys, M.F., Hearnshaw, H. M., and Maguire, D. J. (1990). Object-oriented data modelling for spatial databases, *International Journal of Geographical Information Systems*, 4(4): 369-383.
21. Tomlin, D. (1990). *Geographic Information Systems and Cartographic Modelling*. New Jersey: Prentice-Hall.
22. Al-Taha, K. and R. Barrera (1994). Identities through time. in: M. Ehlers and D. Steiner (Eds.), *International Workshop on Requirements for Integrated Geographic Information Systems*, New Orleans, LA, pp. 1-12.

23. Worboys, M. F. (1994). A Unified Model of Spatial and Temporal Information, *Computer Journal*, 37(1), pp. 26-34.

24. Claramunt, C. and M. Thériault (1996). Toward semantics for modelling spatio-temporal processes within GIS. in: M. J. Kraak and M. Molenaar (Ed.), *7th International Symposium on Spatial Data Handling*, Delft, NL, pp. 47-63.

25. Medak, D. (1999). Lifestyles - an algebraic approach to change in identity. in: M. Böhlen, C. Jensen, and M. Scholl (Ed.), *Spatio-Temporal Database Management, Proceedings of the International Workshop, STDBM'99*, Edinburgh, Scotland, Springer-Verlag, Lecture Notes in Computer Science 1678, pp. 19-38.

26. Hornsby, K. and M. Egenhofer (2000). Identity-based change: a foundation for spatio-temporal knowledge representation. *International Journal of Geographical Information Science* 14(3): 207-224.

27. Casati, R. and Varzi, A.C. (1999). *Parts and Places: Structures in Spatial Representation*, Cambridge, MA: MIT Press/Bradford Books.

28. Grenon, P. and Smith, B. (2004). SNAP and SPAN: Towards dynamic spatial ontology, *Journal of Spatial Cognition and Computation*, 4(1):69-103, 2004.

29. Bédard, Y. (1999). Visual modeling of spatial databases: towards spatial PVL and UML. *Geomatica* 53(2): 169-186.

Author Index

Abdelmoty, Alia I.   125
Agarwal, Pragya   1

Bação, Fernando   22
Béra, Roderic   38
Bittner, Thomas   67

Camara, Gilberto   223
Campbell, James   223
Chakravarthy, Narnindi Sharad   223
Chaudhry, Anjli   51
Claramunt, Christophe   38
Cruz, Isabel F.   51

De Maeyer, Philippe   269
Duckham, Matt   206

Feng, Chen-Chieh   67
Finch, David   125
Flewelling, Douglas M.   67
Frank, Andrew U.   81
Frankel, Andras   94
Fu, Gaihua   125

Grum, Eva   81

Hariharan, Ramaswamy   106
Hornsby, Kathleen   327

Jones, Christopher B.   125

Kazar, Baris M.   140
Kuhn, Werner   299
Kulik, Lars   206

Lee, Jiyeong   162
LeSage, James P.   179

Leyk, Stefan   191
Lilja, David J.   140
Lobo, Victor   22

Malizia, Nicholas R.   251

Nittel, Silvia   206
Nussbaum, Doron   94

Onsrud, Harlan   223

Pace, R. Kelley   140, 179
Painho, Marco   22
Pereira, Gary M.   239
Pontius Jr., Robert Gilmore   251

Rodríguez, Andrea   269

Sack, Jörg-Rudiger   94
Shekhar, Shashi   140
Shirabe, Takeshi   285
Soon, Keanhuat   299
Sunna, William   51

Toyama, Kentaro   106

Vaid, Subodh   125
Van de Weghe, Nico   269
Vasseur, Bérengère   81
Vatsavai, Ranga R.   140

Weissensteiner, Elisabeth   313
Winter, Stephan   313
Worboys, Michael   327

Zimmermann, Niklaus E.   191

Lecture Notes in Computer Science

For information about Vols. 1–3196

please contact your bookseller or Springer

Vol. 3305: P.M.A. Sloot, B. Chopard, A.G. Hoekstra (Eds.), Cellular Automata. XV, 883 pages. 2004.

Vol. 3302: W.-N. Chin (Ed.), Programming Languages and Systems. XIII, 453 pages. 2004.

Vol. 3299: F. Wang (Ed.), Automated Technology for Verification and Analysis. XII, 506 pages. 2004.

Vol. 3294: C.N. Dean, R.T. Boute (Eds.), Teaching Formal Methods. X, 249 pages. 2004.

Vol. 3293: C.-H. Chi, M. van Steen, C. Wills (Eds.), Web Content Caching and Distribution. IX, 283 pages. 2004.

Vol. 3292: R. Meersman, Z. Tari, A. Corsaro (Eds.), On the Move to Meaningful Internet Systems 2004: OTM 2004 Workshops. XXIII, 885 pages. 2004.

Vol. 3291: R. Meersman, Z. Tari (Eds.), On the Move to Meaningful Internet Systems 2004: CoopIS, DOA, and ODBASE. XXV, 824 pages. 2004.

Vol. 3290: R. Meersman, Z. Tari (Eds.), On the Move to Meaningful Internet Systems 2004: CoopIS, DOA, and ODBASE. XXV, 823 pages. 2004.

Vol. 3289: S. Wang, K. Tanaka, S. Zhou, T.W. Ling, J. Guan, D. Yang, F. Grandi, E. Mangina, I.-Y. Song, H.C. Mayr (Eds.), Conceptual Modeling for Advanced Application Domains. XXII, 692 pages. 2004.

Vol. 3287: A. Sanfeliu, J.F.M. Trinidad, J.A. Carrasco Ochoa (Eds.), Progress in Pattern Recognition, Image Analysis and Applications. XVII, 703 pages. 2004.

Vol. 3286: G. Karsai, E. Visser (Eds.), Generative Programming and Component Engineering. XIII, 491 pages. 2004.

Vol. 3284: A. Karmouch, L. Korba, E.R.M. Madeira (Eds.), Mobility Aware Technologies and Applications. XII, 382 pages. 2004.

Vol. 3281: T. Dingsøyr (Ed.), Software Process Improvement. X, 207 pages. 2004.

Vol. 3280: C. Aykanat, T. Dayar, İ. Körpeoğlu (Eds.), Computer and Information Sciences - ISCIS 2004. XVIII, 1009 pages. 2004.

Vol. 3278: A. Sahai, F. Wu (Eds.), Utility Computing. XI, 272 pages. 2004.

Vol. 3274: R. Guerraoui (Ed.), Distributed Computing. XIII, 465 pages. 2004.

Vol. 3273: T. Baar, A. Strohmeier, A. Moreira, S.J. Mellor (Eds.), <<UML>> 2004 - The Unified Modelling Language. XIII, 454 pages. 2004.

Vol. 3271: J. Vicente, D. Hutchison (Eds.), Management of Multimedia Networks and Services. XIII, 335 pages. 2004.

Vol. 3270: M. Jeckle, R. Kowalczyk, P. Braun (Eds.), Grid Services Engineering and Management. X, 165 pages. 2004.

Vol. 3269: J. Lopez, S. Qing, E. Okamoto (Eds.), Information and Communications Security. XI, 564 pages. 2004.

Vol. 3266: J. Solé-Pareta, M. Smirnov, P.V. Mieghem, J. Domingo-Pascual, E. Monteiro, P. Reichl, B. Stiller, R.J. Gibbens (Eds.), Quality of Service in the Emerging Networking Panorama. XVI, 390 pages. 2004.

Vol. 3265: R.E. Frederking, K.B. Taylor (Eds.), Machine Translation: From Real Users to Research. XI, 392 pages. 2004. (Subseries LNAI).

Vol. 3264: G. Paliouras, Y. Sakakibara (Eds.), Grammatical Inference: Algorithms and Applications. XI, 291 pages. 2004. (Subseries LNAI).

Vol. 3263: M. Weske, P. Liggesmeyer (Eds.), Object-Oriented and Internet-Based Technologies. XII, 239 pages. 2004.

Vol. 3262: M.M. Freire, P. Chemouil, P. Lorenz, A. Gravey (Eds.), Universal Multiservice Networks. XIII, 556 pages. 2004.

Vol. 3261: T. Yakhno (Ed.), Advances in Information Systems. XIV, 617 pages. 2004.

Vol. 3260: I.G.M.M. Niemegeers, S.H. de Groot (Eds.), Personal Wireless Communications. XIV, 478 pages. 2004.

Vol. 3258: M. Wallace (Ed.), Principles and Practice of Constraint Programming – CP 2004. XVII, 822 pages. 2004.

Vol. 3257: E. Motta, N.R. Shadbolt, A. Stutt, N. Gibbins (Eds.), Engineering Knowledge in the Age of the Semantic Web. XVII, 517 pages. 2004. (Subseries LNAI).

Vol. 3256: H. Ehrig, G. Engels, F. Parisi-Presicce, G. Rozenberg (Eds.), Graph Transformations. XII, 451 pages. 2004.

Vol. 3255: A. Benczúr, J. Demetrovics, G. Gottlob (Eds.), Advances in Databases and Information Systems. XI, 423 pages. 2004.

Vol. 3254: E. Macii, V. Paliouras, O. Koufopavlou (Eds.), Integrated Circuit and System Design. XVI, 910 pages. 2004.

Vol. 3253: Y. Lakhnech, S. Yovine (Eds.), Formal Techniques, Modelling and Analysis of Timed and Fault-Tolerant Systems. X, 397 pages. 2004.

Vol. 3252: H. Jin, Y. Pan, N. Xiao, J. Sun (Eds.), Grid and Cooperative Computing - GCC 2004 Workshops. XVIII, 785 pages. 2004.

Vol. 3251: H. Jin, Y. Pan, N. Xiao, J. Sun (Eds.), Grid and Cooperative Computing - GCC 2004. XXII, 1025 pages. 2004.

Vol. 3250: L.-J. (LJ) Zhang, M. Jeckle (Eds.), Web Services. X, 301 pages. 2004.

Vol. 3249: B. Buchberger, J.A. Campbell (Eds.), Artificial Intelligence and Symbolic Computation. X, 285 pages. 2004. (Subseries LNAI).

Vol. 3246: A. Apostolico, M. Melucci (Eds.), String Processing and Information Retrieval. XIV, 332 pages. 2004.

Vol. 3245: E. Suzuki, S. Arikawa (Eds.), Discovery Science. XIV, 430 pages. 2004. (Subseries LNAI).

Vol. 3244: S. Ben-David, J. Case, A. Maruoka (Eds.), Algorithmic Learning Theory. XIV, 505 pages. 2004. (Subseries LNAI).

Vol. 3243: S. Leonardi (Ed.), Algorithms and Models for the Web-Graph. VIII, 189 pages. 2004.

Vol. 3242: X. Yao, E. Burke, J.A. Lozano, J. Smith, J.J. Merelo-Guervós, J.A. Bullinaria, J. Rowe, P. Tiňo, A. Kabán, H.-P. Schwefel (Eds.), Parallel Problem Solving from Nature - PPSN VIII. XX, 1185 pages. 2004.

Vol. 3241: D. Kranzlmüller, P. Kacsuk, J.J. Dongarra (Eds.), Recent Advances in Parallel Virtual Machine and Message Passing Interface. XIII, 452 pages. 2004.

Vol. 3240: I. Jonassen, J. Kim (Eds.), Algorithms in Bioinformatics. IX, 476 pages. 2004. (Subseries LNBI).

Vol. 3239: G. Nicosia, V. Cutello, P.J. Bentley, J. Timmis (Eds.), Artificial Immune Systems. XII, 444 pages. 2004.

Vol. 3238: S. Biundo, T. Frühwirth, G. Palm (Eds.), KI 2004: Advances in Artificial Intelligence. XI, 467 pages. 2004. (Subseries LNAI).

Vol. 3236: M. Núñez, Z. Maamar, F.L. Pelayo, K. Pousttchi, F. Rubio (Eds.), Applying Formal Methods: Testing, Performance, and M/E-Commerce. XI, 381 pages. 2004.

Vol. 3235: D. de Frutos-Escrig, M. Nunez (Eds.), Formal Techniques for Networked and Distributed Systems – FORTE 2004. X, 377 pages. 2004.

Vol. 3234: M.J. Egenhofer, C. Freksa, H.J. Miller (Eds.), Geographic Information Science. VIII, 345 pages. 2004.

Vol. 3232: R. Heery, L. Lyon (Eds.), Research and Advanced Technology for Digital Libraries. XV, 528 pages. 2004.

Vol. 3231: H.-A. Jacobsen (Ed.), Middleware 2004. XV, 514 pages. 2004.

Vol. 3230: J.L. Vicedo, P. Martínez-Barco, R. Muñoz, M. Saiz Noeda (Eds.), Advances in Natural Language Processing. XII, 488 pages. 2004. (Subseries LNAI).

Vol. 3229: J.J. Alferes, J. Leite (Eds.), Logics in Artificial Intelligence. XIV, 744 pages. 2004. (Subseries LNAI).

Vol. 3226: M. Bouzeghoub, C. Goble, V. Kashyap, S. Spaccapietra (Eds.), Semantics of a Networked World. XIII, 326 pages. 2004.

Vol. 3225: K. Zhang, Y. Zheng (Eds.), Information Security. XII, 442 pages. 2004.

Vol. 3224: E. Jonsson, A. Valdes, M. Almgren (Eds.), Recent Advances in Intrusion Detection. XII, 315 pages. 2004.

Vol. 3223: K. Slind, A. Bunker, G. Gopalakrishnan (Eds.), Theorem Proving in Higher Order Logics. VIII, 337 pages. 2004.

Vol. 3222: H. Jin, G.R. Gao, Z. Xu, H. Chen (Eds.), Network and Parallel Computing. XX, 694 pages. 2004.

Vol. 3221: S. Albers, T. Radzik (Eds.), Algorithms – ESA 2004. XVIII, 836 pages. 2004.

Vol. 3220: J.C. Lester, R.M. Vicari, F. Paraguaçu (Eds.), Intelligent Tutoring Systems. XXI, 920 pages. 2004.

Vol. 3219: M. Heisel, P. Liggesmeyer, S. Wittmann (Eds.), Computer Safety, Reliability, and Security. XI, 339 pages. 2004.

Vol. 3217: C. Barillot, D.R. Haynor, P. Hellier (Eds.), Medical Image Computing and Computer-Assisted Intervention – MICCAI 2004. XXXVIII, 1114 pages. 2004.

Vol. 3216: C. Barillot, D.R. Haynor, P. Hellier (Eds.), Medical Image Computing and Computer-Assisted Intervention – MICCAI 2004. XXXVIII, 930 pages. 2004.

Vol. 3215: M.G.. Negoita, R.J. Howlett, L.C. Jain (Eds.), Knowledge-Based Intelligent Information and Engineering Systems. LVII, 906 pages. 2004. (Subseries LNAI).

Vol. 3214: M.G.. Negoita, R.J. Howlett, L.C. Jain (Eds.), Knowledge-Based Intelligent Information and Engineering Systems. LVIII, 1302 pages. 2004. (Subseries LNAI).

Vol. 3213: M.G.. Negoita, R.J. Howlett, L.C. Jain (Eds.), Knowledge-Based Intelligent Information and Engineering Systems. LVIII, 1280 pages. 2004. (Subseries LNAI).

Vol. 3212: A. Campilho, M. Kamel (Eds.), Image Analysis and Recognition. XXIX, 862 pages. 2004.

Vol. 3211: A. Campilho, M. Kamel (Eds.), Image Analysis and Recognition. XXIX, 880 pages. 2004.

Vol. 3210: J. Marcinkowski, A. Tarlecki (Eds.), Computer Science Logic. XI, 520 pages. 2004.

Vol. 3209: B. Berendt, A. Hotho, D. Mladenic, M. van Someren, M. Spiliopoulou, G. Stumme (Eds.), Web Mining: From Web to Semantic Web. IX, 201 pages. 2004. (Subseries LNAI).

Vol. 3208: H.J. Ohlbach, S. Schaffert (Eds.), Principles and Practice of Semantic Web Reasoning. VII, 165 pages. 2004.

Vol. 3207: L.T. Yang, M. Guo, G.R. Gao, N.K. Jha (Eds.), Embedded and Ubiquitous Computing. XX, 1116 pages. 2004.

Vol. 3206: P. Sojka, I. Kopecek, K. Pala (Eds.), Text, Speech and Dialogue. XIII, 667 pages. 2004. (Subseries LNAI).

Vol. 3205: N. Davies, E. Mynatt, I. Siio (Eds.), UbiComp 2004: Ubiquitous Computing. XVI, 452 pages. 2004.

Vol. 3204: C.A. Peña Reyes, Coevolutionary Fuzzy Modeling. XIII, 129 pages. 2004.

Vol. 3203: J. Becker, M. Platzner, S. Vernalde (Eds.), Field Programmable Logic and Application. XXX, 1198 pages. 2004.

Vol. 3202: J.-F. Boulicaut, F. Esposito, F. Giannotti, D. Pedreschi (Eds.), Knowledge Discovery in Databases: PKDD 2004. XIX, 560 pages. 2004. (Subseries LNAI).

Vol. 3201: J.-F. Boulicaut, F. Esposito, F. Giannotti, D. Pedreschi (Eds.), Machine Learning: ECML 2004. XVIII, 580 pages. 2004. (Subseries LNAI).

Vol. 3199: H. Schepers (Ed.), Software and Compilers for Embedded Systems. X, 259 pages. 2004.

Vol. 3198: G.-J. de Vreede, L.A. Guerrero, G. Marín Raventós (Eds.), Groupware: Design, Implementation and Use. XI, 378 pages. 2004.